USER'S GUIDEBOOK TO DIGITAL CMOS INTEGRATED CIRCUITS

USER'S GUIDEBOOK TO DIGITAL CMOS INTEGRATED CIRCUITS

Eugene R. Hnatek
Lorlin Industries, Inc.

McGraw-Hill Book Company

New York St. Louis San Francisco Auckland
Bogotá Düsseldorf Johannesburg London Madrid Mexico
Montreal New Delhi Panama Paris São Paulo
Singapore Sydney Tokyo Toronto

Library of Congress Cataloging in Publication Data
  Hnatek, Eugene R
    User's guidebook to digital CMOS integrated circuits.
    Includes index.
    1. Digital integrated circuits. 2. Metal oxide
semiconductors, Complementary. I. Title.
TK7874.H538   621.381'73   80-23677
ISBN 0-07-029067-9

Copyright © 1981 by McGraw-Hill, Inc. All rights reserved.
Printed in the United States of America. No part of this publication
may be reproduced, stored in a retrieval system, or transmitted,
in any form or by any means, electronic, mechanical, photocopying, recording,
or otherwise, without the prior written permission of the publisher.

1234567890 KPKP 8987654321

The editors for this book were Tyler G. Hicks, Barry Richman, and Susan Thomas;
the designer was Blaise Zito Assoc.; and the production supervisor was
Thomas G. Kowalczyk. It was set in Baskerville by Bi-Comp, Incorporated.
Printed and bound by The Kingsport Press.

Contents

Preface, *vii*
Acknowledgments, *ix*

Chapter 1   **An Overview**   1
   The Technology, **2**
   CMOS Logic Circuits, **6**
   Nonlogic Usage, **11**

Chapter 2   **CMOS Logic Performance Characteristics and Circuit Applications**   17
   The Logic--The Inverter, **19**
   The Transmission Gate, **34**
   Memory Cells, **40**
   Dynamic Shift Registers, **42**
   Four-Phase CMOS Logic Circuits, **43**
   Three-State Logic, **45**
   Selected CMOS Logic Circuit Applications, **45**

Chapter 3   **Data Conversion and Telecommunications Circuits**   63
   Data Conversion Circuits, **65**
   Telecommunications Circuits, **151**

Chapter 4   **Memories**   191
What Is a Semiconductor Memory?, **194**
CMOS Memory Cell Architecture, **196**
Memory Organization, **202**
Popular CMOS RAMs, **212**
Popular CMOS ROMs/PROMs, **237**
Selected CMOS Memory Applications, **257**

Chapter 5   **Microprocessors**   265
Microprocessor Architecture, **268**
Commercially Available CMOS Microprocessors, **277**

Index, *331*

Preface

The popularity of CMOS integrated circuits for reasons of low power consumption, high noise immunity, and operation over a wide power supply range—coupled with a variety of new MSI and LSI functions such as memories, microprocessors, A/D and D/A converters, telecommunications circuits, and the like—requires that the practicing engineer keep abreast of state-of-the-art process technology as well as of circuit designs. The increases in circuit density, complexity, and flexibility plus a combination of processing technologies further enhance the usage of CMOS ICs for portable instrumentation, battery-powered, control system, aerospace, and automotive applications.

In this book, practicing design and systems engineers, components engineers, sales and applications engineers, technicians, engineering students, and other interested readers will find a comprehensive overview of the entire topic of digital CMOS integrated circuits, proceeding from a discussion of the basic CMOS process through the CMOS logic functions and their electrical performance peculiarities to LSI functions. The concentration is on commercially available LSI and future VLSI circuits. This is not a book which delves into the details of the design of CMOS ICs; however, it does cover enough processing and monolithic circuit design to enable the user to understand the relationships between the process, the circuit design, and the finished product and how each affects the electrical characteristics. It also describes popular, commercially available circuits for each functional category presented as well as discusses specific applications in sufficient depth to interest the experienced designer. A general background is assumed in semiconductor IC processing terminology, semiconductor physics, and circuit design theory. Engineering students will be able to follow the book if they have been exposed to a course on semiconductor devices.

As with my other books, I wanted this to be a book that was organized and used from a practical viewpoint. As such, it is organized into five sections:

Chapter 1 introduces the subject on a general summary basis. Chapter 2 presents a discussion of the characteristics of CMOS logic circuits to demonstrate the salient features of all CMOS devices and selected logic circuit applications using these ICs. Chapter 3 covers the topics of data conversion and telecommunications circuits, including the latest D/A converter, A/D converter, and IC codec developments. Chapter 4 discusses the topic of CMOS memories—RAMs, ROMs, and PROMs—and their applications; and Chapter 5 deals with microprocessors. The many categories of linear and custom CMOS circuits are not covered. To do so would require an extremely lengthy and cumbersome book. I opted for a readable book.

Throughout the book emphasis is placed on using the popular CMOS ICs of each functional category as illustrative examples, and there are summaries of the salient electrical characteristics of the popular device types contained in each functional category (D/A converters, A/D converters, memories, microprocessors, and the like).

The research that went into putting this book together has convinced me that CMOS, and perhaps CMOS/SOS, is the technology for fabricating LSI circuits in the mid-1980s.

EUGENE R. HNATEK

Acknowledgments

I want to thank all manufacturers of the CMOS ICs discussed for permission to use the material they have published on their devices. Without their cooperation and technological contributions this book would not have been possible. I am also especially grateful to Carol Lopez and Diane Beer for their painstaking efforts in typing the manuscript and its revisions.

USER'S GUIDEBOOK TO DIGITAL CMOS INTEGRATED CIRCUITS

CHAPTER ONE
An Overview

In the past few years, CMOS [complementary (symmetry) metal-oxide semiconductor] technology has taken its place among the most prominent semiconductor technologies. However, this was not always so. Prior to 1977 CMOS was only considered suitable for use as an SSI/MSI (small-scale integration/medium-scale integration) logic-building block technology to be used in low-power portable battery operation where speed was not a critical design-performance parameter. However, from 1977 to the present, advances in both bipolar and MOS (metal-oxide semiconductor) technology [such as device scaling, oxide isolation process, projection alignment, the ability to fabricate both bipolar and MOS devices on the same die, the practical development of SOS (silicon on sapphire) and the like] have changed the design and applications spectrum of CMOS to that of a high-density, high-speed LSI (large-scale integration) technology for use in energy conservation applications. This has been primarily due to the efforts of such companies as RCA Semiconductor, Hewlett Packard Corp., Harris Semiconductor, Intersil Semiconductor, Motorola Semiconductor, Analog Devices, National Semiconductor, and Hughes Aircraft Corp., as well as work done by many research facilities.

CMOS is used to fabricate gates, flip-flops, buffers, multiplexers, and many other logic forms, as well as shift registers, D/A (digital-to-analog) and A/D (analog-to-digital) converters, codecs, memories, and microprocessors. Moreover, it has been widely used in watches, clocks, pocket and desk calculators, communications controllers, CB (citizen's band) synthesizers, cross-point switches, and the like. Table 1-1 summarizes the major areas of CMOS applications.

THE TECHNOLOGY

As a technology, CMOS combines the low-power characteristics of I^2L (integrated injection logic) with the high-speed characteristics of NMOS (N-channel metal-oxide semiconductor). As such, CMOS occupies a position midway between the two on the classically used figure of merit for semiconductor technologies—the speed-power curve—as shown in Fig. 1-1. That illustration also compares the speed-power characteristics of CMOS with TTL (transistor-transistor logic): 54L, 54LS and 54H.

The principle features that make CMOS an attractive technology are the following:

1. High noise immunity
2. Operation from unregulated power supply
3. Microwatt power dissipation for portable controls, instruments, telecommunications systems where battery-supported backup is re-

TABLE 1-1 Dominant CMOS Applications

Time Keeping	Electronic Watches, Clocks, Control Timers
Automobiles	Autoclocks, Seat Belts, Controls for Pollution and Drive Train, Ignition Timing and Control, Fuel Injection
Industrial Controls	Process Control Equipment for Mills, Refineries, and Printers
Computer	Peripherals, Communication Buffer, Scratchpad Memories
Telephone	Tone Synthesizing and Detection, Crosspoints, Dialers, Codecs, Speech Synthesizers
Medical	Heart Pacers, Fluid Analyzers, Patient Monitoring
Communication	RF Digital Tuning, CATV Converters, Pocket Pagers
Military	Fuses, Airborne Computers, ECM Equipment, Controllers
Satellites	Data Acquisition, Lower Power Consumption Logic, Computers
Miscellaneous	Utility Meter Reading, Traffic Controllers, Smoke Detectors, Video Games

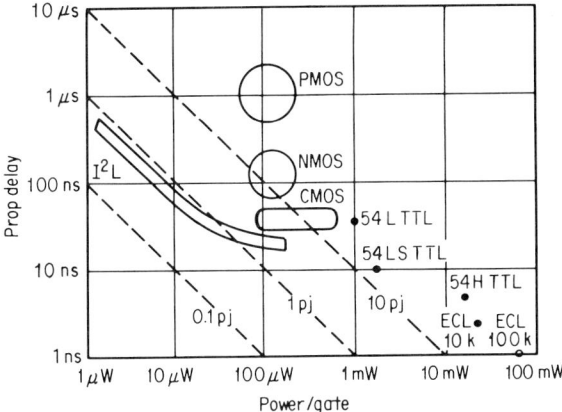

Fig. 1-1 Comparison of Speed and Power Capabilities of Various Logic Families.

quired, and battery-operated medical electronic devices such as heart pacemakers

4. Complex functions that are not available in other high-noise-immunity families at a speed comparable to NMOS
5. Low noise generation, which appeals particularly to manufacturers of medical electronics equipment where noise-free operation is highly important

From a circuit construction viewpoint, there is no such thing as a discrete CMOS transistor; the simplest CMOS device is an inverter consisting of an N- and P-channel transistor on one chip. CMOS, then, must be explained in terms of basic MOSFET (metal-oxide semiconductor field-effect transistor) theory.

Unlike bipolar or conventional transistors, whose operation depends on two kinds of charge carriers (holes and electrons), FETs (field-effect transistors) depend on only one. Thus, a P-channel FET depends on holes and an N-channel FET depends on electrons as carriers.

FETs can be explained in terms of charge control and the charge can be considered as a kind of "vehicle" which moves in accordance with very strict rules. Each FET consists of three elements: the source, the drain, and the gate. The source may be considered the "entrance" to an underground transit system and the drain the "exit." The flow of charge "traffic" is regulated from the surface by the gate, which acts as a kind of "traffic cop" or "controller."

Any charge placed on the gate creates an equal, but opposite, charge in the semiconductor layer or path beneath existing between the source and drain. The charge on the gate at the surface, therefore, influences the "underground traffic" charges in much the same way as traffic lights influence the flow of vehicular traffic.

In MOSFETs, a metal-gate electrode is separated from the conduction layer by an insulator. This permits the gate voltage to increase conductivity in the channel without drawing input power. It is this feature that keeps power consumption so low in MOS and CMOS.

The simplest CMOS device is an inverter circuit having both an N-channel and P-channel MOSFET on the same substrate and connected in parallel as shown in Fig. 1-2. In this network, power is consumed only during the transition from a logic low output (Q_1 is saturated, but Q_2 is off) to a logic high output (Q_1 is off, but Q_2 is saturated).

Under normal static conditions, therefore, power is consumed by the device only because of the existence of leakage current through the transistor which is off. This explains the minute magnitude of current and power consumption during the quiescent stage.

But even when operating, both transistors are only partially on for only a fraction of the operating interval, so that the current drawn is still in the microampere region. At moderate speeds in the 10 to 11-kHz range, power dissipation is less than 1 μW per gate. The power dissipation does, of course, rise at higher frequencies, so that at TTL speeds metal-gate CMOS structures may dissipate as much as 20 mA per gate.

The transition from P to N conduction (and vice versa) and the overlap current-condition transition is shown in Fig. 1-3. In that scheme, the inverter threshold is centered around the midpoint of the input V_{CC} voltage (power supply voltage) [which is also widely labeled as V_{DD} (drain supply voltage)]. This factor accounts for the high-voltage noise immunity of CMOS for both logic low or logic high inputs.

Additionally, unlike every other logic family, CMOS circuits do not

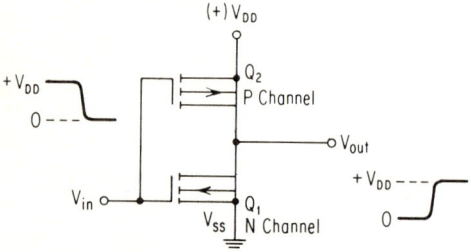

Fig. 1-2 The Basic CMOS Inverter Circuit. An Input High Causes Q_1 to Conduct and Q_2 to Cut Off. An Input Low Reverses the Output.

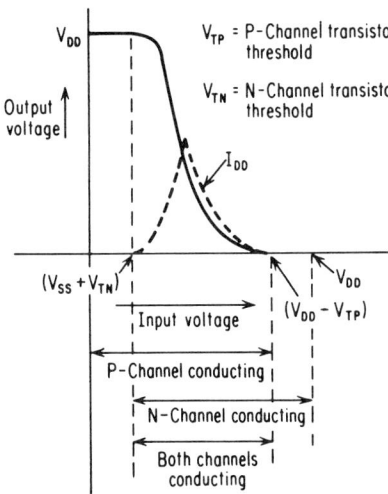

Fig. 1-3 CMOS Voltage Transfer Curve. Current through the Device Flows Only During the Transition When Q_1 and Q_2 Are Simultaneously Conductive.

need expensive close tolerance power supplies or expensive on-card regulation. Because an inverter configuration is so insensitive to voltage variations, the circuits function from 3 to 18 V.

Because CMOS has a nearly ideal logic transfer characteristic, it possesses good temperature stability and is extremely immune to noise (Fig. 1-4). This immunity makes CMOS a powerful logic technology for such environments as the automobile and factory. Its guaranteed noise margin is approximately 1.5 V, compared to TTL systems with noise margins of only 0.4 V, so it is obvious why CMOS logic circuits for electronic ignition and injection fuel systems are already being built. In manufacturing process control equipment, too, standard CMOS logic circuits are rapidly replacing TTL packages.

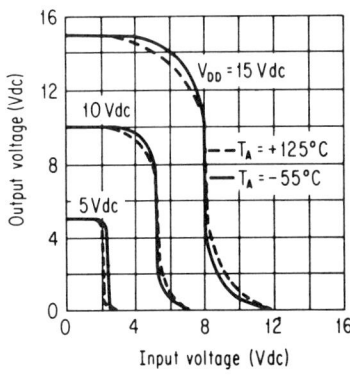

Fig. 1-4 The Transfer Characteristics of a CMOS Inverter. Notice the Outstanding Temperature Stability of the Device and the Sharpness of the Knee.

Unfortunately, CMOS devices are not as fast as the more common bipolar TTL circuits. The most frequently used CMOS manufacturing process, standard metal gate, results in device capacitance levels which limit practical switching speeds to less than 5 MHz. As a result, CMOS is not likely to be used in high-speed counters or for data processing, for example. With the advent of oxide-isolated CMOS, high-speed operation is being realized, especially in the area of LSI.

CMOS LOGIC CIRCUITS

The most popular CMOS logic family is the 4000 series, introduced by RCA in 1968. It contains a very wide selection of circuit functions and has many alternate sources.

The leading alternative to the 4000 series is the 54C/74C family, introduced by National Semiconductor. Both the 4000 and 54C/74C circuits use standard metal-gate fabrication techniques and silicon substrates. Consequently, they have comparable performance characteristics.

But unlike the 4000 units, 54C/74C ICs (integrated circuits)—promoted as the CMOS logic counterpart of the 7400 TTL circuits—exhibit the same key interface characteristics of equivalent TTL units. Pinouts are the same, too. Hence, designers familiar with popular TTL circuits can transfer that experience to a 54C/74C design.

Several manufacturers have introduced proprietary 4000-type ICs to answer the call for more complex CMOS circuits. For example, Motorola in its 14500 series is offering a growing number of MSI functions, while its 14400 series provides monolithic MSI and LSI communications circuits optimized toward particular applications.

While 54C/74C ICs are designed for TTL sockets, other alternatives to the leading series stay with the 4000 series pinouts. These circuits are sole-source lines offered by Fairchild and Motorola and they permit upgrading of existing 4000 circuit designs. Improved performance results from the manufacturers' use of either isolation techniques with silicon substrates or SOS (silicon-on-sapphire) fabrication methods and silicon-gate and ion-implant processes to achieve benefits like improved speeds, lowered thresholds, and increased circuit densities. Standard products using these processes are available.

CMOS has improved steadily over the last 5 yr as ways have been found to increase its output drive capabilities and decrease its sensitivity to input and output patterns.

As a supplement to the 4000 series, the upgraded B series of standard CMOS chips is now available from most CMOS manufacturers. Its superior input, output, and internal transfer characteristics all make it easier to use than the earlier series.

In standard, unbuffered CMOS circuits, pattern sensitivity at the output was a problem. For example, in the unbuffered two-input NOR (NOT/OR) gate (Fig. 1-5a) either of the N-channel transistors connected to ground [V_{SS} (source supply voltage)] conducts when either input is high, causing the output to go low through the ON resistance of the conducting device. If both inputs are high, both N-channel devices are on and this has the effect of halving the ON resistance by making the output impedance (and hence fall time) a function of input variables.

Similarly, the P-channel devices are switched on by low signals; that is, when both inputs are low, conduction will occur from V_{DD} to the output. Since the P-channel devices are in series, their ON resistance must be decreased (by enlarging their chip area) if the output impedance is to be held within specification. As the number of gate inputs increases, even larger P-channel devices are required, and the output impedance to V_{SS} becomes even more pattern-sensitive.

A conventional, unbuffered, CMOS two-input NAND (NOT/AND) gate interchanges the parallel and serial transistor gating (Figure 1-5b). The changes in output resistance then move to the P-channel transistors connected to V_{DD}, while the N-channel devices must be increased in size since they are now the ones connected in series.

The solution to this pattern sensitivity is the buffered gates of Fig. 1-5c and 1-5d, which show fully buffered NOR and NAND gates.

Since there are two buffered inverters at the output of each gate, high gain is available to minimize switching threshold variations, and the output drive requirements are isolated from all input conditions.

In constructing the buffered gate, most companies use standard, small-geometry CMOS transistors to generate the required logic function which, in turn, is used to drive the low-impedance buffer stages. Consequently, process complexity is not much increased and the layout size is kept minimal since only two large output transistors are required in any circuit configuration. Meanwhile, the rise and fall times at the output of the gate are independent of input pattern.

There is even a bonus or two. One advantage is increased system speed, since the internal logic gates can be driven harder and the output buffer stages can be used to minimize the resulting increase in distortion.

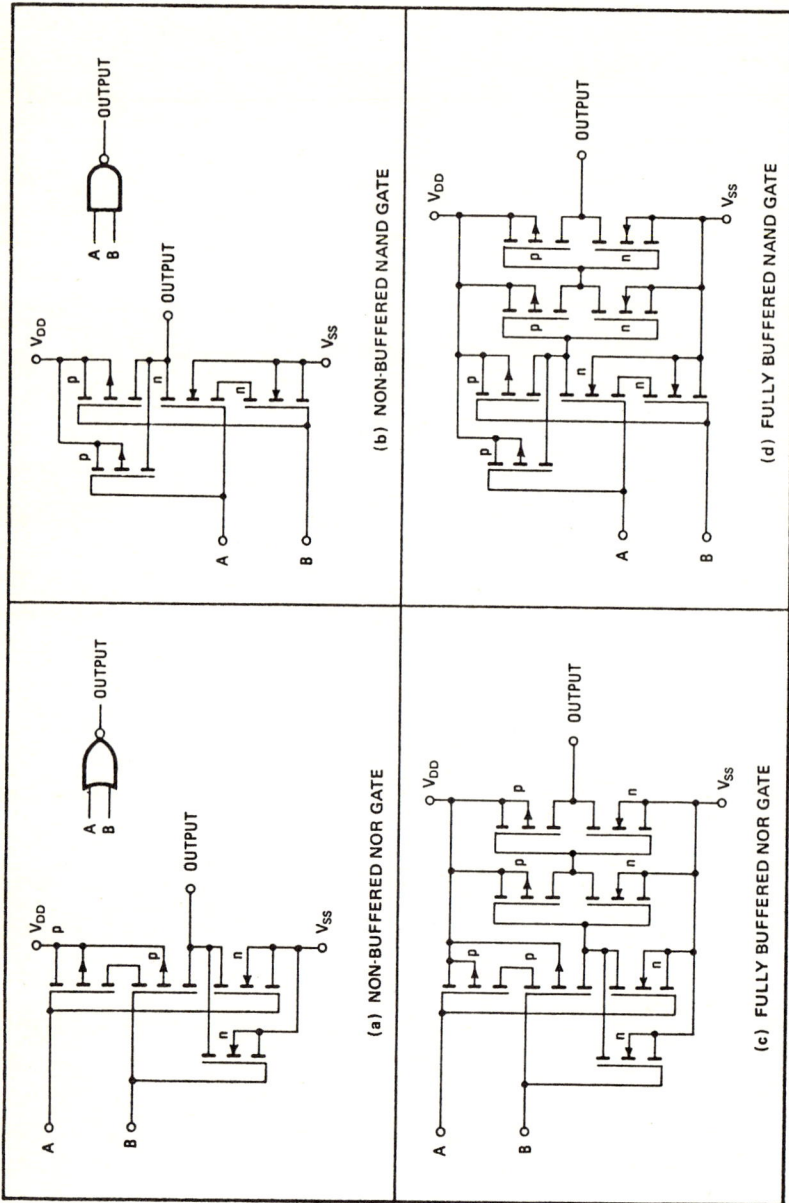

Fig. 1-5 B-Series CMOS Circuits Containing Buffered Gates Increase Drive Outputs and Minimize the Pattern Sensitivity that Could Otherwise Trouble Them.

CMOS Logic Circuits 9

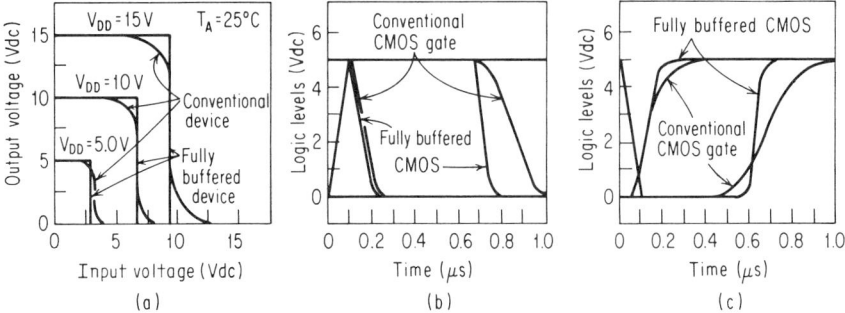

Fig. 1-6 Buffered Outputs Offer Better Noise Immunity, as Shown by Transfer Characteristics (a), and Better Gain Resulting in Cleaner Pulse Shapes, as Shown in the Output Traces (b) and (c).

This means that the designer now has a means of making propagation delay less sensitive to output load capacitance.

Another advantage of the buffered gate is improved noise immunity. Nearly ideal transfer characteristics are realized (Fig. 1-6a) because of the increased voltage gain, which is greater than 10,000.

This high gain also provides significant pulse shaping (Fig. 1-6b and 1-6c). For input transition times of 100 ns or less, the output waveforms of both conventional and buffered CMOS gates are similar. When the input transitions are stretched to 1 μs, the transition times of the conventional gates are increased while those of the buffered gate remain unchanged. This feature prevents the pulse characteristics in a system from progressively deteriorating. Table 1-2 compares the salient features of buffered and unbuffered CMOS.*

CMOS designers, when faced with the demand for more complex MSI and LSI functions, had to overcome three design factors that limited CMOS density: metal interconnections, gate alignments, and guarding structures that separated the N- and P-channel elements in the complementary configuration.

The problem was especially acute in standard metal-gate CMOS structures (Fig. 1-7a) where each N- or P-channel device required a metal gate that had to overlap the source and drain regions. To eliminate conductive channels between devices, each transistor needed to be surrounded first by an undiffused spacing and then by a diffused region whose polarity was opposite to that of the source and drain. These guard rings

---

* For more details on output buffers, refer to A. Nguyen-Huu, "Advances in CMOS Device Technology," *Comp. Des.*, January 1975.

10   An Overview

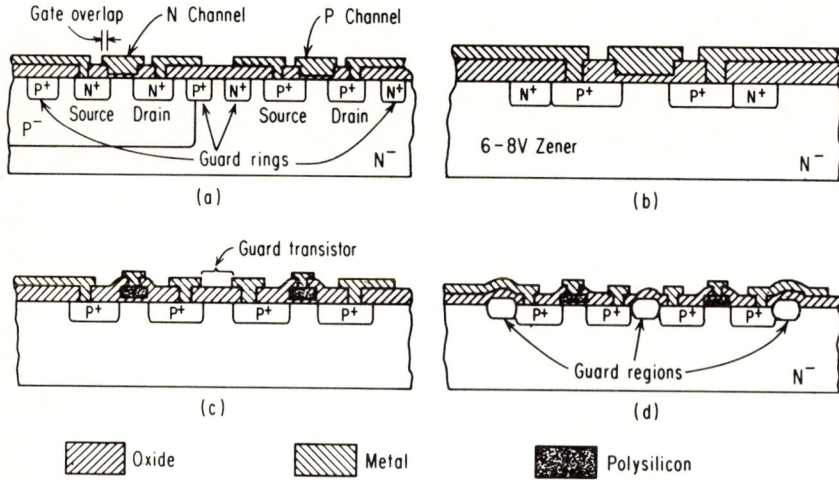

Fig. 1-7 (a) Metal-Gate CMOS, (b) Guard Area Reduction, (c) Oxide Guarding, and (d) Full-Voltage Oxide Guarding. For Higher Device Complexity, Conventional Guard Rings (a) Occupy Too Much Space. A Zener Guard Ring (b) Is Smaller, but Restricts Voltage to 8 V.

wasted so much space as to preclude the use of standard metal-gate CMOS structures in high-density circuit designs.

The interconnect limitation could be solved by the use of a two-layer metal system or with the use of polysilicon as the gate material. The polysilicon can be used as an interconnect crossunder and, being a self-aligning structure, it eliminates the need for gate alignment tolerances as well.

A straightforward way of reducing the area required for guard rings is to eliminate the spacing between source/drain and guard diffusions (Fig. 1-7b). The result is the formation of N^+P^+ zener diodes with breakdown

TABLE 1-2 Comparison of Buffered and Unbuffered CMOS

Characteristic	Buffered Version ("B")	Unbuffered Version ("UB")
Propagation Delay (Speed)	Moderate	Fast
Noise Immunity/Margin	Excellent	Good
Output Impedance and Output Transition Time	Constant	Variable
AC Gain	High	Low
Output Oscillation for Slow Inputs	Yes	No
Input Capacitance	Low	High

voltages of about 6 to 8 V. Such a technique should be used on a limited basis since it is, of course, only suitable for products with a maximum operating limit of 5 V.

A better method for reducing guard ring area is the use of oxide in the guard region. These oxide guards may be thermally grown at the beginning of the process and then selectively etched. In the surface oxide guarded transistors shown in Fig. 1-7c, the guard is actually a FET with a much thicker gate oxide, and hence with a higher threshold voltage than the switching device. The guard device will thus be turned off (even though interconnect metal may cross its channel region) and will prevent device crosstalk at operating voltage levels. The requirement that "steps" on the IC surface be minimized limits the thickness of the guard oxide, however, and again restricts it to a lower operating supply range.

To overcome this limitation, a thicker gate oxide is possible if part of it is grown below the surrounding silicon through the use of a nitride technique. The cross section resulting from this approach is shown in Figure 1-7d.

A third technique for obtaining thinner guard rings is to use silicon on an insulating substrate (SOS). The switching transistors are isolated by selective etching of the silicon film.

The combination of the polysilicon gate structure with either of the oxide isolation schemes or with SOS produces a very-high-density structure. The significant fact is that the polysilicon gates may cross over the guard rings, enabling both the N- and P-channel devices of an inverter, for example, to be fabricated with a single run of polysilicon and with only one gate contact. When combined with a multilayer metal system, this approach could provide a density three to four times that presently available with metal-gate CMOS.

Alternatively, higher density can be realized with the existing process utilizing circuit techniques. One of these techniques involves the implementation of dynamic logic with CMOS—a technique that is most efficient in multilevel logic organizations like decoders. In some applications, dynamic CMOS may be on concurrently with the chip's static circuitry. In a decoder, however, only one decode path is on at a time, and the power increase is negligible.

NONLOGIC USAGE

As mentioned previously, the real potential of CMOS is in nonlogic MSI and LSI circuit implementations. Consequently these will be emphasized in this book. The myriad of available logic circuits and applications will not be covered, nor will analog circuits such as CMOS op amps, analog switches and multipliers, and phase-locked loops and timers.

Data Processing/Communications Circuits

The data processing/communications marketplace represents the fastest-growing application for CMOS technology, followed closely by the memories and microprocessors field. CMOS D/A converters, codecs (encoders-decoders), PLLs (phase-locked loops) fast-bit rate generator circuits, tone decoders, and BCD (binary-coded decimal) rate multipliers, to name a few devices, are rapidly appearing in answer to the telecommunication industry's need for extremely-low power dissipation and high-speed data transmission circuits. What is more, such interface circuits as IC modems, byte-sized bit and line drivers, and UARTs (universal asynchronous receiver-transmitters) are all becoming available.

The myriad of available data converters are presented in detail in Chap. 3. Suppliers of CMOS data converters include Analog Devices, National Semiconductor, Intersil Semiconductor, Siliconix, Teledyne Semiconductor, and others.

A new development in the communications industry is the single-chip codec. An encoder contains all the elements needed to convert analog voice into digital data for transmission over a telephone while the decoder performs the reverse function—that is, performs the D/A conversion needed to restore the sound of the voice transmission at the other end. Previously these conversions were achieved by high-speed A/D (analog-to digital) and D/A converters.

Recently the advance of LSI technology, especially in the CMOS area, has made it economically feasible and technically possible to encode each channel separately and multiplex the resulting digital signals on a single chip.

CMOS codecs are supplied by Mostek (MK5151, MK5116), Motorola (MC14406, MC14407), AMI (3501, 3502 two-chip circuit), National Semiconductor Corp. (58100, LF3700—a CMOS/BIFET bipolar 2-chip circuit), and Siliconix (DF341, DF342 two-chip circuit).

Motorola's MC14400 series CMOS communications circuits are medium- and large-scale circuits, each a subsystem in itself, with some of them optimized toward a particular application. They are completely compatible with the rest of the CMOS family, which gives the system designer an enormous choice of peripheral logic functions to use in conjunction with the subsystem chips.

Offered in the 14400 series are such functional blocks as a 2-of-8 tone encoder, a bit-rate generator and multiplier, a quad-precision time-driver, synthesizers (145155 and 145156), and a contact bounce eliminator. All are monolithic implementations of complex functions that previously required tens of packages of TTL.

Sharp Corporation's CMOS speech synthesizer for use in consumer products contains all necessary components except a 6-bit DAC. Its 4-kilobyte ROM has sufficient capacity for 13,622 seconds of speech (in the form of a male voice, a female voice, or a tone) and can be expanded by the addition of up to 1 megabyte of external ROM.

Other companies actively building CMOS telecommunications circuits such as tone encoders, tone generators, frequency receivers and synthesizers, cross-point switches, relay drivers, contact bounce eliminators, telephone dialers, and the like include Intersil, American Microsystems, Inc., Mostek, RCA, National Semiconductor, and Nippon Electric.

Memory/LSI Products

With an increased focus on energy conservation, one can expect the number of commercially available memories and microprocessors to increase. RCA pioneered CMOS microprocessors with the 1802 single-chip 8-b bulk CMOS processor with an instruction set that fits a wide variety of I/O (input/output) requirements. However, this family was not widely accepted because of weaknesses in both hardware (availability of second sources and peripheral circuits) and software (versatility, sophistication). Thus RCA developed the 1804 single-chip CMOS/SOS register-oriented 8-b microcomputer to provide easy I/O access.

Harris Semiconductor's HM6100 and Intersil's IM6100 (12-b CMOS microprocessors) are based on Digital Equipment Corporation's PDP-1/E minicomputer's instruction language. This allows a designer to take advantage of Digital Equipment Corp.'s existing minicomputer software, and not only frees Intersil from supplying an expensive software package for its system, but also means that a designer can emulate existing systems in CMOS and take advantage of the low-power features of that technology without a major investment in software.

Both Harris Semiconductor and Intersil offer complete microcomputer support chips for these processors—RAMs (random-access memories), ROMs (read-only memories), UARTs, and other specialized interface circuits.

In addition Intersil has developed CMOS pin-compatible versions of

Intel's industry standard 8049, 8741, 8748, and 8035 NMOS 8-b microprocessors. These devices do not sacrifice performance by using CMOS technology. In many cases the CMOS products outperform the NMOS ones. For example, Mitel's CMOS version of Motorola's 6802—the MP46802—runs off of a 5-MHz clock, whereas the original NMOS part has a maximum clock of 2 MHz. And though power dissipation of CMOS increases with clock speed, the MP46802 dissipates only 15 mW at 1 MHz, whereas the NMOS part dissipates more than 20 times that figure.

Other CMOS microprocessors have been introduced by Motorola (the MC146805 CMOS version of the MC6805), by Texas Instruments (the 4-b TMS1000C and the 8-b TMS1200C), and by National Semiconductor (the 8-b NSC 800). However, the prime mover for CMOS LSI has been Hewlett-Packard's decision to develop a 16-b microcomputer chip set using CMOS/SOS technology in-house and to use the resulting devices for their microcomputer. This singular decision and its implications, coupled with the dire need for energy conservation, has caused a tremendous resurgence of interest in CMOS technology by most of the major semiconductor manufacturers.

Complementing these microprocessor chips are a host of memories and I/O and interface circuits that promise complete families of CMOS computing elements. Today's 4096-b CMOS memories built on both bulk and sapphire film substrates are already capable of 35 to 300 ns of access time at practically zero quiescent power dissipation—and thus are devices that virtually solve the volatile memory problem. 1024-, 4096-, and 8192-b ROMs and EPROMs (erasable programmable read-only memories) are available to supply the storage needed in microcomputer-based designs.

4096- and 16,384-b RAMs, built on bulk silicon using a silicon-gate (Si Gate) process and developed by Intersil, Harris, Hitachi, and Intel, and as well as those built on SOS and developed by RCA, Hughes Aircraft, and Hewlett-Packard, have electrical characteristics that rival today's static NMOS RAMs. They offer nanowatt standby power operation, access times under 100 ns, and single 5-V static operation. Along with the 64-, 256-, and 512-b CMOS memories already available, they make up a full complement of micropower memories for use in low-power applications like point-of-sale terminals, remote sensing, and communications systems.

What makes these CMOS RAMs so attractive is their usefulness in a system that must retain its memory when power is turned off. With such low standby power, they need only one small battery on the memory board to emulate a nonvolatile memory system.

Two of the popular CMOS memories are Intel's 5101 and Intersil's 6508 1024-b RAMs. The 5101 is organized in the 256- by 4-b format popular for microprocessor systems and the 6508 is a 1024- by 1-b device. Both are silicon-gate CMOS devices that, being static, eliminate clocks, interface circuits, and special power supplies while minimizing package size. Gaining in popularity are Harris' 6504 (4-kb × 1-b) and 6514 (1-kb × 4-b) RAMs and Hitachi's HM6147, which are being widely second-sourced.

Matsushita Electric Industrial Company has developed an 80-ns (t_{AA}), 300-mW (15-mW standby), 8-kb × 8-b CMOS RAM that is packaged in a 28-pin DIP. The chip contains 271,400 MOSFETs and 131,000 load resistors for a total of 402,500 individual elements.

A detailed discussion of these and other CMOS memories is presented in Chap. 4, and Chap. 5 is concerned with CMOS microprocessors.

CHAPTER TWO
CMOS Logic Performance Characteristics and Circuit Applications

In order to provide you with a book that covers the primary impact of CMOS ICs (large-scale integrated circuits) and yet remain affordable in price, it was necessary to limit the discussion of certain topics. As such, the topic of CMOS logic circuits, which has been the mainstay of CMOS technology prior to mid-1977, is discussed only from the viewpoint of demonstrating the performance characteristics obtained from CMOS technology. It has been projected that the real application for CMOS will be VLSI (very-large-scale integration), as by 1982, it is estimated, 20 percent of all VLSI circuits will be fabricated using CMOS technology because of the power dissipation of large arrays.

Designed in the mid-1960s, the 4000A series CMOS circuits have by now become widely accepted as a viable production technology rather than as merely a research and development technology. There are three reasons for this:

1. Number of functions available
2. Usage history (reliability data)
3. Development of several second sources

As a comparison, the total number of all families of 54/74 series TTL circuits (more than 300 different types), along with their flexibility and ease of use, still overwhelms the available number of 4000 series CMOS circuits, of which there are more than 110 different types. However, the number of functions available using CMOS technology is rapidly increasing and, due to design and processing advances, CMOS possesses tremendous potential for LSI circuit implementation.

Compared with most bipolar devices, MOS circuits consume less power. The reason for this is the utilization of MOS transistors as load elements. This not only results in space savings, but yields higher values of load resistance; hence, lower operating current is required. In the on mode, current flow through the load causes dc power dissipation only on the order of 0.5 to 2.0 mW.

Even this small amount of dc power dissipation can be reduced significantly through CMOS circuitry. Indeed, by combining PMOS (P-channel metal-oxide semiconductor) and NMOS on the same chip in a complementary-symmetry configuration, dc power dissipation can be reduced to virtually zero. This type of micropower operation is particularly desirable for battery-operated equipment where unattended operation for long periods is required.

CMOS circuits can be designed for the low-power extreme if frequency is unimportant—for example, 100 kHz operation can be achieved at only a few microwatts of power drain. On the other hand, CMOS circuits can be designed to operate at speeds in excess of 10 MHz, at the expense of greater power dissipation due to the increased rate of signal transitions. In the standby state, power dissipation is virtually zero. Previously, the advantage of CMOS circuits was offset somewhat by their increased processing complexity. However, the advent of scaled NMOS has resulted in comparable levels of processing complexity between CMOS and NMOS, negating the prior disadvantage of CMOS circuits. Thus the high speed and low power drain of new CMOS processes and circuits have given them the edge over NMOS.

CMOS can be combined with other technologies and processes to provide more refined electrical characteristics. Ion implantation can be used in conjunction with the basic CMOS technology to provide the following:

1. Low-threshold voltage shifts
2. Depletion load pull-ups that are insensitive to supply voltage variations
3. High-value implanted resistors (10 kΩ/□) and greater tolerance resistors

Additionally, use of the SOS process in conjunction with CMOS can dramatically increase operating speed from 45 to 5 ns per gate.

In the following discussion, basic CMOS logic circuits are used to demonstrate the salient features of all CMOS devices in a simplistic manner.

THE LOGIC—THE INVERTER*

The basic building blocks for all CMOS logic functions are N- and P-channel MOS transistors connected to form an inverter (Fig. 2-1). These devices function as voltage-controlled switches and are capable of bilateral current flow between source and drain.

The heavily doped source and drain diffusions are separated by a narrow gap over which lies a thin gate oxide and aluminum metallization. In order for the transistors to conduct current from source to drain, one must apply voltage in excess of the threshold from gate to source. The threshold is the voltage that must be exceeded in order to invert the silicon between the source and drain to form a conducting channel. Increasing the gate-to-source bias beyond the threshold voltage further inverts the material under the gate electrode, increasing conductivity.

The MOS characteristic is represented by two curve segments (regions of operation), an inverted parabola for nonsaturated operations $(V_{GS} - V_T) \geq V_{DS}$, and a straight line for saturated operation $(V_{GS} - V_T) \leq V_{DS}$ (where V_{GS} stands for gate-to-source voltage, V_T for threshold voltage, and V_{DS} for drain-to-source voltage).

In the nonsaturated region, the characteristics of the MOSFET are similar to a resistor; the impedance of the channel is approximated by the slope of the curves. The current in this region is given by the equation

$$I_D = 2K(V_{GS} - V_T)V_{DS} - KV_{DS}^2 \tag{2-1}$$

where K is a constant dependent upon processing parameters and the channel geometry and I_D is the drain current.

In the saturated region, the MOSFET behaves similarly to a current source, as illustrated by the constant drain current independent of the drain-to-source voltage. The currents in the saturated region are given by the equation

$$I_D = K(V_{GS} - V_T)^2 \tag{2-2}$$

---

* Portions of this section are taken with permission from J. McCullen, "Motorola Complementary MOS ICs," Motorola Application Note AN-538A, Motorola, Phoenix, Arizona, 1972.

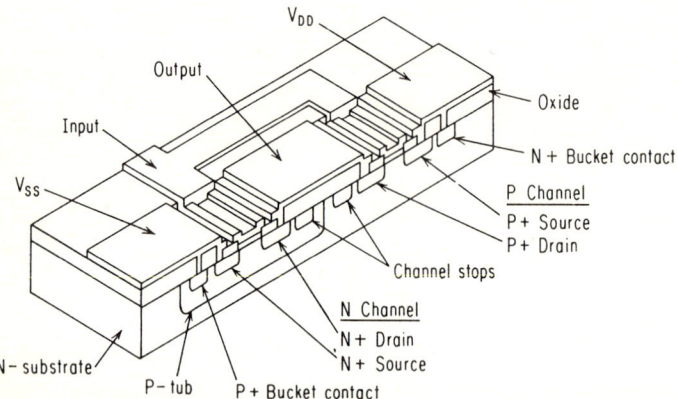

Fig. 2-1 The P- and N-Channel Devices Are Connected to Form an Inverter. *(Courtesy Motorola Semiconductor Products, Inc. Used with permission.)*

The maximum drain current I_D is almost proportional to the square of the gate-to-source voltage, V_{GS}. And since V_{GS} is limited to the power supply voltage ($V_{DD} - V_{SS}$), the drive capability of CMOS is proportional to the square of the power supply voltage.

Figure 2-2 depicts a schematic diagram of the inverter cross section of Fig. 2-1. In this circuit, when the input signal is low (ground), the N-channel transistor, Q_2, is off and the P-channel device is on. The output, therefore, is "shorted" to the positive supply voltage. If the load resistance is a MOS gate, which has a very high input resistance, virtually no current is drawn from the supply and the output voltage approaches $+V_{DD}$. When the input signal goes high, Q_2 is turned on and Q_1 is turned off. Again, no dc current can flow from the supply, but the output is drawn to ground through the low ON resistance of Q_2. The output voltage varies from almost $+V_{DD}$ to zero.

During the on portion of the input signal, the CMOS inverter draws no dc current at all. The only power dissipation occurs during the transi-

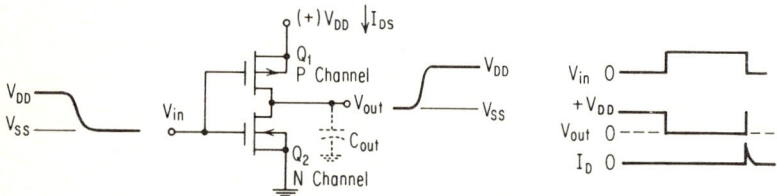

Fig. 2-2 CMOS Inverter Schematic Diagram. *(Courtesy Motorola Semiconductor Products, Inc. Used with permission.)*

tions of the input signal. In either logic state, one MOS transistor is on while the other is off. Because one transistor is always turned off, the quiescent power consumption of the CMOS inverter is extremely low; more precisely, it is equal to the product of the supply voltage and the leakage current.

Looking at the characteristic curves of MOS transistors, one can see how rise and fall times, propagation delays, and power dissipation will vary with power supply voltage and capacitive loading. Figure 2-3 shows the characteristic curves of N- and P-channel enhancement mode transistors.

There are a number of important observations to be made from these curves. Refer to the curve of $V_{GS} = 15$ V for the N-channel transistor. Note that for a constant drive voltage, V_{GS}, the transistor behaves like a current source for V_{DS}s greater than $V_{GS} - V_T$. For V_{DS}s below $V_{GS} - V_T$, the transistor behaves essentially like a resistor. Note also that, for lower V_{GS}s, there are similar curves except that the magnitude of the I_{DS}s (drain-to-source currents) are significantly smaller and that, in fact, I_{DS} increases approximately as the square of increasing V_{GS}. The P-channel transistor exhibits essentially identical, but complemented, characteristics.

The output drain characteristics also vary as a function of temperature, as shown in Fig. 2-4. The typical characteristics for a standard CMOS output N-channel device, operating with a supply of 10 V, is shown in Figure 2-4a. The complementary P-channel characteristics are shown in Fig. 2-4b. For device temperatures above 25°C, the decrease in drain current can be approximated by a negative temperature coefficient of approximately $-0.3\%/°C$.

Figure 2-5 shows the voltage transfer curve for the basic inverter at a power supply voltage of 10 V. When V_{in} is between 8 and 10 V, the potential difference between the gate and substrate of Q_1 is less than the threshold voltage and Q_1 is off. The potential difference between the gate

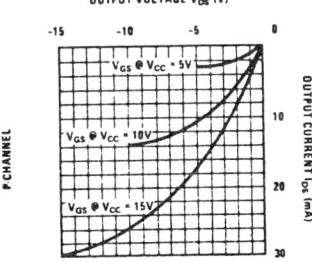

Fig. 2-3 Logical 1 Output Voltage versus Source Current.

22 CMOS Logic Performance Characteristics and Circuit Applications

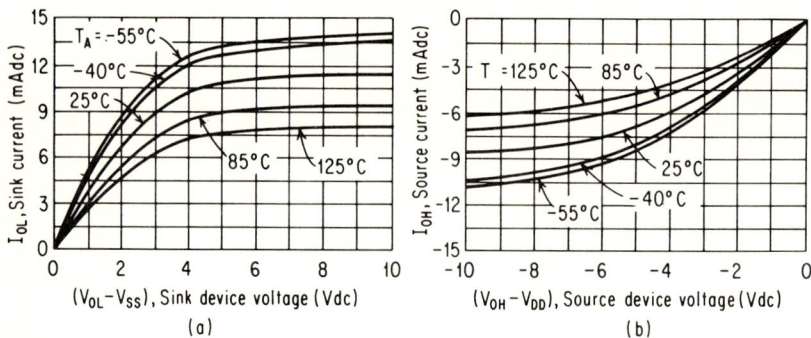

Fig. 2-4 Typical Output Characteristics for V_{DD} = 10 V as a Function of Temperature. (a) N Channel, (b) P Channel.

of Q_2 and its substrate is 8 to 10 V; thus Q_2 is on. At the output, the inverter appears as 500-Ω to 1-kΩ resistance to ground (V_{SS}), and extremely high ($\geq 10^9$-Ω) resistance to +10 V (V_{DD}).

The current drain from the 10-V supply is less than 15 nA, which results in very low power dissipation. When V_{in} is between 0 and 2 V, Q_2 is off and Q_1 is on. In this case, the output appears as a low resistance to +10 V and a high resistance to ground. As V_{in} makes the transition from 2 to 8 V, both Q_1 and Q_2 are in an on condition, resulting in current flow from V_{DD} to ground. Figure 2-6 illustrates this current as a function of V_{in}. This graph was generated by applying various dc voltages at V_{in} and measuring I_D. In a situation involving the application of a switching

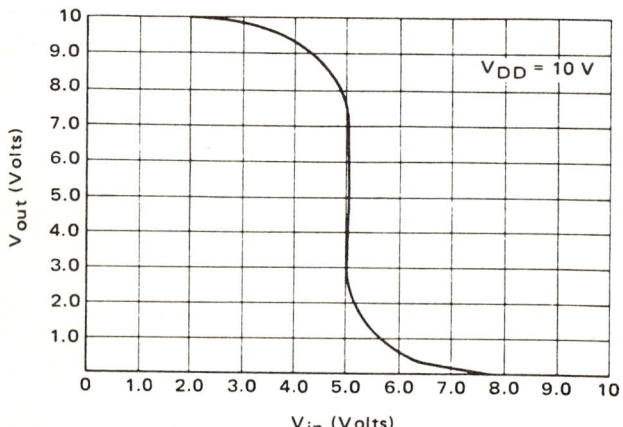

Fig. 2-5 CMOS Inverter Transfer Curve. *(Courtesy Motorola Semiconductor Products, Inc. Used with permission.)*

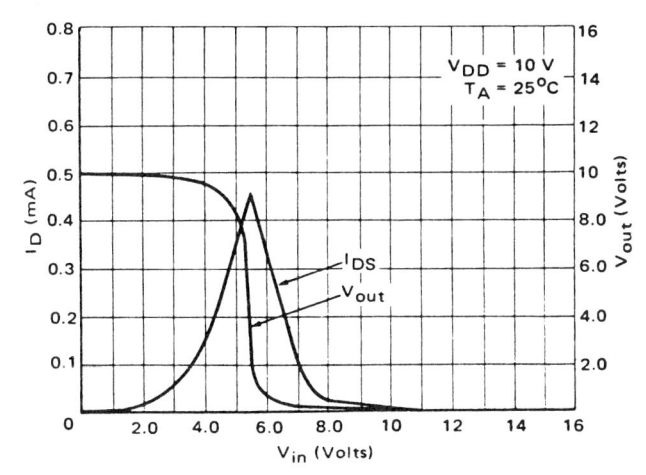

Fig. 2-6 CMOS Inverter—Power Supply Current versus Input Voltage. *(Courtesy Motorola Semiconductor Products, Inc. Used with permission.)*

waveform to V_{in}, the current I_D does not flow through both Q_1 and Q_2. The reason for this may be explained as follows:

Assume Q_1 is off and Q_2 is on. As V_{in} switches from V_{DD} to ground, there is a period of time when both Q_1 and Q_2 are on. However, because of stray and load capacitance on the output, the output voltage is still nearly 0 V when Q_1 and Q_2 are both on. Thus, little current flows through Q_2 to ground. Because of this fact, the power dissipation of CMOS circuits, under switching conditions, is due almost entirely to capacitive loading.

Because the dynamic power dissipation results from capacitive loading, it is also a function of the frequency at which the capacitance is charged and discharged. Figure 2-7 illustrates this relationship for a basic inverter. It can be seen that the power dissipation is linear with frequency.

Leakage current plays an important role in quiescent power supply and three-state loading considerations. When designing a circuit which must operate over wide temperature ranges, the effect of temperature on leakage must be considered in worst-case design.

Leakage is due primarily to internal reverse-biased PN junction leakage. As such, the leakage increases exponentially with increasing temperature, as indicated by the formula

$$I_L = (T_I) = I_L(T_0)e^{\Delta T/K} \tag{2-3}$$

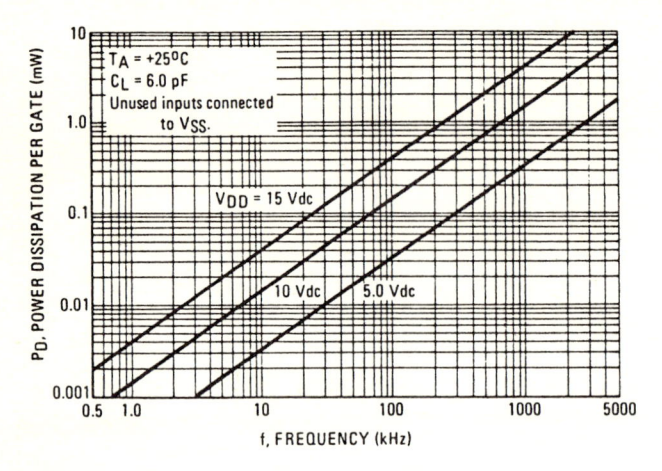

24 CMOS Logic Performance Characteristics and Circuit Applications

Fig. 2-7 Power Dissipation of CMOS Inverter. *(Courtesy Motorola Semiconductor Products, Inc. Used with permission.)*

where $I_L(T_0)$ is the leakage current measured at temperature T_0, K is the constant of the rate of increase, and the temperature change is $\Delta T = T_1 - T_0$.

For silicon junctions, I_L doubles approximately every 10°C, thus making $K_{Si} = 14°C$. A CMOS device that has 5×10^{-9} A leakage at room temperature (25°C) may be expected to have 5×10^{-6} A leakage at 125°C.

If one tries to drive a capacitive load with the basic inverter of Fig. 2-2, the initial voltage change across the load will be ramplike due to the current source characteristic followed by a rounding off due to the resistive characteristic dominating as V_{DS} approaches zero. Referring to the basic CMOS inverter in Fig. 2-2, as V_{DS} approaches zero, V_{out} will approach V_{DD} or ground, depending on whether the P- or the N-channel transistor is conducting.

Now if V_{DD} is increased, and therefore V_{GS}, the inverter must drive the capacitor through a larger voltage swing. However, for this same voltage increase, the drive capability (I_{DS}) has increased roughly as the square of V_{GS} and, therefore, the rise times and the propagation delays through the inverter have decreased.

Thus, for a given design, and therefore for a fixed capacitive load, increasing the power supply voltage will increase the speed of the system at the expense of power dissipation. Figure 2-8 shows the effect of capacitive loading and power supply voltage on propagation delays.

The input capacitance is dependent upon the input voltage, as shown in Fig. 2-9. The input capacitance increases when the output is in the

The Logic—The Inverter   25

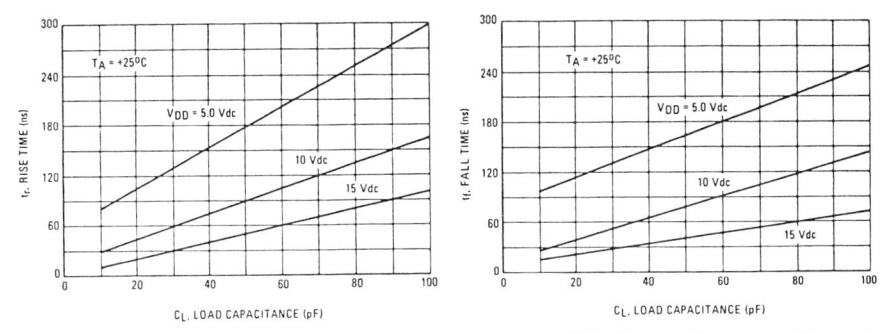

Fig. 2-8 Typical Delay Characteristics of two Input NOR Gates. *(Courtesy Motorola Semiconductor Products, Inc. Used with permission.)*

transition region. This increase is due to internal feedback. Under normal operating conditions, this effect is not present because the input transition is complete before the feedback takes place due to the propagation delay of the gate.

An inherent advantage of the generic complementary MOS process is the tendency of the N- and P-channel thresholds to "track" together over wide temperature variations in such a manner that the input threshold to a CMOS device remains quite constant.

Figure 2-10 shows the typical transfer characteristic curves of a CMOS gate. The high input impedance of the gate causes the input and output signals to swing completely from 0 V (logic 0) to V_{DD} (logic 1) when sufficient settling time is allowed. The switching point is shown to be typically 45 to 50 percent of the magnitude of the power supply voltage, and it varies directly with that voltage over the entire range.

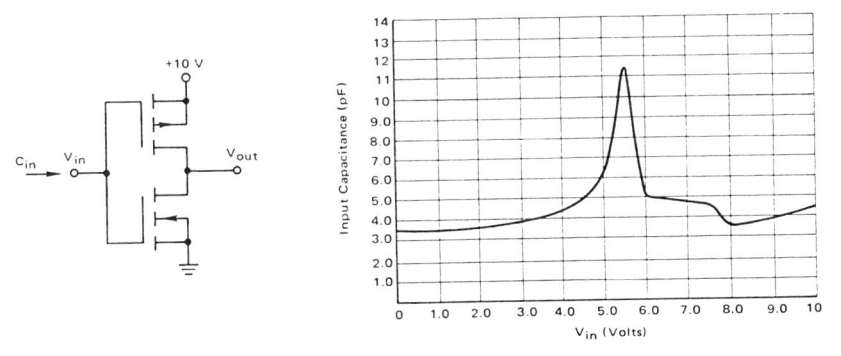

Fig. 2-9 Input Capacitance versus Input Voltage. *(Courtesy Motorola Semiconductor Products, Inc. Used with permission.)*

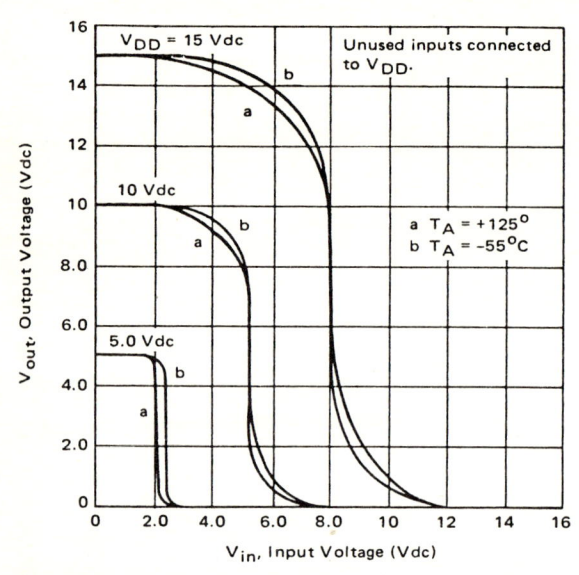

26 CMOS Logic Performance Characteristics and Circuit Applications

Fig. 2-10 Typical Voltage Transfer Characteristics versus Temperature. *(Courtesy Motorola Semiconductor Products, Inc. Used with permission.)*

The transfer characteristics for a TTL gate, a buffered CMOS gate, and an unbuffered single-stage CMOS gate, as illustrated in Fig. 2-11, show the high noise immunity of CMOS devices.

The exceptionally high ac and dc noise immunity of CMOS ICs is attributable to the relatively low output impedances (ca. 600 Ω), moderate-speed operation, steep transfer characteristics (above), and output voltages within 10 mW of V_{DD} when driving other CMOS circuits. The high values of noise immunity are achieved with power consumption in the microwatt range, as compared with much higher (milliwatt) consumptions and lower noise immunities (ca. 1 V) for saturated bipolar logic. Because of their high cross-talk noise immunity, CMOS circuits are useful in applications requiring long lines or in circuits with closely coupled wiring. Moreover, CMOS circuits are not susceptible to ground-line noise or to noise produced by improper ground returns; therefore, it is not necessary to use large ground planes. Similarly, the high-power-supply noise immunity eliminates the need for complex filtering circuits.

However, a primary source of misunderstanding concerning the voltage noise immunity specifications lies in the way that this parameter is defined. Historically, CMOS manufacturers have followed the unity-gain, worst-case voltage noise immunity definitions, as depicted in Figure

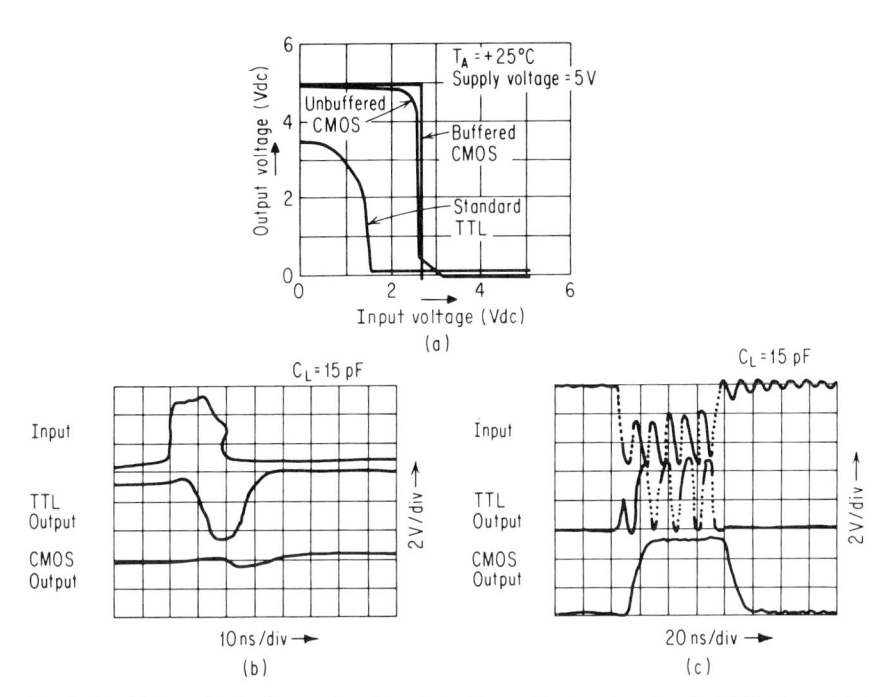

Fig. 2-11 Voltage Noise Immunity. *(Reprinted from Electronics, Jan. 9, 1975. Copyright © 1975 by McGraw-Hill Inc. Used with permission.)*

2-12a. While CMOS devices typically reject voltage noise pulses of 45 percent of the supply voltage, the standard guaranteed value across the industry is 30 percent.

Because CMOS logic gates typically change state near 50 percent of the supply voltage, and because of the steep transfer characteristics exhibited during transitions, exceptionally high input voltages are required

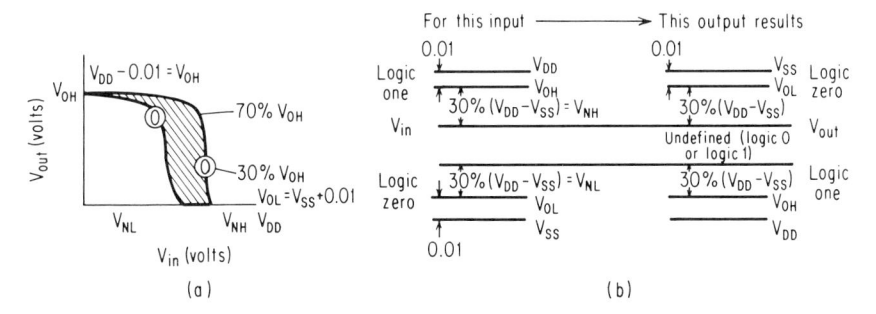

Fig. 2-12 Noise-Voltage Margin Definitions Are Given in (a) and Guaranteed Input/Output Relationships in (b). *(From EDN, Jan. 5, 1973. Reprinted with permission.)*

to significantly change or falsely switch the output logic state. Typically, then, the input to a CMOS logic gate operating at a dc supply voltage of 10 V may change by 4.5 V before the output begins to change state. In addition, ac noise immunity is extremely high because of the high static or dc noise immunity and the moderate-speed operation of standard CMOS circuits. The high ac noise immunity also implies high immunity to cross-talk noise.

Figures 2-13 and 2-14 illustrate CMOS logic gate noise immunity. These definitions assure that the logic level at the output of the driving device is recognized as the same logic level at the input to the load device. The dc noise level at the junction is equal to or less than the difference in maximum magnitudes of the output and input logic levels.

An analysis of noise immunity involves consideration of immunity to both ac and dc noise. Whereas ac noise is usually thought to be made up of those noise spikes with pulse widths shorter than the propagation delay of a logic gate, dc noise spikes are considered to have pulse widths longer than the propagation delay of one gate. Since ac noise immunity, which varies in direct proportion to dc noise immunity, is largely a function of the propagation delays and output transition times of logic gates, it is a function of input and output capacitances.

Table 2-1 illustrates typical noise immunity voltages and output impedances for TTL and CMOS logic circuits. With these parameters, the noise current sufficient to cause false operation can be determined. The noise current from external sources required to switch a CMOS device is actually quite low, making TTL superior in this respect. In regard to noise immunity, then, CMOS is unbeatable from a voltage and ac noise standpoint, but susceptible to noise from outside sources.

In addition to illustrating the high noise immunity of CMOS devices, Fig. 2-10 also reveals the negligible change in threshold voltage as tem-

TABLE 2-1 Noise Immunity Comparison: CMOS versus TTL

Parameter	Standard TTL	Low-Power TTL	CMOS ($V_{DD} = 5$ V)	CMOS ($V_{DD} = 10$ V)
Typical Output Impedance	100	10	400	200
Typical Noise Immunity	1.5 V	1 V	2.5 V	5 V
Noise Current Required to Induce Noise	15 mA	100 mA	6.25 mA	25 mA

SOURCE: *Electronics*, Jan. 9, 1975. Copyright © 1975 by McGraw-Hill, Inc. Reprinted with permission.

The Logic—The Inverter   29

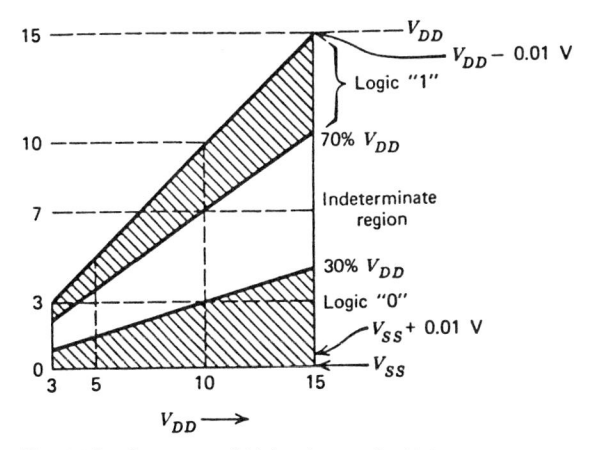

Fig. 2-13 Guaranteed Noise-Immunity Values.

perature varies from −55°C to +125°C, typically less than 5 percent. By comparison, a bipolar threshold may vary as much as 40 percent, as shown in Fig. 2-15.

Threshold variation over temperature becomes an important factor when determining worst-case noise margins, but it is of greater concern in "one-shot" multivibrators. The threshold levels in these types of circuits directly affect frequency, duty cycle, and time-out. Therefore, the

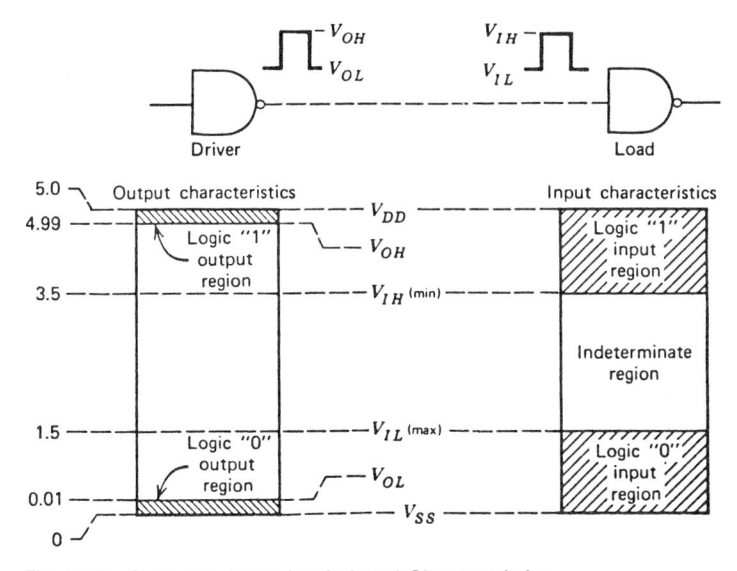

Fig. 2-14 Output-to-Input Logic-Level Characteristics.

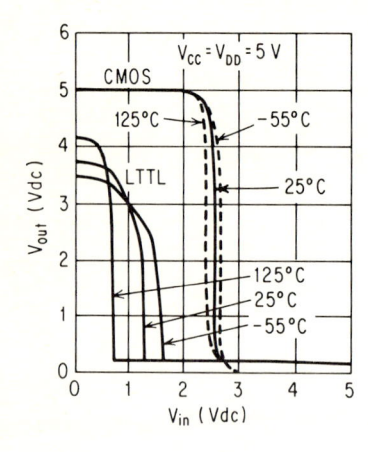

Fig. 2-15 Transfer Characteristics of the CMOS Inverter Show Much Less Dependence on Temperature than Does TTL.

stability of MOS devices over temperature is a definite advantage in these special applications.

Because of the ideal nature of these switching characteristics, CMOS devices operate reliably over a much wider range of voltage than other forms of logic circuits.

Since the output is equally isolated from both V_{DD} and V_{SS} terminals, CMOS can operate with negative as well as positive supplies. The only requirement is that V_{DD} be more positive than V_{SS}.

Variations in the channel resistance of the P- and N-type MOSFETs in a CMOS circuit affect several important device characteristics—namely, current sinking and sourcing capability (I_{OL}, I_{OH}), switching through-current (I_{TC}), propagation delay (t_{PHL}, t_{PLH}), and output voltage rise and fall times (t_r, t_f).

The channel resistance of an MOS device is inversely proportional to the surface mobility of the majority carriers (holes in a P-type device and electrons in an N-type device). Since the mobility is a function of temperature, it is not surprising to find that channel resistance and thus the current and switching time parameters change with variations in temperature.

The normalized graph of the current and switching parameters versus ambient temperature (Fig. 2-16) should prove a useful tool to the designer in predicting the typical performance to be expected of a CMOS circuit at different temperatures.

During switching, the node capacitances, within a given circuit, and the load capacitances external to the circuit, are charged and discharged through the P- or N-type device conducting channel. As the magnitude

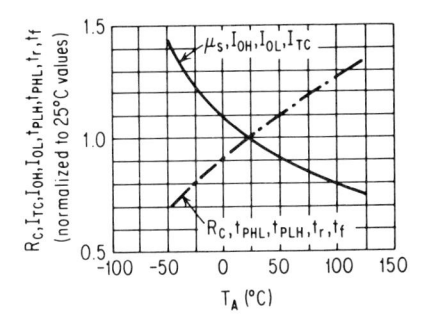

Fig. 2-16 Normalized Plot of Typical Sink Current (I_{OL}), Source Current (I_{OH}), Through-Current (I_{TC}), Surface Mobility (μs), Channel Resistance (R_C), Propagation Delay Times (t_{PHL}, t_{PLH}), and Rise and Fall Times (t_r, t_f) versus Ambient Temperature (T_A).

of V_{DD} increases, the impedance of the conducting channel decreases accordingly. This lower impedance results in a shorter RC (resistance-capacitance) time constant (this nonlinear property of MOS devices is due to current saturation at large values of drain-to-source voltage). The result is that the maximum switching frequency of a CMOS device increases with increasing supply voltage. See Fig. 2-17a.

Figure 2-17b shows curves of propagation delay as a function of supply voltage for a typical gate device. However, the tradeoff for low supply voltage (i.e., lower output current to drive a load) is lower speed of operation.

CMOS propagation delay is usually specified at drain-voltage levels of 5 and 10 V with a 20 ns input transition, a 15 pF output-load capacitance, and an ambient temperature of 25°C; each CMOS gate input is typically specified at 5 pF (Fig. 2-17).

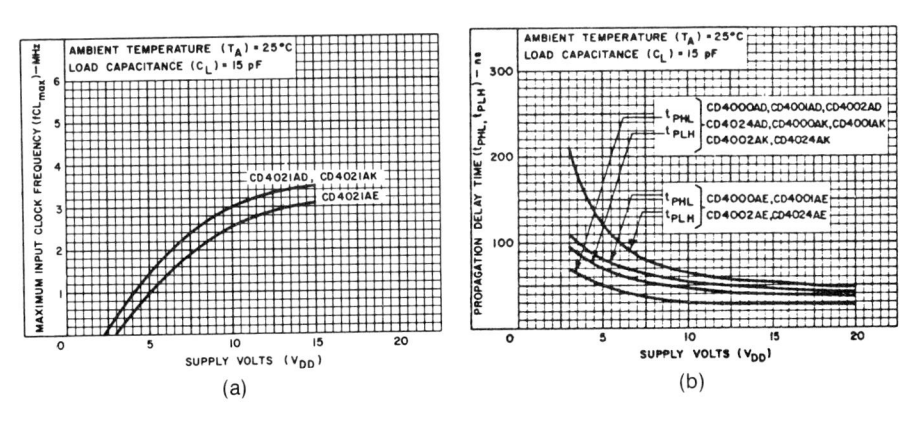

Fig. 2-17 Operating Frequency and Propagation Delay as a Function of Power Supply Voltage. (a) Maximum Guaranteed Operating Frequency as a Function of Power Supply Voltage, (b) Propagation Delay as a Function of Power Supply Voltage for the Basic Gate. (*Reprinted from RCA Application Note ICAN6576, RCA Solid-State Division, Somerville, NJ, 19XX. Used with permission.*)

32 CMOS Logic Performance Characteristics and Circuit Applications

A load of 15 pF corresponds to a fan-out of 1 or 2—hardly a worst-case condition. Furthermore, as Fig. 2-18a illustrates, CMOS is significantly more sensitive to capacitive loading than TTL: the larger the load capacitance, the longer the propagation delay and the slower the system.

To obtain a true estimate of worst-case propagation delays, the designer must first determine the typical capacitance per fan-out for the system. Then the correct propagation delays for that factor may be found from the curves of delay time versus the capacitance, supplied with every CMOS device. When the design is complete, the capacitance estimates should be double-checked and final worst-case propagation delays calculated. (Fairchild specifies propagation delay at the more realistic load-capacitance value of 50 pF as well as at 15 pF.)

No matter who the manufacturer, CMOS propagation delay is consistently specified for an input transition of 20 ns. Naturally, in a real system, transition times are almost always greater, causing propagation delays to be different than those indicated on the data sheet. There are several possible reasons for this difference. Figure 2-18b, for example, shows how the delay of a CMOS gate depends on its driving source. The delay of a gate driven by a pulse generator is smaller than that of a gate driven by another similar gate.

Temperature and power supply voltage also influence propagation delay. Curves for determining the effects of these two factors are included with each CMOS device. Generally, propagation delay increases

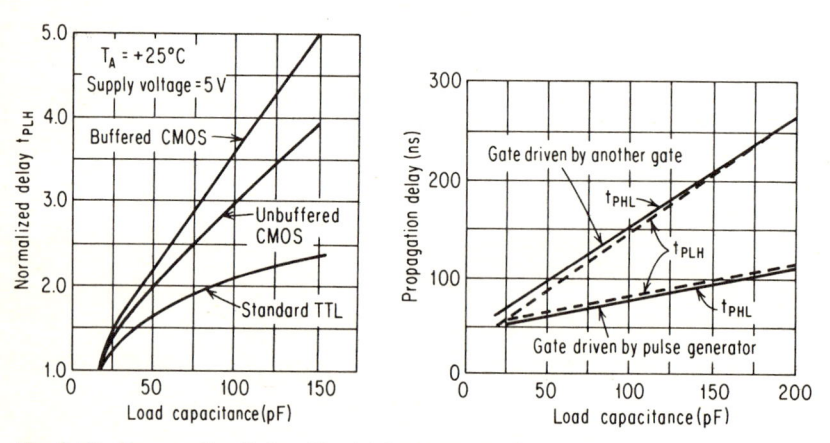

Fig. 2-18 Propagation Delay. Plot (a) Shows that CMOS Delay Time Is More Sensitive to Load Capacitance than TTL. Besides Load Capacitance, Delay Time Also Depends on the Driving Source (b). *(Reprinted from Electronics, Jan. 9, 1975. Copyright © 1975 by McGraw-Hill, Inc. Used with permission.)*

with increasing temperature and/or decreasing power supply voltage, as was shown in Figures 2-8 and 2-17.

To create logic gates, the CMOS P and N transistors of the basic inverter are combined in series-parallel arrangements. CMOS gates are logically just the opposite of DTL (diode-transistor logic) and TTL in one important respect. While DTL has large, expandable fan-in and limited fan-out, CMOS has limited fan-in and almost unlimited fan-out. The fan-in to a CMOS gate is limited because a complete PN pair must be added for each input. This is why most of the CD4000A gates have no more than four inputs and these cannot be expanded by diodes as in DTL. The fan-out, on the other hand, is limited only by the effect of added wiring capacitance on operating speed. From the dc standpoint, fan-outs much larger than any designer is ever likely to want—in the hundreds—are possible.

SET/RESET Flip-Flops

Two NOR gates, when connected as in Fig. 2-19, form a SET/RESET flip-flop. When the SET and RESET inputs are low, one amplifier output is low and the other is high. If the SET input is raised to a higher level, the associated N-type unit is turned on, so that the output of the SET stage goes high and becomes logic 1 or $\bar{Q}$. Under these conditions, the flip-flop is said to be in the SET state. Raising the RESET input level causes the other output to go high, and places the flip-flop in the RESET

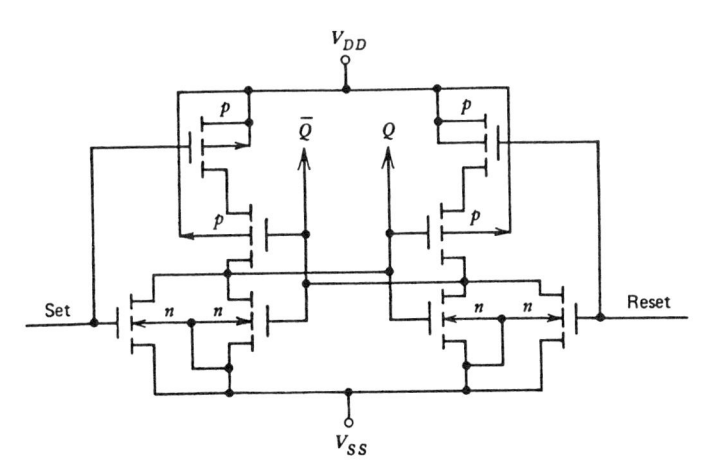

Fig. 2-19 A SET/RESET Flip-Flop in which All P-Unit Substrates Are Connected to V_{DD} and All N-Unit Substrates Are Connected to V_{SS}.

state (Q represents the low-output state). Thus, the circuit of Fig. 2-19 represents a static flip-flop with SET and RESET capability.

THE TRANSMISSION GATE

A second important building block for the construction of CMOS integrated circuits is the transmission gate shown in Fig. 2-20. The CMOS transmission gate is an SPST (single-pole, single-throw) switch formed when the P and N transistors are arranged in-parallel to the power supply.

This switch expands the versatility of CMOS circuits in both digital and linear applications. The perfect transmission gate or switch may be characterized as having zero forward and reverse resistance when closed and infinite resistance when open (i.e., it has an infinite on/off impedance ratio). The CMOS transmission gate approaches these ideal conditions.

When the transmission gate is on, a low resistance exists between the input and the output, which allows current flow in either direction through the gate. The voltage on the input line must always be positive with respect to the substrate (V_{SS}) of the N-channel device and negative with respect to the substrate (V_{DD}) of the P-channel device. The gate is on when the gate (G_1) of the P-channel device is at V_{SS} and gate G_2 of the N-channel device is at V_{DD}. When G_2 is at V_{SS} and G_1 is at V_{DD}, the transmission gate is off and a resistance greater than 10^9 Ω exists between input and output.

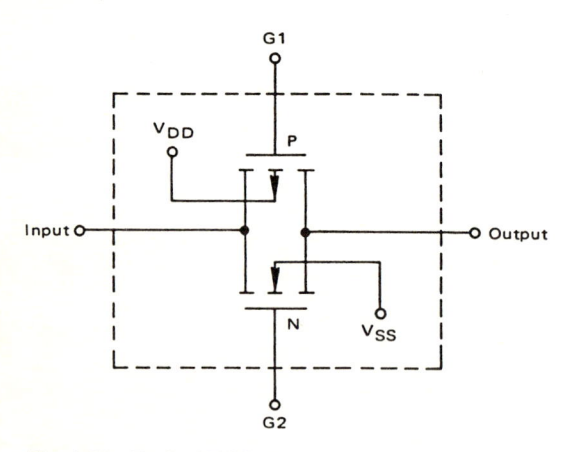

Fig. 2-20 Basic CMOS Transmission Gate. *(Courtesy Motorola Semiconductor Products, Inc. Used with permission.)*

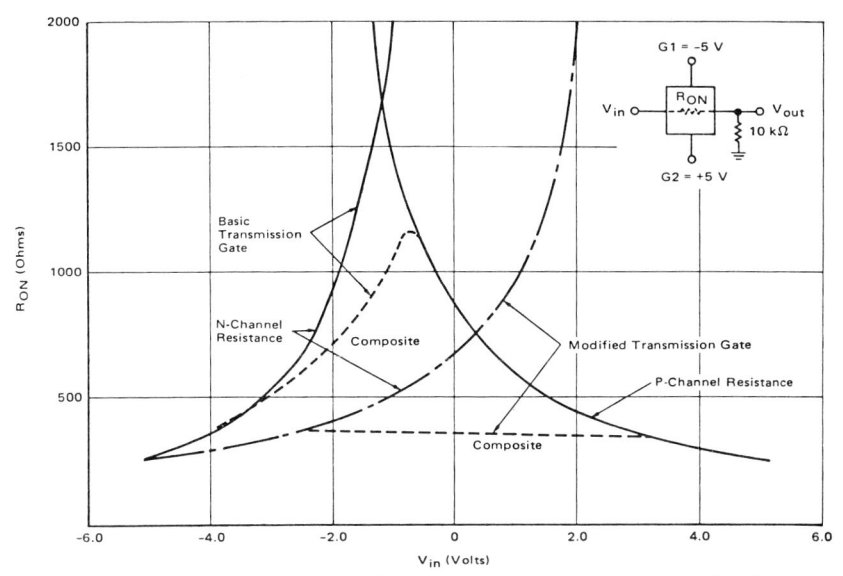

Fig. 2-21 CMOS Transmission Gate R_{on} versus V_{in}. (*Courtesy Motorola Semiconductor Products, Inc. Used with permission.*)

The control signals to the gates of this pair are made opposite to each other rather than of the same polarity as in the case of the basic inverter pair. A built-in inverter pair element is usually used along with the transmission pairs to provide the control signal complementation. Therefore, the pair is either turned on or off together to act like an SPST switch in the signal path.

The advantage of having the opposite-polarity devices in-parallel is that neither signal switch swing will be limited by the gate thresholds and the signal can swing across the whole CMOS supply range. As the sources and drains of these MOS devices are interchangeable, it does not matter which way the signal flows.

The resistance between the input and output of a basic transmission gate in the on condition is dependent upon the voltage applied at the input, the potential difference between the two substrates ($V_{DD} - V_{SS}$), and the load on the output. R_{on} is defined as the input-to-output resistance with a 10-kΩ load resistor connected from the output to ground. Figure 2-21 illustrates an interesting peaking effect which occurs in the R_{on}-versus-V_{in} curves of the basic transmission gate in Fig. 2-20. When V_{in} is at or near V_{DD}, the P-channel device provides the low resistance. The N-channel device is off since the potential difference between G_2 and the drain or source of the N-channel device is less than the threshold voltage. When V_{in} is at or near V_{SS}, the N-channel device is conducting and the P-channel device is off. At voltages between the two extremes, both devices are partially on and the value of R_{on} is due to the parallel resistance of the P- and N-channel devices. The different slopes of the curve on either side of the peak are due to the greater sensitivity of the N-channel resistance to the substrate degeneration (or substrate bias). Thus, the rate of increase in R_{on} with respect to V_{in} is greater for input voltages between V_{SS} and the "peaking voltage" than for input voltages greater than the "peaking voltage."

Figure 2-22 shows a modification to the basic transmission gate: the addition of a third device to control the substrate bias of the N-channel device. The effect of this third device is to delay the turn off of the N-channel device, which results in a much flatter R_{on}-versus-V_{in} curve. This concept is used in the CD4016 quad analog switch.

When used to pass or impede digital signals in CMOS logic systems, two or more of these switches are often used to select which of several signals is to be fed to a gate. Only one transmission switch is turned on at a time, and its low impedance—several hundred ohms—will connect one of the signals. The other off gates, meanwhile, have such high impedances—100 MΩ—that they might as well be open circuits.

The transmission gate is a valuable tool used for accomplishing many CMOS technology MSI and LSI designs. The transmission gate can be combined with a basic inverter circuit to form a single switch, as in Fig. 2-23. Only one control voltage is required because the inverter provides the control voltage necessary for the complementary unit. The circuit

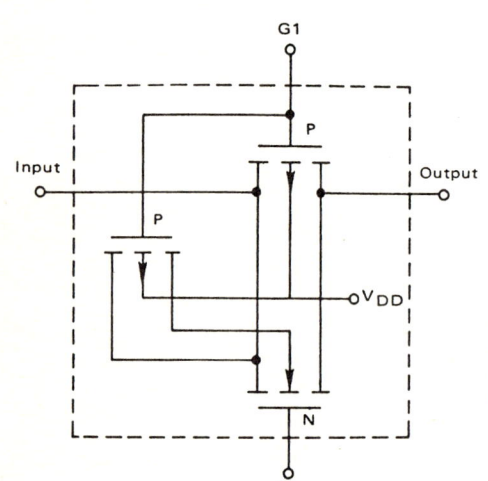

Fig. 2-22 Modified CMOS Transmission Gate. *(Courtesy Motorola Semiconductor Products, Inc. Used with permission.)*

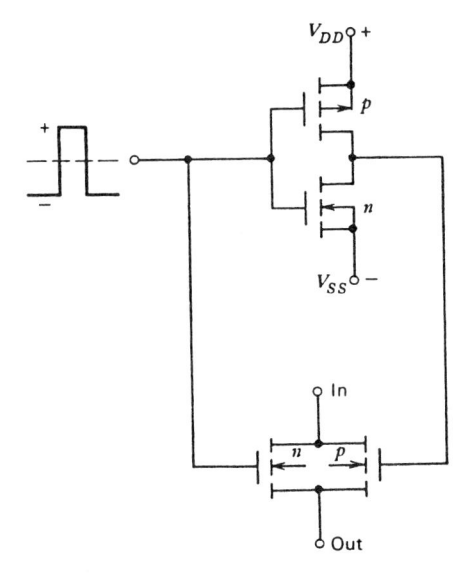

Fig. 2-23 Combination of Transmission Gate and Basic Inverter to Form a Switch.

shown in Fig. 2-23 is useful in a variety of analog and digital multiplexing applications.

D-Type Flip-Flops

An example of how the transmission gate is used inside the CD4000 series of flip-flops is shown in Fig. 2-24. The block diagram shows a master flip-flop formed from two inverters and two transmission gates (shown as switches); the master feeds a slave flip-flop having a similar configuration. When the input signal is at a low level, the TG_1 transmission gates are closed and the TG_2 gates open. This configuration allows the master flip-flop to sample incoming data and the slave to hold the data from the previous input and feed them to the output. When the control is high, the TG_1 transmission gates open and the TG_2 transmission gates close, so that the master holds the data entered and feeds them to the slave. The D flip-flop is static and holds its state indefinitely if no pulses are applied (i.e., it stores the state of the input prior to the last clocked input pulse). Both the "clock" CL and "inverted clock" $\overline{CL}$ (Fig. 2-24c) are required; clock inversion is accomplished by an inverter internal to each D flip-flop.

JK Flip-Flops

The circuit diagram for a JK flip-flop appears in Fig. 2-25, while the truth table for the flip-flop is shown below the diagram. The JK flip-flop is similar in some respects to the D flip-flop, but it has additional circuitry

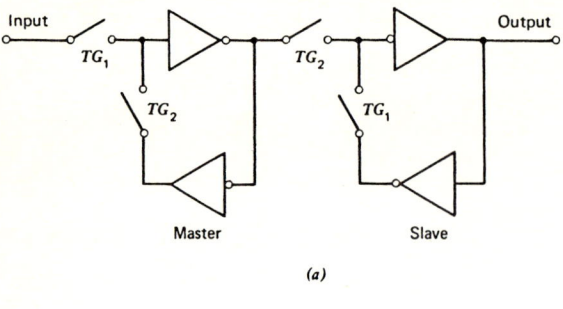

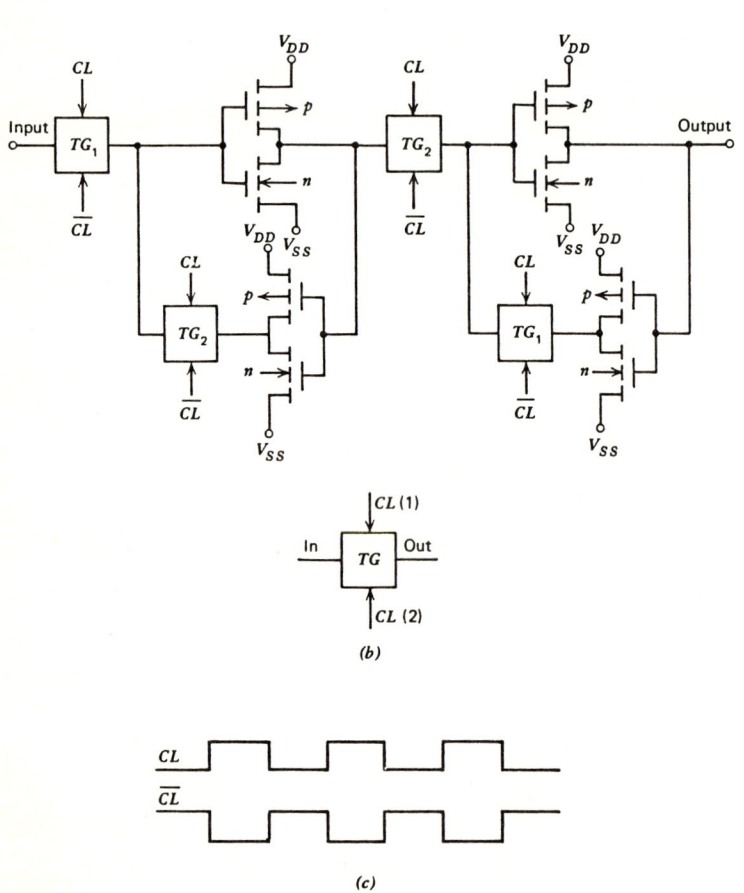

38 CMOS Logic Performance Characteristics and Circuit Applications

Fig. 2-24 Diagrams for a D-Type Flip-Flop: (a) Block, (b) Schematic, (c) Clock Pulse. Here TG Represents "Transmission Gate"; All P-Unit Substrates Are Connected to V_{DD}, and All N-Unit Substrates Are Connected to V_{SS}. Input-to-Output is a Bidirectional Short Circuit when Control Input 1 is Low and Control Input 2 is High; it is an Open Circuit when Control Input 1 is High and Control Input 2 is low.

● t_{n-1} INPUTS						+ t_n OUTPUTS	
CL△	J	K	S	R	Q	Q	Q̄
╱	1	X	0	0	0	1	0
╱	X	0	0	0	1	1	0
╱	0	X	0	0	0	0	1
╱	X	1	0	0	1	0	1
╲	X	X	0	0	X		← (NO CHANGE)
X	X	X	1	0	X	1	0
X	X	X	0	1	X	0	1
X	X	X	1	1	X	∗	∗

△ – LEVEL CHANGE
× – DON'T CARE
∗ – INVALID CONDITION
● – t_{n-1} REFERS TO THE TIME INTERVAL PRIOR TO THE POSITIVE CLOCK PULSE TRANSITION.
+ – t_n REFERS TO THE TIME INTERVALS AFTER THE POSITIVE CLOCK PULSE TRANSITION.

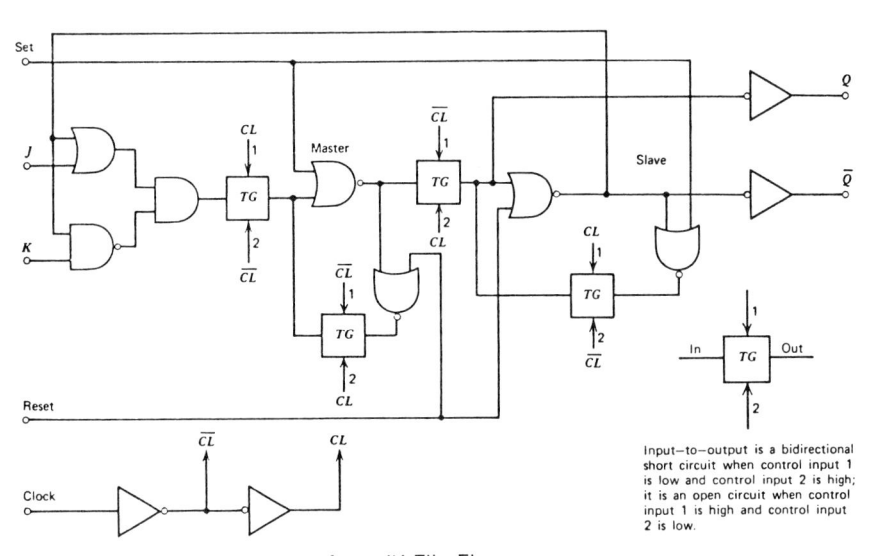

Fig. 2-25 Schematic Diagram for a JK Flip-Flop.

to accommodate the J and K inputs. The J and K inputs provide separate clocked SET and RESET inputs, and allow the flip-flop to change state on successive clock pulses. In the truth table t_{n-1} and t_n refer to the time intervals before and after the positive clock pulse transition, respectively; CL△ indicates level change, × indicates "don't care," and asterisks represent an invalid condition. The JK flip-flop circuit also has SET and RESET capability; the inverters in the master and slave flip-flops each have an added OR input for direct (unclocked) setting and resetting of the flip-flop.

MEMORY CELLS

The basic storage element common to CMOS memories consists of two CMOS inverters cross-coupled to form a flip-flop as in Fig. 2-26. Single-transistor transmission gates are employed as a simple and efficient means of performing the logic functions associated with storage-cell selection (i.e., the sensing and storing operations). The resulting word-organized storage cell (Fig. 2-27) is composed of six transistors, one word line W, and two digit-sense lines D_1 and D_2. Addressing is accomplished by energizing a word line; this turns on the transmission gates on both sides of the selected flip-flop. Because the cell in Fig. 2-27 has P-channel transmission gates, a ground level voltage is required for selection. Figure 2-28 shows an eight-transistor bit-organized memory cell employing X-Y selection. A modification of this circuit in which the Y-select transistors are common for each column of storage elements is used in large memory arrays.

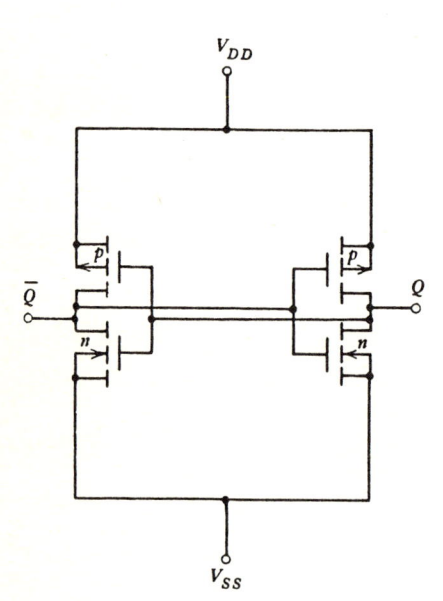

Fig. 2-26 The Basic Storage Element Common to CMOS Memories.

Memory Cells 41

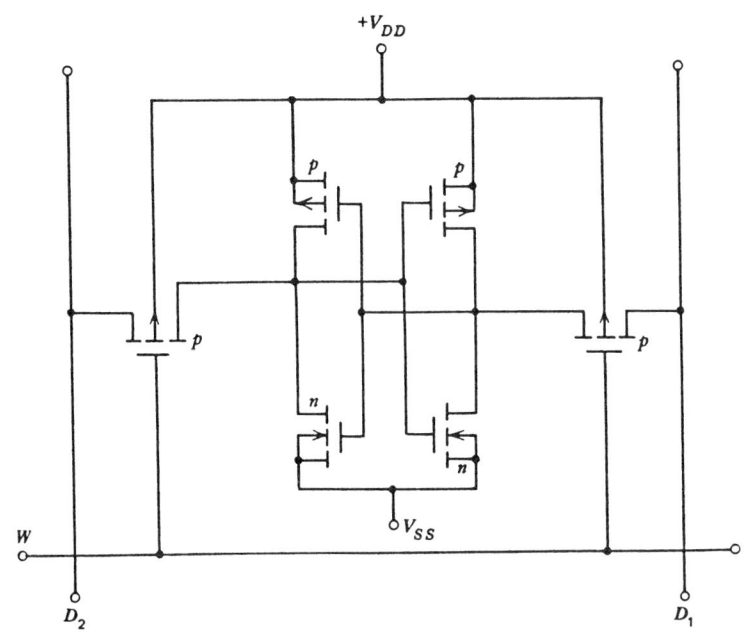

Fig. 2-27 A Word-Organized Storage Cell: W is the Word Line, D_1 and D_2 Are Data Lines.

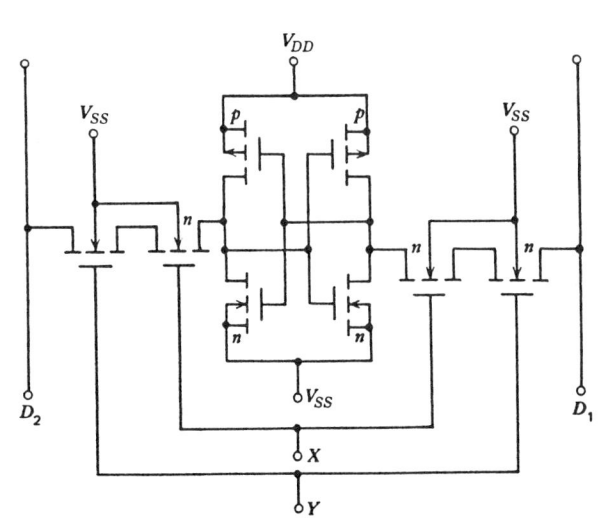

Fig. 2-28 An Eight-Transistor Bit-Organized Memory Cell with X-Y Selection.

42 CMOS Logic Performance Characteristics and Circuit Applications

DYNAMIC SHIFT REGISTERS

Figure 2-29a diagrams a two-stage shift register; each stage consists of two inverters and two transmission gates. Each transmission gate is driven by two out-of-phase clock signals, arranged as in Fig. 2-29b, so that, when alternate transmission gates are turned on, the others are turned off. When the first transmission gate in each stage is turned on, it couples the signal from the previous stage to the inverter and causes the signal to be stored on the input capacitance to the inverter. The shift register utilizes the input of the inverter for temporary storage. When the transmission gate is turned off on the next half-cycle of the clock, the signal is stored on this input capacitance, and the signal remains at the output of the inverter where it is available to the next transmission gate which is turned on. Again, this signal is applied to the input of the next inverter where it is stored on the input capacitance of the inverter, making the signal available at the output of the stage. Thus, a signal progresses to the right by one half-stage on each half-cycle of the clock, or by one stage per clock cycle.

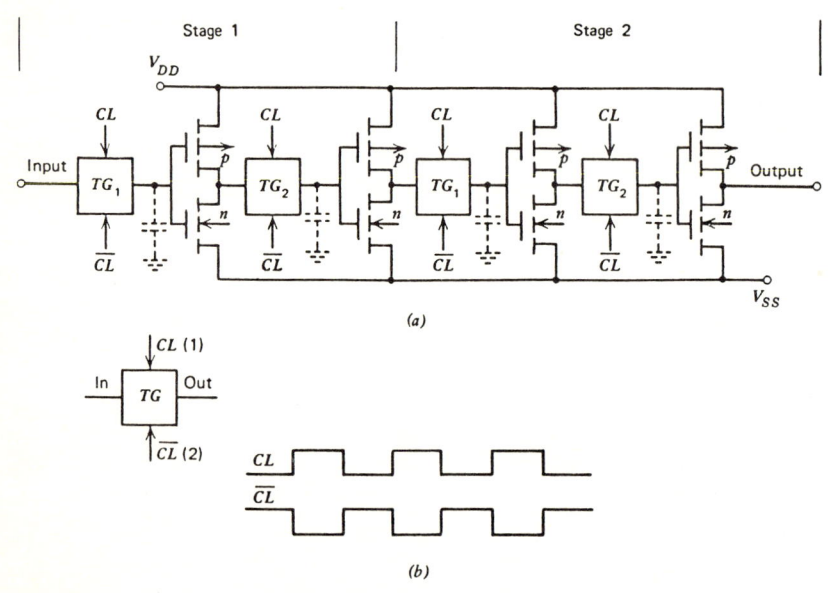

Fig. 2-29 (a) Two-Stage Shift Register, (b) Clock-Pulse Diagram. All P-Unit Substrates Are Connected to V_{DD}; All N-Unit Substrates Are Connected to V_{SS}. Input to Output Is a Bidirectional Short Circuit when Control Input 1 is Low and Control Input 2 is High; It Is an Open Circuit when Control Input 1 is High and Control Input 2 is Low.

Because the shift register is dependent on stored charge, which is subject to slow decay, there is a minimum frequency at which it will operate; reliable operation can be expected at frequencies as low as 5 kHz.

CMOS dynamic shift registers have all the advantages of other CMOS devices, including low power dissipation, high noise immunity, and wide operating voltage range; in addition, they are superior in two important ways to the single-channel (PMOS and NMOS) dynamic shift registers. First, the CMOS device easily generates the two-phase clock signals required, internal to itself, with just one supply voltage. Second, TTL and DTL logic compatibility is maintained on all inputs and outputs with one supply voltage.

FOUR-PHASE CMOS LOGIC CIRCUITS*

Logic circuits often use a four-phase clocking technique to charge and discharge capacitors without providing a direct path from the power supply to ground. This arrangement saves considerable power compared to circuits having a load resistor between the power supply and ground.

Four-phase logic can be implemented using CMOS to provide a four-phase clock signal from a single-phase clock, and it requires only one power supply voltage. CMOS devices are used as switches which charge and discharge capacitors in accordance with circuit requirements but without a direct connection between power supply and ground. Figure 2-30 shows clock waveforms. Figure 2-31 is a schematic of several four-phase circuit implementations.

The circuit shown in Fig. 2-31b will be described in detail for illustrative purposes and operates as follows: During clock time t_1 to t_2 (Fig. 2-30), ϕ_1 turns transistor Q_3 on and holds transistor Q_2 off. This charges capacitor C_2 to $+V_{DD}$. From time t_3 to t_4, ϕ_1 turns Q_3 off and ϕ_2 turns Q_2 on. If at this time the input to Q_1 (LOGIC$_2$) is a logic 1 ($+V_{DD}$), Q_2 and Q_1 will form a conductive path which discharges C_2 to ground or logic 0. If, however, the input to Q_1 is a logic 0 (ground is 0 V), then C_2 remains charged to $+V_{DD}$ (logic 1). The voltage on C_2 or the output of this circuit then remains unchanged and is valid during time t_4 to t_9. This circuit thus operates as an inverter which is precharged during t_1 to t_2, evaluated during t_3 to t_4, and valid during t_4 to t_9.

* Complementary MOS Four-Phase Logic Circuits," *Computer Design* Tech. Brief, September 1974, p. 110. Used with permission.

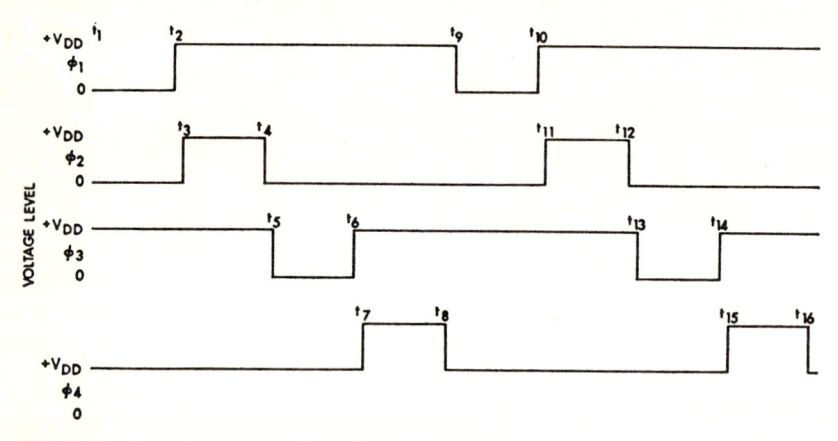

44 CMOS Logic Performance Characteristics and Circuit Applications

Fig. 2-30 Complementary Four-Phase Clock Waveforms. *(Reprinted from Computer Design, September 1974. Used with permission.)*

The circuit shown in Fig. 2-31 operates in a similar manner and is also an inverter; it is precharged during t_3 to t_4, evaluated during t_5 to t_6, and valid during t_4 to t_9. The circuit in Fig. 2-31d is similar to that in Fig. 2-31b, while that in Fig. 2-31a is similar to the one in Fig. 2-31c.

Because the complementary inverter has a finite and predictable propagation delay, the four-phase clock waveform can be generated by the circuit from a one phase input. Thus, the output of the inverter is the inverse of the input, but is delayed by a small time. The output of a number of inverters can be joined in series and connected to logic gates to produce the required four-phase clock pulses shown in Fig. 2-30.

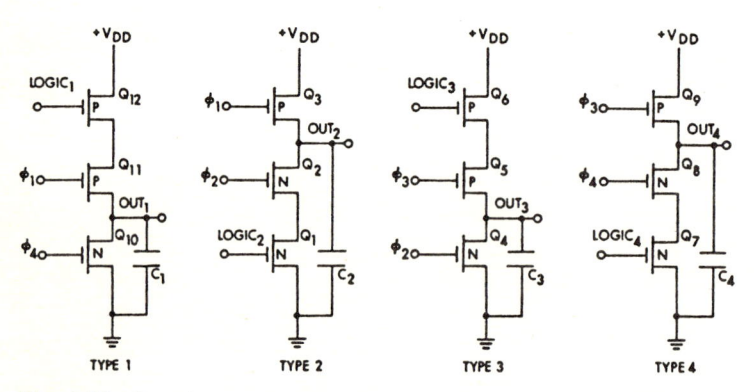

Fig. 2-31 Complementary Four-Phase Circuits. *(Reprinted from Computer Design, September 1974. Used with permission.)*

THREE-STATE LOGIC

In many systems, the wire-ORing technique of the passive pull-up outputs has been used in bussing-type applications. This technique greatly reduces system wiring and package count by the sharing of common input/output lines.

To eliminate the problems of wire ORing in TTL, the concept of three-state logic was introduced. This allows the system designer to develop multiplexing schemes which select a single element to drive a common line while disabling all other drivers on that line. The concept also allows more drivers to be attached to a common input/output bus line.

CMOS, like active pull-up TTL, cannot be wire-ORed, since both the current sinking and sourcing devices in a CMOS output are MOS transistors. Thus, to eliminate the problems of wire ORing, the three-state logic concept is provided in many CMOS logic functions because it can be very easily implemented.

SELECTED CMOS LOGIC CIRCUIT APPLICATIONS

The available literature (predominately from RCA Semiconductor) lists a veritable trove of practical CMOS applications. Because of this fact plus the practical need to provide a manageable manuscript and to focus on the real growth of CMOS—in LSI and VLSI—it was decided to limit the number of applications of CMOS logic circuits to a bare minimum—to those concerned with display circuits and data acquisition, entry, and manipulation.

Display Circuits

Where data is being processed, the user normally wants a readout. As such, counters often go hand in hand with digital display circuits.

When driving LED (light-emitting diode) displays, a greater current drive is required of the CMOS circuit. As such, several CMOS circuits have been specifically designed to provide LED drive capabilities for CMOS systems. Some of these are Motorola's MC 14511 BCD-to-7-segment latch-decoder-driver (Fig. 2-32), Solid State Scientific's SCL 4426 and 4433 decade counters with 7-segment display outputs, Siliconix's DF412 LCD (liquid crystal device) driver, and Intersil's ICM

46 CMOS Logic Performance Characteristics and Circuit Applications

7216/7226, 7217/7227, and 7218 CMOS LED display driver systems to name a few.

The MC 14511 shown in Fig. 2-32 can source up to 25 mA of current per segment, which means that it cannot drive MAN_1-type displays, but could easily handle MAN_3 size or smaller readouts.

There are several ways to implement a visual display. For example, a 3-digit display "counter" (Fig. 2-33) combines the counter, latches, and

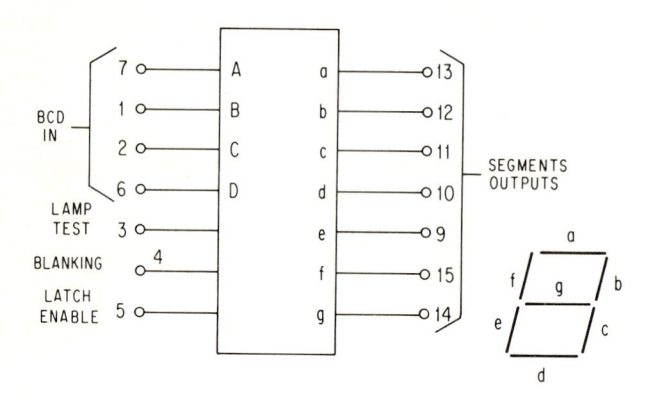

LE	BI	LT	D	C	B	A	a	b	c	d	e	f	g	DISPLAY
X	X	0	X	X	X	X	1	1	1	1	1	1	1	8
X	0	1	X	X	X	X	0	0	0	0	0	0	0	Blank
0	1	1	0	0	0	0	1	1	1	1	1	1	0	0
0	1	1	0	0	0	1	0	1	1	0	0	0	0	1
0	1	1	0	0	1	0	1	1	0	1	1	0	1	2
0	1	1	0	0	1	1	1	1	1	1	0	0	1	3
0	1	1	0	1	0	0	0	1	1	0	0	1	1	4
0	1	1	0	1	0	1	1	0	1	1	0	1	1	5
0	1	1	0	1	1	0	0	0	1	1	1	1	1	6
0	1	1	0	1	1	1	1	1	1	0	0	0	0	7
0	1	1	1	0	0	0	1	1	1	1	1	1	1	8
0	1	1	1	0	0	1	1	1	1	0	0	1	1	9
0	1	1	1	0	1	0	0	0	0	0	0	0	0	Blank
0	1	1	1	0	1	1	0	0	0	0	0	0	0	Blank
0	1	1	1	1	0	0	0	0	0	0	0	0	0	Blank
0	1	1	1	1	0	1	0	0	0	0	0	0	0	Blank
0	1	1	1	1	1	0	0	0	0	0	0	0	0	Blank
0	1	1	1	1	1	1	0	0	0	0	0	0	0	Blank
1	1	1	X	X	X	X	•							•

X = Don't care   • Depends upon the BCD code applied during the 0 to 1 transition of LE.

Fig. 2-32 LED Displays Can Be Driven Directly from CMOS Systems with Circuits Such as the 14511.

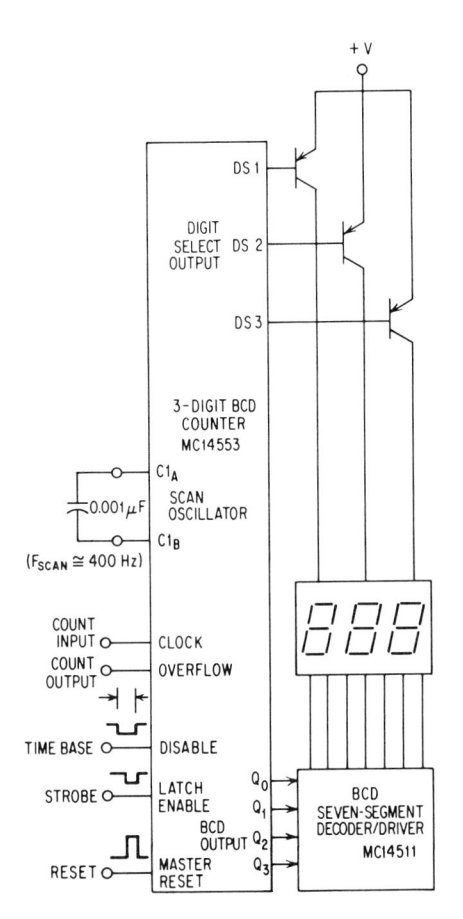

Fig. 2-33 An MSI/CMOS 3-Digit Counter-Display System Requires Only Two ICs—a Counter Subsystem and a Display Driver.

multiplex driver. This device also includes a multiplex scan-oscillator and an input trigger to accommodate slow rise-time inputs. As a 3-digit device, it lends itself to systems that require multiples of 3 digits. If a system requires exactly 4 digits, however, it would be more economical to use two MSI 3-digit counters and forget the extra digits, rather than use the previous counter design.

Intersil offers several counter circuits that were designed to drive digital displays. Three of these are the ICM7216/7226, the ICM7217/7227, and the ICM7218.

The Intersil ICM7216/7226 is a universal counter IC that contains the functions of a frequency counter, period counter, unit event counter, frequency ratio counter, and time interval counter all on a single chip. Furthermore, it has 8-digit multiplexed LED display outputs capable of switching up to 250 mA per digit, typically. These outputs will directly drive both digits and segments of large LED displays. Maximum digit output current capability is 400 mA, and maximum segment current output is 60 mA. The counter operates from a single +5-V supply, and can function at frequencies up to 10 MHz. Besides replacing as many as 100 discrete components, the ICM7216 is compatible with most microprocessors.

In addition to its multicounter capability, other important features of the ICM7216 include

- Four internal gate times (0.01, 0.1, 1, and 10 s) in the frequency counter mode
- 1-, 10-, 100-, or 1000-cycle gate time in the period, frequency ratio, and time interval modes.
- Frequency measurements from direct current to 10 MHz; period measurements from 0.5 μs to 10 s
- Stable on-chip oscillator capable of operating with either a 1- or a 10-MHz crystal
- Internally generated multiplex timing with interdigit blanking, leading zero blanking, and overflow indication
- Decimal point and leading zero blanking controlled directly on-chip or externally
- Standby mode that turns off the LED display and puts the chip into a low-power (10-mW) mode

The universal counter comes in two versions, depending on how the LED display is driven: common anode (ICM7216A) or common cathode (ICM7216B).

Figure 2-34 shows a typical counter using the ICM7216 universal counter IC.

The ICM7217 and ICM7227 are 4-digit, presetable UP/DOWN counters with an on-board prestable register the contents of which are continuously compared to the counter. These devices operate from a single +5-V supply and provide three main outputs: a CARRY/BORROW output which allows for direct cascading of counters, a ZERO output which indicates when the count is zero, and an EQUAL output which indicates when the count is equal to the value contained in the register. Data is multiplexed into and out of the device by means of a Tri-State® BCD I/O port, which acts as a high-impedance input when loading and

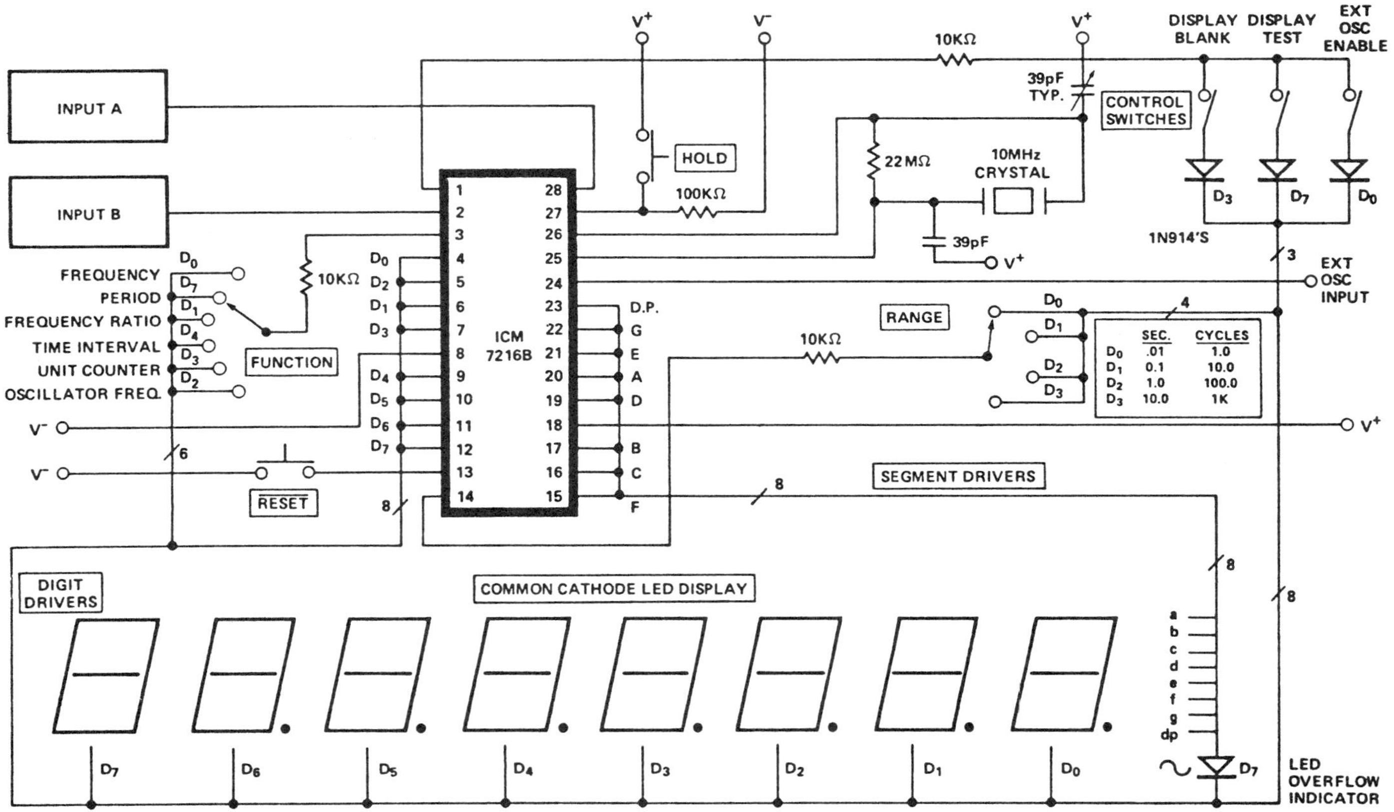

Fig. 2-34 Display Counter Using the ICM7216 Counter IC. *(Courtesy of Intersil.)*

provides a multiplexed BCD output. The CARRY/BORROW, EQUAL, and ZERO outputs, and the BCD port functioning as an output, will drive one standard TTL load. The quiescent power dissipation is less than 5 mW.

The ICM7217 versions are intended for use in hardwired applications where thumbwheel switches are used for loading data and simple SPDT (single-pole double-throw) switches are used for chip control. The ICM7227 versions are intended for use in processor-based systems where presetting and control functions are performed under processor control.

These circuits provide multiplexed 7-segment LED display outputs, with common anode or common cathode configurations available. Digit and segment drivers are provided to directly drive displays of up to 1-in character height at a 25 percent DUTY cycle. The frequency of the on-board oscillator (and thus the multiplex frequency) may be controlled with a single capacitor, or the oscillator may be allowed to free run. Leading zeros are blanked, and the display drivers may be disabled, allowing the display to be used for other purposes. The data appearing at the 7-segment and BCD outputs is latched; the contents of the counter are transferred into the latches under external control by means of the STORE pin.

The ICM7217/7227 (common anode) and ICM7217A/7227A (common cathode) versions are decade counters, providing a maximum count of 9999, while the ICM7217B/7227B (common anode) and ICM7217C/7227C (common cathode) are intended for timing purposes, providing a maximum count of 5959. Several applications examples are presented to show the versatility of the device.

A simple implementation of a 4-digit frequency counter is shown in Fig. 2-35. In that illustration, an ICM7207A is used to provide the 1-s gating window and the $\overline{\text{STORE}}$ and $\overline{\text{RESET}}$ signals. In this configuration, the display reads hertz directly. With pin 11 of the ICM7207A connected to V^+, the gating time will be 0.1 s, which will give tens of hertz in the least-significant digit. For shorter gating times, an ICM7207 may be used (with a 6.5536-MHz crystal), giving a 0.01-s gating with pin 11 connected to V^+ and a 0.1-s gating with pin 11 open.

To implement a 4-digit tachometer, the ICM7207A with a 1-s gating should be used. In order to get the display to read directly in r/min (revolutions per minute), the rotational frequency of the object to be measured must be multiplied by 60 (or 600 using a 0.1-s gating for faster update). This can be done electronically using a phase-locked loop, or

Selected CMOS Logic Circuit Applications   51

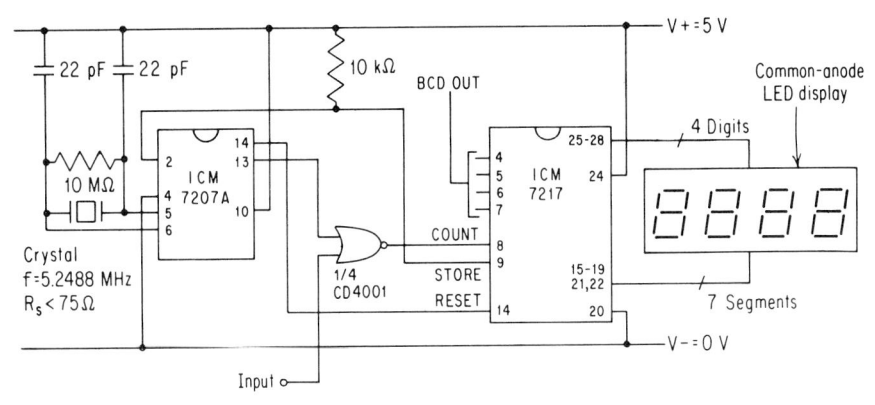

Fig. 2-35 Precision Frequency Counter. (*Courtesy of Intersil.*)

mechanically by using a disk, rotating with the object, that has the appropriate number of holes drilled around its edge to interrupt the light from an LED to a photodetector.

The circuit shown in Fig. 2-36 uses a 556 dual timer to generate the gating, $\overline{\text{STORE}}$ and $\overline{\text{SET}}$ signals instead of an ICM7207A. One timer is configured as an astable multivibrator, using R_A, R_B, and C to provide an output (556, pin 5) that is positive for approximately 1 s and negative for approximately 300 to 500 μs to serve as the gating signal. The gating positive time is given by $G_H = 0.693\,(R_A + R_B)C$, while the gating low

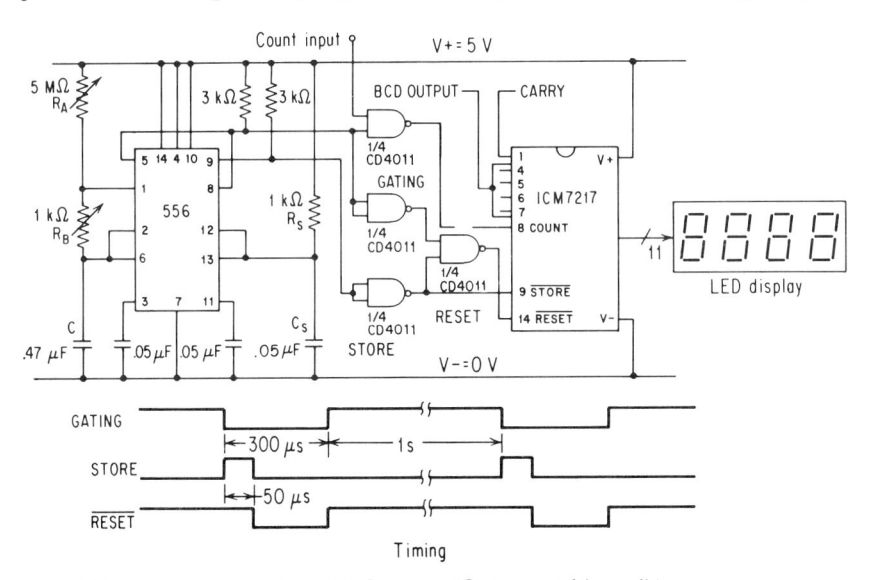

Fig. 2-36 Inexpensive Frequency Counter. (*Courtesy of Intersil.*)

time is $G_L = 0.693\, R_B C$. The system is calibrated by using a 5-MΩ potentiometer for R_A as a "coarse" control and a 1-kΩ potentiometer for R_B as a "fine" control. The other timer in the 556 is configured as a one-shot triggered by the negative-going edge of the gating. This one-shot output (556, pin 9) is inverted to serve as the $\overline{\text{STORE}}$ pulse and to hold $\overline{\text{RESET}}$ high. When the one-shot times out and $\overline{\text{STORE}}$ goes high, $\overline{\text{RESET}}$ goes low, resetting the counter for the next measurement. The one-shot pulse width will be approximately 50 μs with the component values shown. When fine-trimming the gating signal with R_B, care should be taken to keep the gating low time ($=0.693 R_B C$) at least twice as long as the one-shot pulse width.

The low-power operation of the ICM7217 makes its use as an LCD interface highly desirable, as shown in Fig. 2-37.

The Siliconix DP411 4-digit BCD-to-LCD display driver easily interfaces to the ICM7217A with one CD4000-series package to provide a total system power consumption of less than 5 mW. The common-cathode devices should be used since in these versions the digit drivers are CMOS, while in the common-anode devices the digit drivers are NPN devices and will not provide full logic swing.

The Intersil ICM7218 A/E series consists of LED-driver ICs, capable of interfacing with microprocessors or other digital systems, and an 8-digit display.

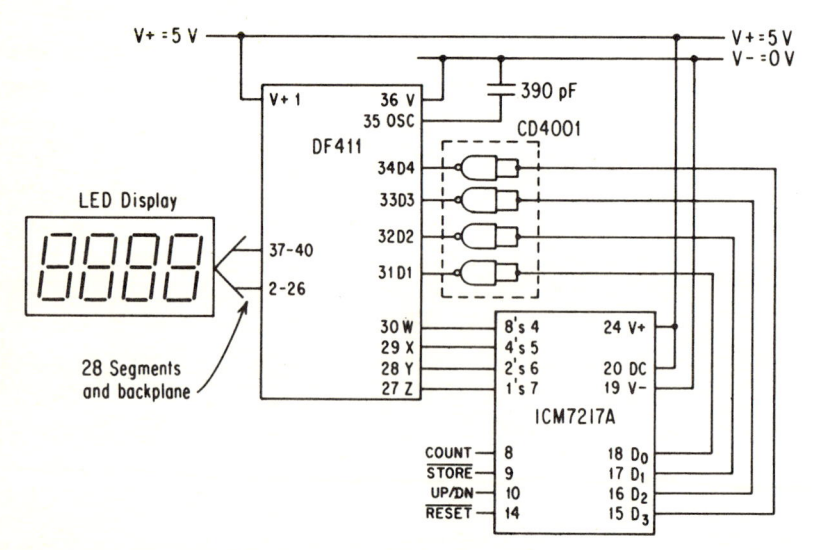

Fig. 2-37 LED Display Interface. (*Courtesy of Intersil.*)

Selected CMOS Logic Circuit Applications 53

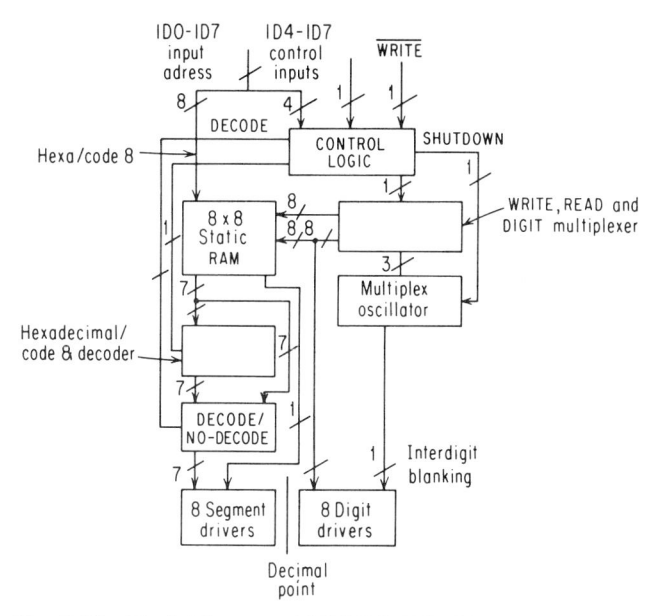

Fig. 2-38 Block Diagram of ICM7218 LED Display Driver System. (*Courtesy of Intersil.*)

Each device (see Fig. 2-38) incorporates an 8 × 8-b memory array, two types of 7-segment decoders (hexadecimal or code B), all multiplexing scan circuitry, and digit/segment drivers on a single chip. Both common-anode and common-cathode LED drive versions are available. The common-anode output drive is approximately 200 mA per digit at a 12 percent delay cycle (approx. 40 mA per segment with 5 segments being driven). The common cathode version has approximately 20-mA/segment peak drive. The drive will be correspondingly less if high-impedance LED displays are used.

A typical display example is shown in Fig. 2-39, in which the ICM7218 display system is combined with an 87C48 microprocessor, an IM6604 CMOS EPROM and an IM6507 CMOS RAM. The P_{21} I/O port and the $\overline{WR}$ ($\overline{WRITE}$) port of the microprocessor provide the MODE and $\overline{WRITE}$ signals for the ICM7218.

The control information (DATE COMING—ID_7; $\overline{SHUTDOWN}$—ID_4; DECODE—ID_6 and HEXA or CODE B—ID_5) is written in the ICM7218 when P_{21} (MODE) is high and a $\overline{WR}$ pulse occurs.

Display data is written into each digit location in memory on eight successive $\overline{WR}$ enables only if the control information contains a DATA ENABLE and a P_{21} (MODE) is low. If the DATA COMING signal has

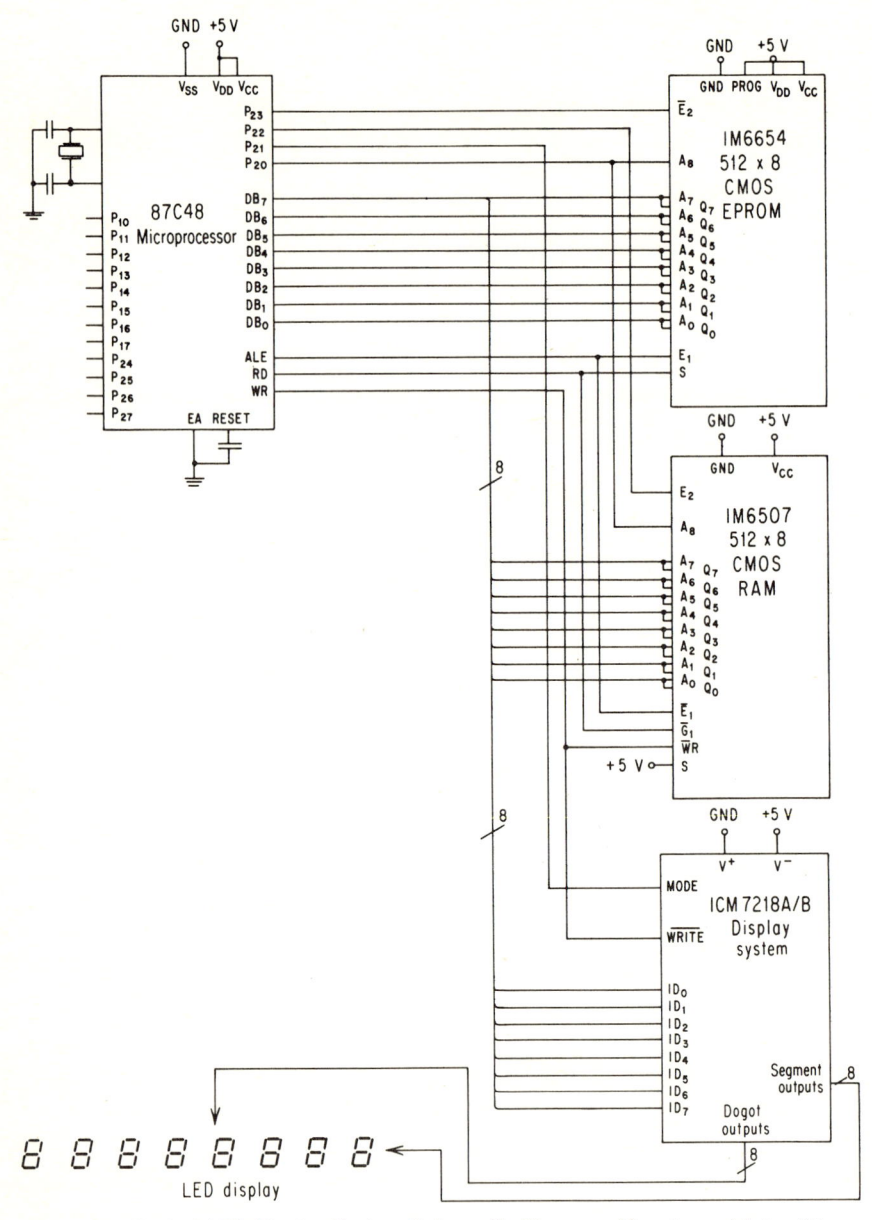

54 CMOS Logic Performance Characteristics and Circuit Applications

Fig. 2-39 Typical LED Display System Schematic Diagram. (*Courtesy of Intersil.*)

Selected CMOS Logic Circuit Applications   55

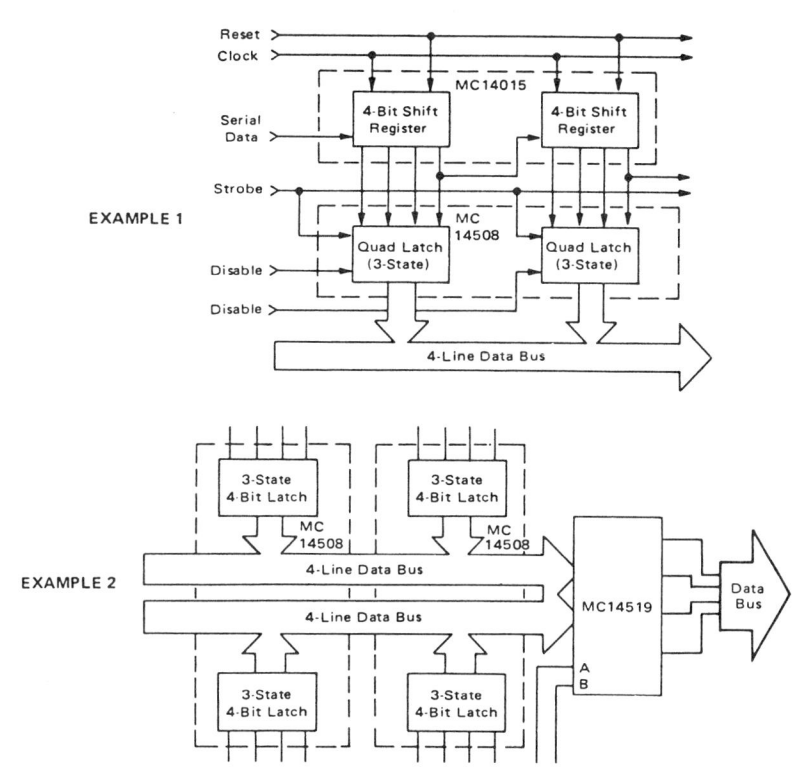

Fig. 2-40 Typical Three-State Applications. *(Courtesy Motorola Semiconductor Products, Inc. Used with permission.)*

not been enabled, the system will continue normal display operation based on the instruction given in the control information.

Data Acquisition, Entry, and Manipulation*

DATA BUSSING

Figure 2-40 shows two examples of the MC14508 dual 4-bit latch used for data bussing. In example 1, an 8-b data word is shifted into the MC14015 shift register and then strobed into the dual latch. Either 4-b byte may be selected and placed onto a common 4-line output data bus. Example 2 shows a second level of multiplexing by using the MC14519

---

* Portions of this section taken from J. Tonn, "Introduction to CMOS Integrated Circuits with Three State Output," Motorola Semiconductor Products, Inc., Application Note AN-715, Motorola, Phoenix, Arizona, 1974. Used with permission.

56 CMOS Logic Performance Characteristics and Circuit Applications

(quad two-channel data selector) to select either one of two input buses to be connected to a third output bus.

STORAGE REGISTERS

A dual 4-b storage register, as shown in Fig. 2-41, can be implemented by connecting the data outputs of the MC14508 back to the data inputs and controlling that state of the strobe and disable lines. Simple NAND gates are used to control the READ and WRITE modes and also to synchronize the data flow with respect to a common system clock. The two 4-input cross-coupled gates prevent the outputs of both registers from simultaneously driving the common bus lines. When in the READ mode (READ = 1, WRITE = 0), the register contents are placed on the bus lines on the positive edge of the clock. In the WRITE mode (READ = 0, WRITE = 1), data enters the register on the positive clock transition and is latched on the falling edge. If READ = WRITE = 0, the register is disabled, and if READ = WRITE = 1, the register will be in the WRITE mode.

The MC14508 shown in Fig. 2-42 serves as an 8-bit storage register. The operation is similar to that previously discussed for Fig. 2-41 with a minor exception. That is: If the READ and the WRITE controls are

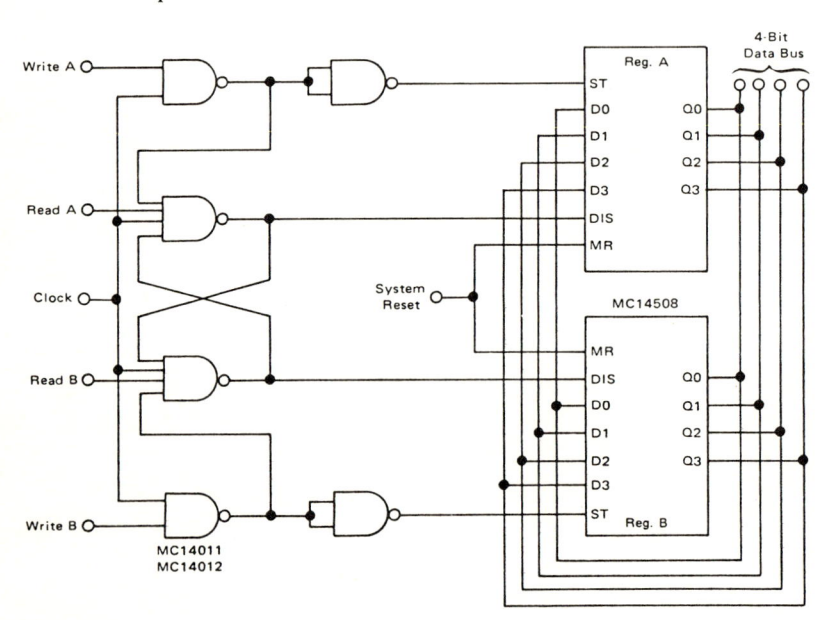

Fig. 2-41 Dual 4-b Storage Register. (*Courtesy Motorola Semiconductor Products, Inc. Used with permission.*)

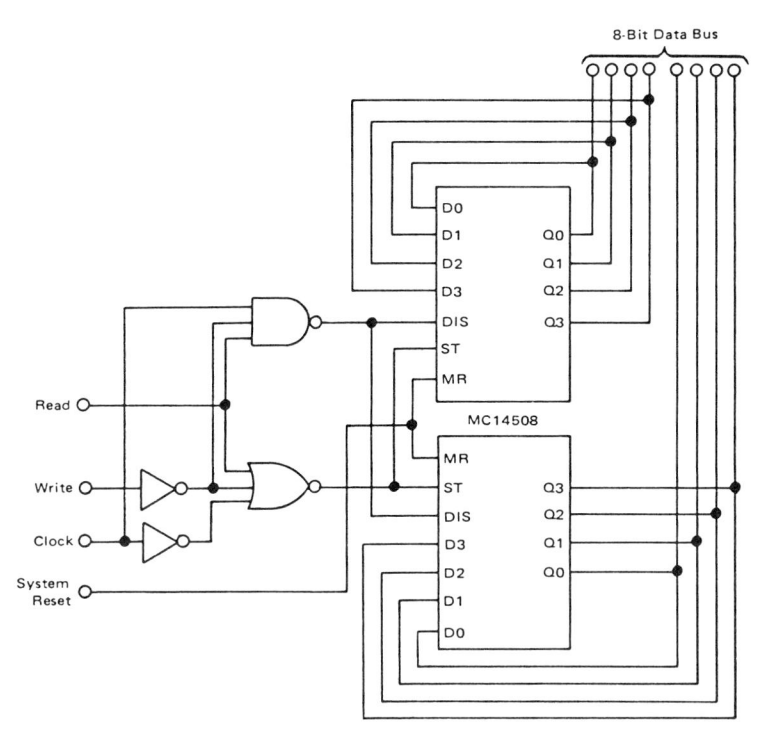

Fig. 2-42 An 8-b Storage Register. *(Courtesy Motorola Semiconductor Products, Inc. Used with permission.)*

simultaneously in a logic 1 state or in the logic 0 state, the register is disabled from the data bus and the contents are unchanged.

Figure 2-43 shows the interconnection of 4034 bus registers to allow shift right or shift left with parallel inputs.

DATA ROUTING

Two MC14512 8-channel data selectors are used in Fig. 2-44 with the MC14514 4-b latched decoder to effect a complex data routing system. A total of 16 inputs from data registers are selected and transferred via a three-state data bus to a data distributor for rearrangement and entry into 16 output registers. In this way, sequential data can be rerouted or intermixed according to patterns determined by DATA SELECT and DISTRIBUTION inputs.

Data is placed into the routing scheme via the 8 inputs on both MC14512 data selectors. One register is assigned to each input. The signals on A_0, A_1, and A_2 choose one of eight inputs for transfer out to

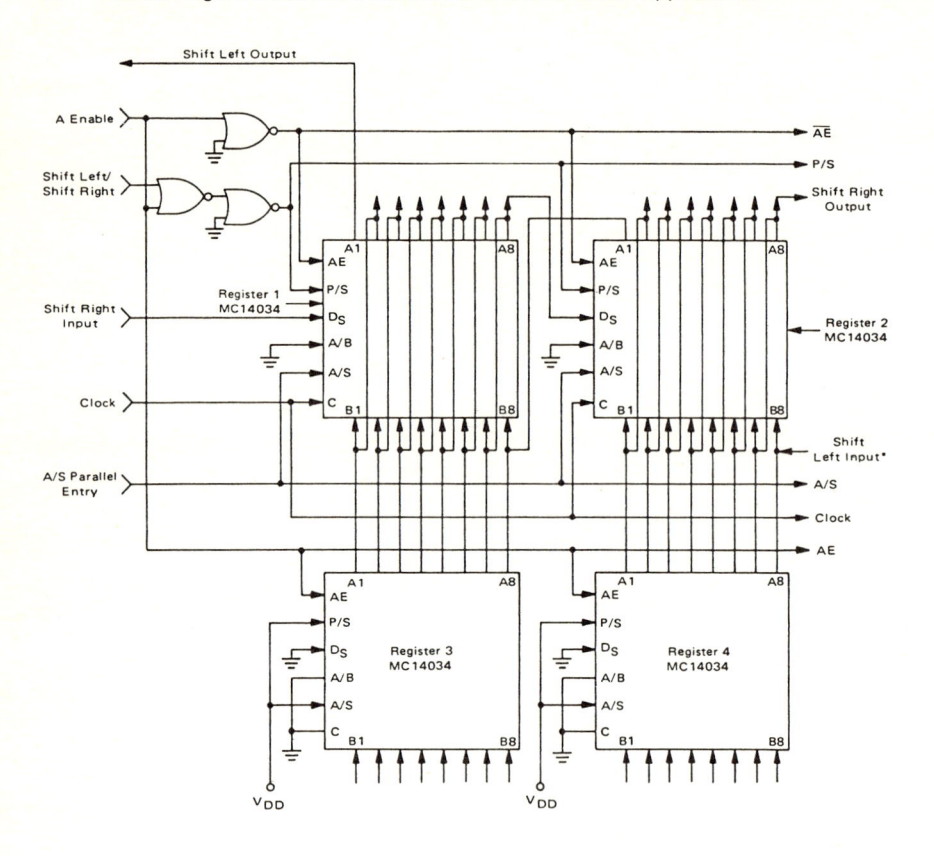

58 CMOS Logic Performance Characteristics and Circuit Applications

A "High" ("Low") on the Shift Left/Shift Right input allows serial data on the Shift Left Input (Shift Right Input) to enter the register on the positive transition of the clock signal. A "high" on the "A" Enable Input disables the "A" parallel data lines on Reg. 1 and 2 and enables the "A" data lines on registers 3 and 4 and allows parallel data into registers 1 and 2. Other logic schemes may be used in place of registers 3 and 4 for parallel loading.

When parallel inputs are not used Reg. 3 and 4 and associated logic are not required.

*Shift left input must be disabled during parallel entry.

Fig. 2-43 Shift Right/Shift Left with Parallel Inputs. *(Courtesy Motorola Semiconductor Products, Inc. Used with permission.)*

the three-state data bus. A fourth signal, labeled DIS, disables one of the MC14512 selectors, assuring transfer of data from only one register.

In addition to a choice of input registers 1 through 16, the rate of transfer of the sequential information can also be varied. That is, if the MC14512 were addressed at a rate that is eight times faster than the shift

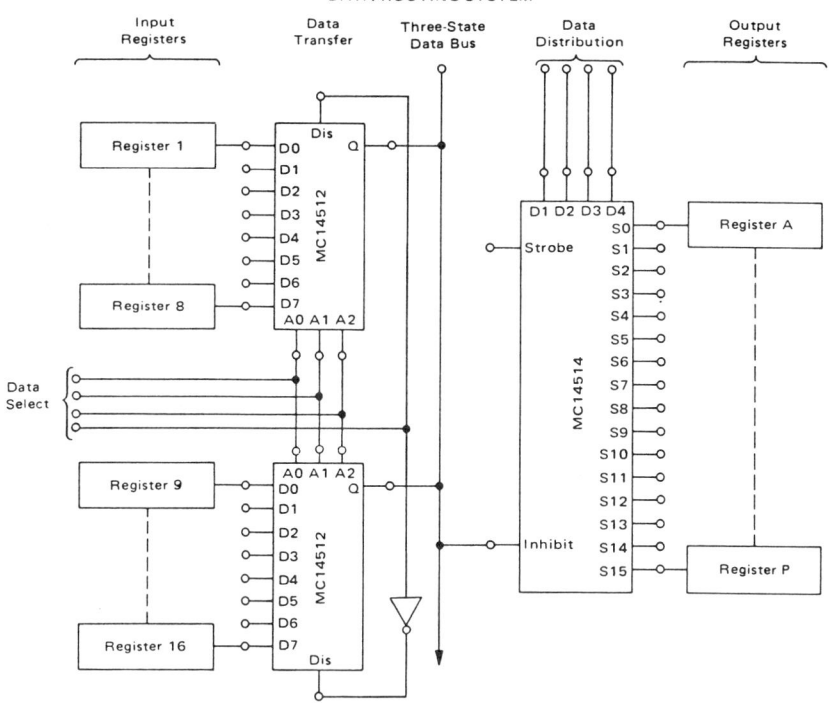

Fig. 2-44 Data Routing System. (*Courtesy Motorola Semiconductor Products, Inc. Used with permission.*)

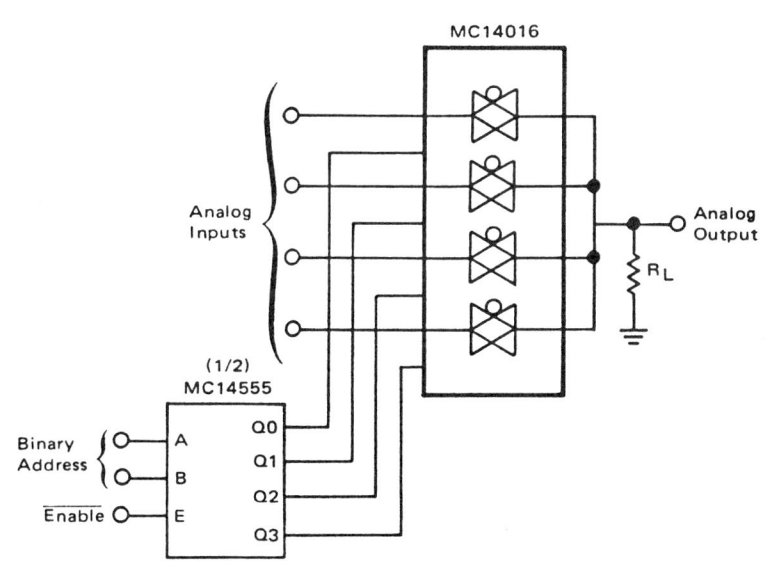

Fig. 2-45 A Four-Channel Analog Data Selector. (*Courtesy Motorola Semiconductor Products, Inc. Used with permission.*)

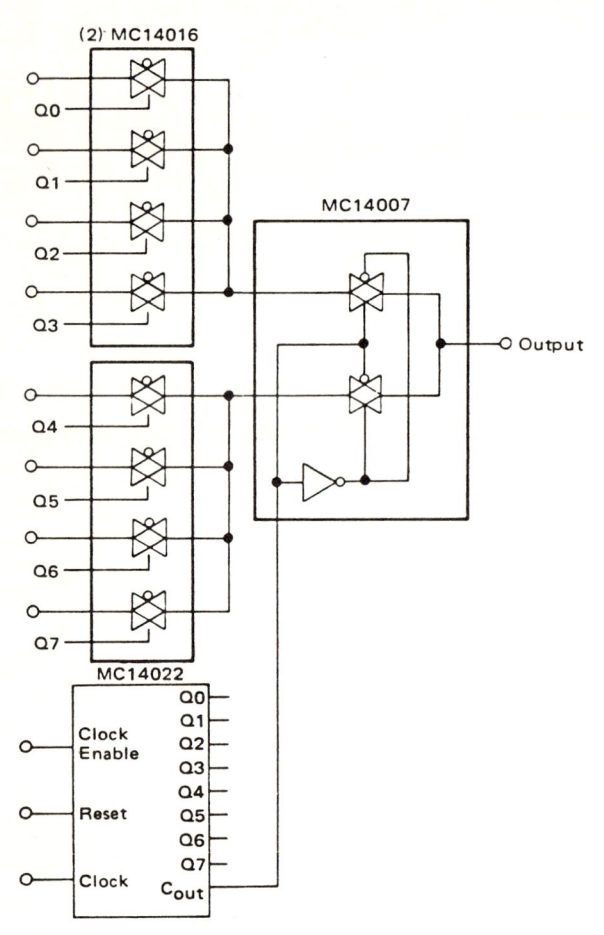

Fig. 2-46 Two-Level 8-Channel Multiplexer. *(Courtesy Motorola Semiconductor Products, Inc. Used with permission.)*

frequency of the input registers, the MSB (most-significant bit) from each register could be selected for transfer to the data bus. Therefore, all of the MSBs from all of the registers can be transferred to the data bus before the next-most-significant bit is presented for transfer by the input registers.

Information from the three-state bus is redistributed by the MC14514 4-b latch decoder. Using the 4-b address, D_1 through D_4, the information on the INHIBIT line can be transferred to the ADDRESSED OUTPUT line to the desired output registers, A through P. This distribution of

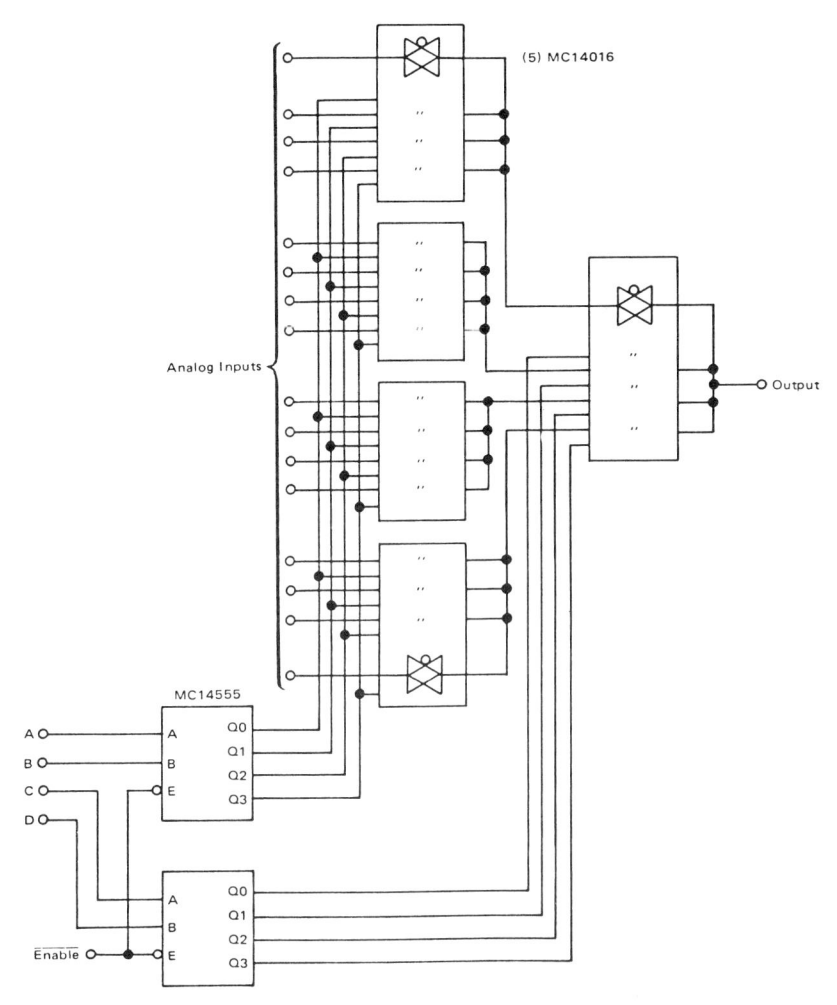

Fig. 2-47 Two-Level 16-Channel Multiplexer. *(Courtesy Motorola Semiconductor Products, Inc. Used with permission.)*

data bits to the output registers can be made in many complex patterns. For example, all of the MSBs from the input registers can be routed into output register A, all of the next-most-significant bits into register B, etc. In this way, horizontal, vertical, or other methods of data slicing can be implemented.

In addition, Harris Semiconductor provides a number of interface circuits for use in data bus applications: the HD6431 three-state latching bus driver; the HD6432 bidirectional bus driver; the HD6433 bus

62 CMOS Logic Performance Characteristics and Circuit Applications

separator-driver; the HD6440 latched decoder-driver, and the 6495 three-state buffer driver.

Harris offers a programmable bit-rate generator, the HD6405, and a universal asynchronous receiver transmitter (UART) circuit, the HD6402; Intersil offers the IM6402/6403 UART.

MULTIPLEXING/DEMULTIPLEXING

In applications requiring 1-of-4 channel selection, the 4016 and ½ of the MC14555 for control may be used, as shown in Fig. 2-45. The input channels are selected with a 2-b binary address (A, B). The $\overline{\text{ENABLE}}$ control input when high will switch all channels off independently of the binary address.

Figure 2-46 shows a two-level 8-channel multiplexer in which the second level of multiplexing is performed by the 14007. The 14022 is used for time division multiplexing with C_{out} ($C_{out} = 1$ if the count is less than 4) used to control the 4007.

The two-level multiplexer shown in Fig. 2-47 has a slightly different address decoding scheme than any previously shown. One-half of the MC14555 (A and B input bits) simultaneously selects four inputs in the first level (one in each of four 4016 packages). The second half of the 14555 (C and D input bits) selects one of the four input branches of the tree to be connected to the output. The $\overline{\text{ENABLE}}$ input is active low and will disable all analog switches in the tree, and it can be used for further levels of multiplex decoding.

Intersil, SGS, RCA, Motorola, and others offer single-chip 4-channel-differential, 8-channel-differential, and 16-channel multiplexers and cross-point switches that further ease the design of data acquisition, entry, and manipulation circuits.

CHAPTER THREE
Data Conversion and Telecommunications Circuits

The most exciting aspect of the full utilization of CMOS characteristics is its application to LSI circuits—memories, microprocessors, D/A and A/D converters, codecs, UARTs, complete data acquisition circuits, and the like—both by itself (bulk and/or oxide-isolated) and in conjunction with SOS.

LSI circuit implementation using CMOS technology provides greater speed and better performance at a lower cost than other technologies due to circuit innovations possible only with CMOS. Many circuit components can be eliminated by using the parasitic capacitances and overdrive characteristics of digital CMOS circuits properly during chip design. For example, a counting function requiring 24 transistors in a straightforward design requires only 12 transistors, or even six transistors and six diodes when implemented in CMOS. Figure 3-1 shows three versions of a divide-by-3 counting circuit that are easy to make in CMOS and virtually impossible any other way. All three do the same job and require a single-phase clock instead of the conventional two-phase clock. All three can be expanded by adding more sections to divide by 5 and 7 without forbidden states. The truth table is the same for all three circuits, except that, for the version shown in Fig. 3-1c, the clock phase must be reversed.

The dynamic divider shown in Fig. 3-1a is the fastest CMOS divider configuration now possible. Implemented with a 5-V supply and a low-threshold CMOS process, it operates to nearly 100 MHz. Its four-transistor structure—two P- and two N-channel transistors in series—is one of the most useful CMOS configurations.

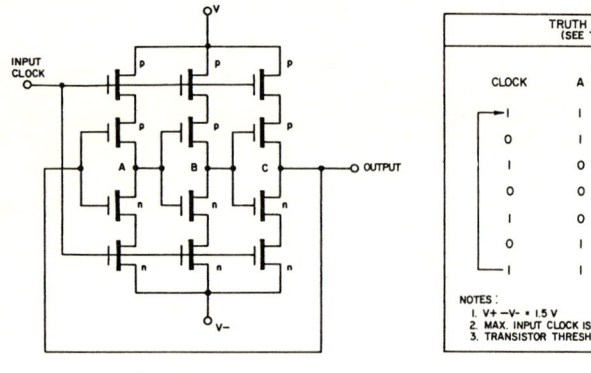

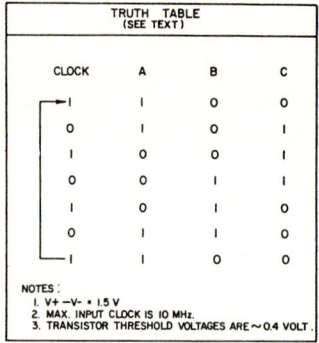

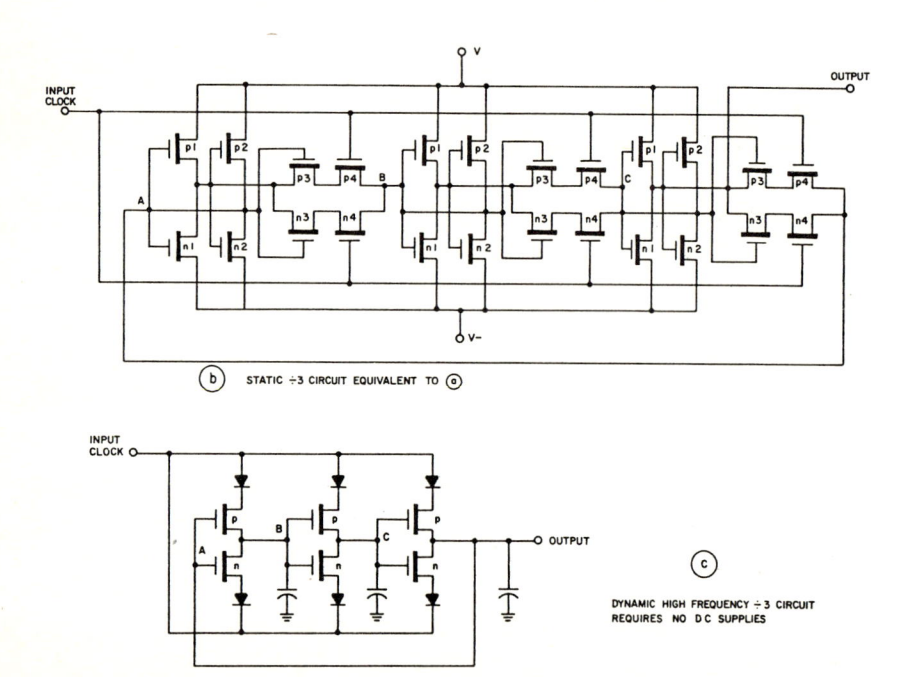

Fig. 3-1 Divide-by-3 Circuits Illustrate CMOS Circuit Innovations. *(Taken from CMOS: Higher Speeds, More Drive and Analog Capability Expand Its Horizons, Electronic Design 23, November 8, 1978. Reprinted with permission.)*

Figure 3-1b shows a static equivalent of the circuit shown in Fig. 3-1a. Three latches are used to store data instead of the parasitic capacitances at nodes A, B, and C. Each latch consists of two inverters, $P_1 N_1$ and $P_2 N_2$. Inverter elements P_2 and N_2 are tiny devices easily overdriven by the preceding complex transmission gate.

Figure 3-1c can be used only in a totally oxide-isolated or SOS CMOS technology. For power, it uses no dc supply, but works off the input clock. It is this type of innovative implementation of CMOS circuit design that has stirred the imagination of IC designers and system users alike regarding potential LSI circuits and applications for CMOS. It is with this in mind that the following three chapters will be devoted to CMOS LSI circuits. This chapter concerns itself with the topic of data conversion and telecommunications circuits.

DATA CONVERSION CIRCUITS

D/A Converters

D/A converters (DACs) are a key electronic data acquisition and manipulation system. Basically DACs contain a resistor network (either of the current ladder, R/2R current ladder, or voltage ladder types) in combination with current or voltage switches, a voltage reference, and a voltage follower. The use of CMOS switches and CMOS op amp followers is increasing. For detailed design information on DACs, see Eugene R. Hnatek, *A User's Handbook of Digital-to-Analog and Analog-to-Digital Converters*, John Wiley & Sons, Inc., New York, 1976.

DISCRETE D/A CONVERTERS*

D/A converter designs can be implemented by using standard CMOS logic-building blocks, as shown in Fig. 3-2 where three CD4007As perform the switch function using a 10-V logic level. A change in input logic level causes the output to swing to either the positive or the negative supply voltage. Power consumption is low, typically a few microwatts to a few milliwatts, depending on the ladder resistance and voltage choice. Details of the circuit shown in Fig. 3-2 can be found in RCA Application Note ICAN6080.

* Portions of this section excerpted from (1) "Digital-to-Analog Conversion Using the RCA CD4007A COS/MOS IC," RCA Application Note ICAN6080, RCA, Somerville, NJ, 1974; (2) "Transmission and Multiplexing of Analog or Digital Signals Utilizing the CD4016A Quad Bilateral Switch," RCA Application Note ICAN6601, RCA, Somerville, NJ, 1975. Used with permission.

66 Data Conversion and Telecommunications Circuits

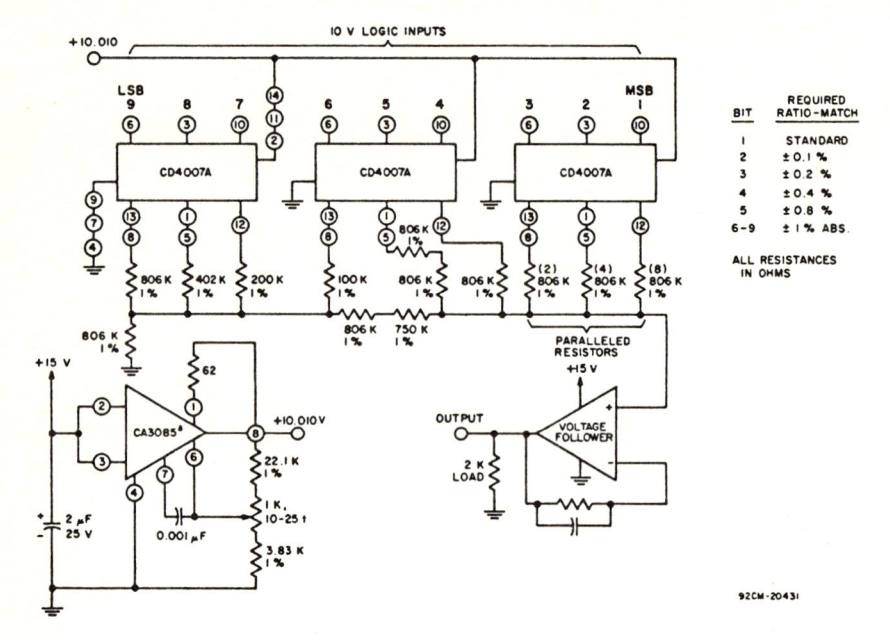

Fig. 3-2 A 9-b DAC Using the CD4007A Dual Complementary Pair Plus Inverter. *(Courtesy RCA Solid State Division.)*

Figure 3-3 shows a circuit diagram of a voltage-fed D/A converter that uses two CD4016A Qual bilateral switches and an R/2R network. The circuit provides 2^n equal voltage steps, where n is the number of legs used. V_R is the dc reference input (voltage). The R/2R ladder network functions as a successive voltage divider. Each CD4016A connects or disconnects its leg to the reference voltage or to ground, and the output voltage is proportional to the input reference. Thus, for the 4-b system shown, the LSB (least-significant bit) contributes $V_R/2^4$ or $V_R/16$ to the total output voltage. A CD4004A binary counter is used to provide a staircase output from the D/A converter. An analog reference voltage may be used as the input. Laboratory data indicates less than a 1 percent error in accuracy. The overall accuracy is dependent upon the switch ON resistance (R_{on}), the leakage current, and the discrete component accuracy.

The current-fed D/A converter is shown in Figure 3-4. This circuit is similar in performance to the voltage-fed R/2R, except that the current-fed circuit has a higher accuracy. Higher accuracy is obtained because the same current is supplied from a constant-current source over a wide variety of load values (including R_{on} of the CD4016As). The accuracy of

Data Conversion Circuits   67

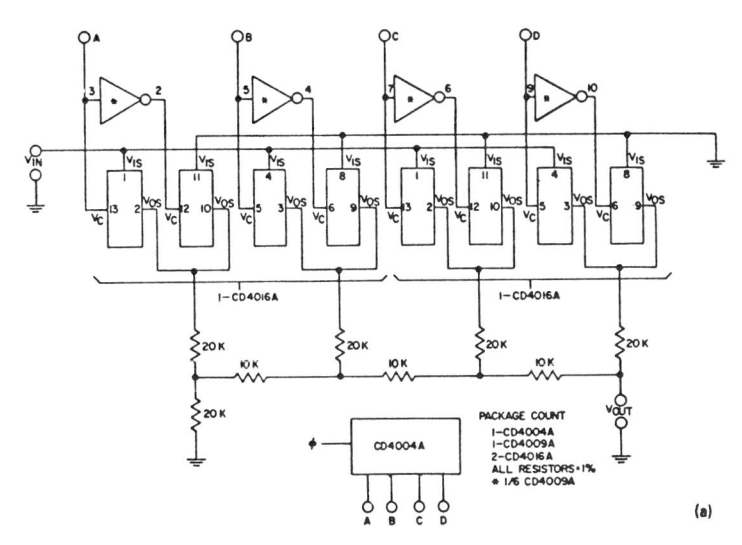

Fig. 3-3 Voltage-Fed R/2R Resistive Ladder Network D/A Converter. *(Courtesy RCA Solid State Division.)*

the circuit is limited only by the accuracy of the resistors used. In this circuit, a constant-current source feeds the R/2R network through the CD4016A. The current is divided by each branch in the same manner as voltage is divided in the voltage-fed circuit.

These examples demonstrate the accuracy and simplicity available

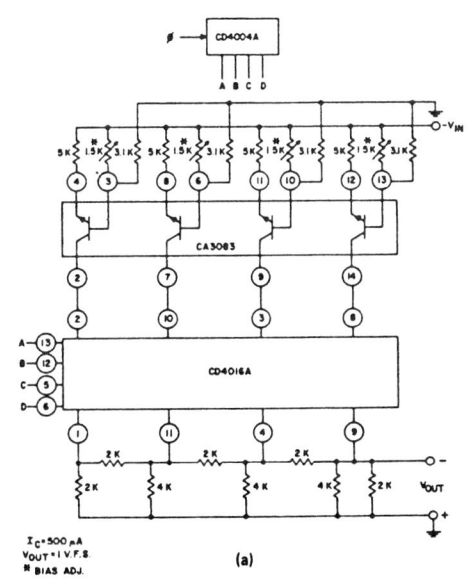

Fig. 3-4 Current-Fed R/2R Resistive Ladder Network D/A Converter. *(Courtesy RCA Solid State Division.)*

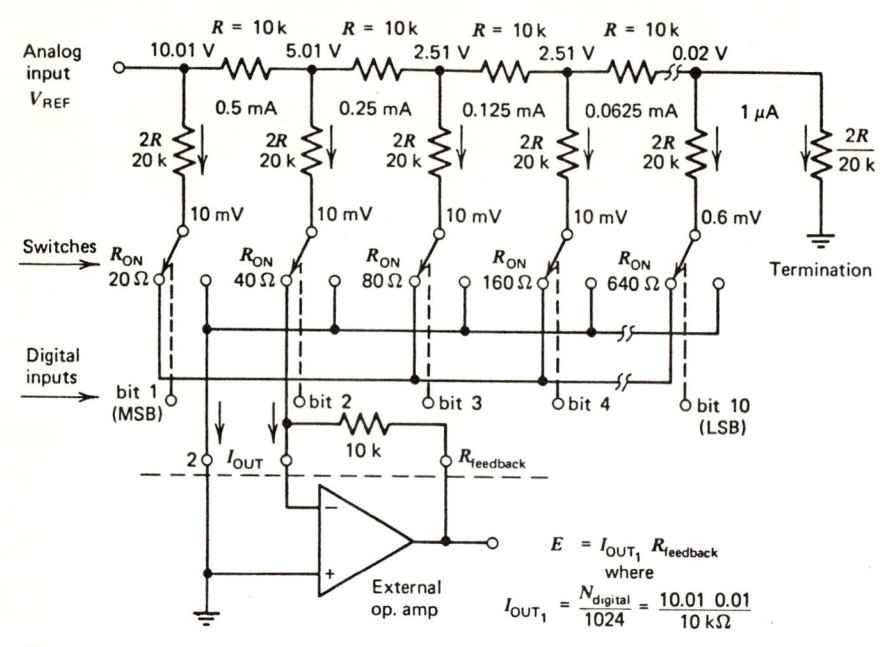

68 Data Conversion and Telecommunications Circuits

Fig. 3-5 Functional Diagram of the AD7520 D/A Converter with $V_{REF} = 10.01$ V. Bits 5 through 9 Are Omitted for Clarity. *(Courtesy Analog Devices.)*

with economical components and modest power supply requirements. In addition, the design flexibility afforded by the CMOS building blocks simplifies the generation of DAC systems tailored to individual needs. CMOS switches used in conjunction with CMOS counters also find application in A/D conversion systems. The low-power and high-noise-immunity features of these devices make them attractive A/D system components.

COMMERCIALLY AVAILABLE MONOLITHIC CMOS D/A CONVERTERS

The AD7520 10-b Multiplying CMOS D/A Converter*

Figure 3-5 depicts the functional diagram of Analog Devices' AD7520 10-b multiplying CMOS D/A converter. It consists of 10 CMOS SPDT current-steering switches and their driving logic and a thin-film–on–CMOS inverted R/2R ladder network. The digital input, which responds to the wide voltage swings of CMOS logic, is also compatible with TTL

---

* J. Cecil and J. Whitmore, "A 10-bit Monolithic CMOS D/A Converter," *Analog Dialogue*, Vol. 8-1, Analog Devices, Norwood, MA, 1974. Used with permission.

Data Conversion Circuits   69

and DTL logic levels. Two complementary current outputs are available for use with inverting operational amplifiers.

Besides the 10-b resolution, the AD7520 family has maximum nonlinearities as low as ±0.05 percent of V_{REF}, a nonlinearity temperature coefficient of 2 ppm/°C, and a maximum feedthrough error of $\frac{1}{2}$ LSB (0.1 percent) at 100 kHz. Typical settling time following a full-scale digital input change is 500 ns.

While monolithic 6- and 8-b converters have been easily achievable, 10-b conversion has been more difficult to obtain with good yields (and low cost) because of the finite beta (β) of the switching devices, the V_{BE}-matching requirement, the matching and tracking requirements of the resistance ladders, and the tracking limitations caused by the thermal gradients produced by high internal power dissipation.

All of these problems can be solved or avoided with CMOS devices. They have nearly infinite current gain, eliminating β problems. There is no equivalent in CMOS circuitry to a bipolar transistor's V_{BE} drop; instead, a CMOS switch in the on condition is almost purely resistive, with the resistance value controllable by device geometry.

The R/2R ladder is composed of 2-kΩ/□ silicon-chromium resistors (a 10-kΩ resistor has a very manageable length/width ratio of 5 : 1) deposited on the CMOS die. While the absolute temperature coefficient of these resistors is 150 ppm/°C, their tracking with temperature is better than 1 ppm/°C. The feedback resistor for the output amplifier is also provided on the chip, to ensure that the DACs gain temperature coefficient is better than 10 ppm/°C by compensating for the absolute temperature coefficient of the network.

Finally, the low on-chip dissipation of only 20 mW (including the dissipation of the ladder network), in conjunction with the excellent tracking capabilities of the thin-film resistors, minimizes linearity drift problems caused by internally generated thermal gradients. It also helps to minimize the power and cooling requirements for circuitry in which the AD7520 is used.

Referring to Fig. 3-5, binary weighted currents flow continuously in the shunt arms of the network, with 10 V applied at the reference input: 0.5 mA flows in the first, 0.25 mA in the second, 0.125 mA in the third, and so on. The $I_{out,1}$ and $I_{out,2}$ output buses are maintained at ground potential, either by operational amplifier feedback or by a direct connection to common.

The switches steer the current to the appropriate output lines in response to the individually applied logic levels. For example, a high digital input to SW₁ will cause the 0.5 mA of the MSB to flow through $I_{out,1}$. When the digital input is low, the current will flow through $I_{out,2}$. If $I_{out,1}$ flows through the summing point of an operational amplifier and $I_{out,2}$ flows to ground, then "high" logic will cause the nominal output voltage of the op amp to be -0.5 mA $\times$ 10 kΩ = -5 V, for a positive reference voltage of 10 V, while "low" logic will make the contribution of bit 1 zero. With all bits on (i.e., "high"), the nominal output will be -9.99 V. With all bits off, the output will be zero.

Linearity errors and—more important—their variation with temperature are affected by variations of resistance in both the resistors and the switches. The resistor-network tracking is excellent. However, the switches, while tracking one another, will not track the resistance network. With identical switches having realistic resistance values (say, 100 Ω), one would expect that, as temperature changed, the variation of resistance in the series legs would transform the network into an R/nR network, with n sufficiently different from two to destroy the binary character of the network and cause the converter to become non-monotonic.

The key to the linearity of the AD7520 is that the geometries of the switches are tapered so as to obtain ON resistances that are related in binary fashion, for the first 6 b. Thus, the nominal values of switch resistance range from 20 Ω for the first bit and 40 Ω for the second bit through 640 Ω for the last 5 b. The effect is, as can be seen in Fig. 3-5, to provide equal voltages at the ends of the six most-significant arms of the ladder. (0.5 mA $\times$ 20 Ω = 0.25 mA $\times$ 40 Ω, etc. = 10 mV.) Since this drop is, in effect, in-series with the reference, it causes an initial 0.1 percent scale-factor ("gain") error, which is well within the specifications but does not affect the linearity. Since the switches tend to track one another with temperature, linearity is essentially unaffected by temperature changes, and the gain error is held to within the 10-ppm/°C specification.

A 10-b linearity could have been obtained by scaling the ON resistance of all the switches to a negligible value, say 10 Ω, but the switches would have required very large geometries, which would result in a 30 to 50 percent larger chip, at a substantial increase in cost.

Figure 3-6 illustrates one of the 10 current switches and its associated internal drive circuitry. The geometries of the input devices 1 and 2 are scaled to provide a switching threshold of 1.4 V, which permits the digital inputs to be compatible with TTL, DTL, and CMOS. The input

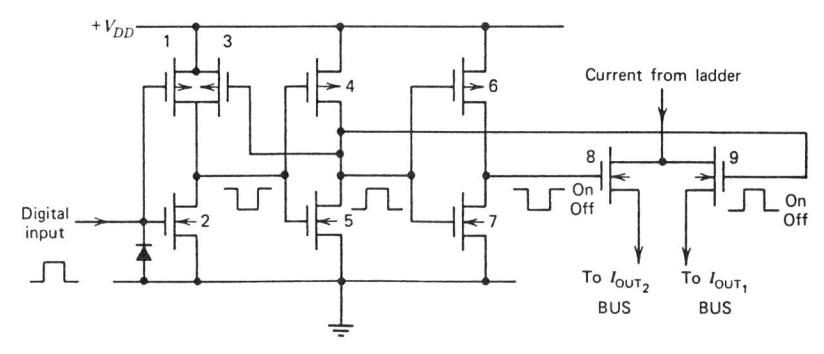

Fig. 3-6 CMOS Switch Used in the AD7520. Digital Input Levels May Be DTL, TTL, or CMOS. *(Courtesy Analog Devices.)*

stage drives two inverters (4, 5, 6, and 7), which in turn drive the N-channel output switches.

Figure 3-7 shows the equivalent circuit of the AD7520 at the two extremes of input, (*a*) all inputs high and (*b*) all inputs low. V_{REF} (or I_{REF}, if a current reference is used) sees a nominal 10-kΩ resistance, regardless of the switch states. The current source $I_{REF}/1024$ represents a 1-LSB current loss through the 20-kΩ ladder termination resistor. R_{on}, in this case, is the equivalent resistance of all 10 switches connected to (1) the $I_{out,1}$ bus or (2) the $I_{out,2}$ bus. Current source I_{lkg} represents junction and surface leakage to the substrate. Capacitors $C_{out,1}$ and $C_{out,2}$ are the output capacities to ground for the on and off switches. C_{SD} is the open-switch capacitance.

The 1000:1 ratio between R_{ladder} and R_{on} provides a number of benefits, all related to the small voltage drop across R_{on}:

1. V_{REF} can assume values exceeding the absolute maximum CMOS rating, V_{DD}. For example, V_{REF} could be as large as ± 25 V, even if the AD7520's V_{DD} were only $+17$ V.

2. The nonlinearity temperature coefficient depends primarily on how well the ladder resistances track. Since R_{on} is only a small fraction of R_{ladder}, R_{on} tracking errors will be felt only as second- and third-order effects.

3. The same argument holds true for power supply variations. Any change of switch ON resistance, as the power supply changes, will be swamped by the 1000:1 attenuation factor. Power supply rejection is better than $\frac{1}{3}$ LSB per volt.

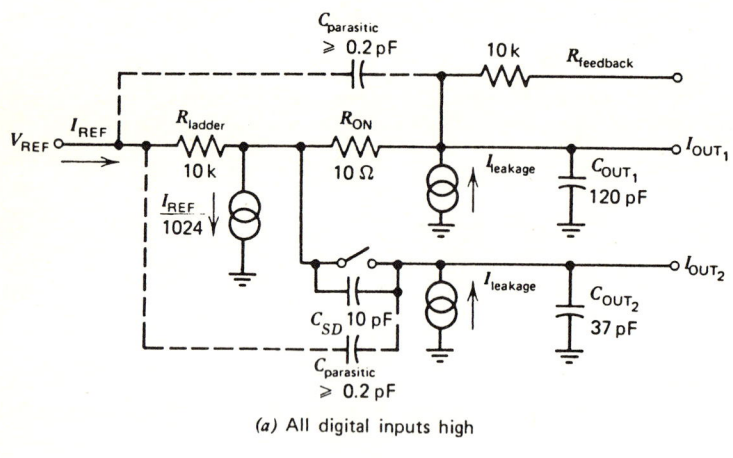

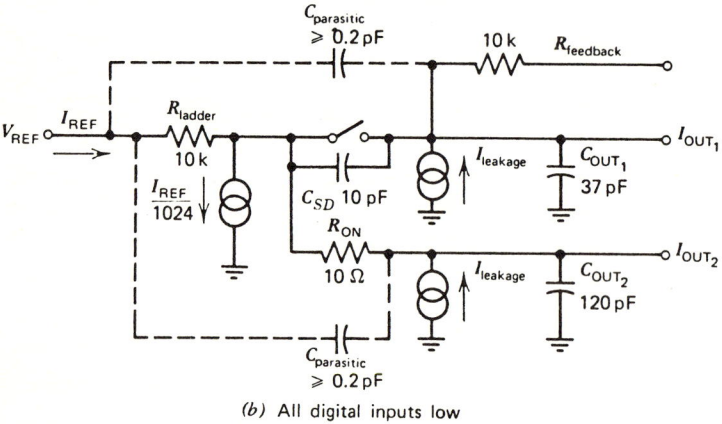

72 Data Conversion and Telecommunications Circuits

Fig. 3-7 Equivalent Circuits of the AD7520 D/A Converter. *(Courtesy Analog Devices.)*

4. If V_{REF} is a fast ac signal, the feedthrough coupling via CSD, the open-switch capacitance, will be negligible, again because of the 1000:1 voltage stepdown. The parasitic capacitances from V_{REF} to $I_{out,1}$ and $I_{out,2}$ comprise the major source of ac feedthrough. Careful board layout by the user can result in less than ½ LSB of ac feedthrough at 100 kHz.

Since the ON resistance depends only on the value of V_{DD}, not the current through the switch, and the resistance network is unaffected by V_{REF}, the full-scale output current (all bits high) is nominally $V_{REF}/10.01$ kΩ less the constant-current losses shown in Fig. 3-7. This means that I_{out} is almost perfectly proportional to V_{REF} over the entire range from -10

Data Conversion Circuits   73

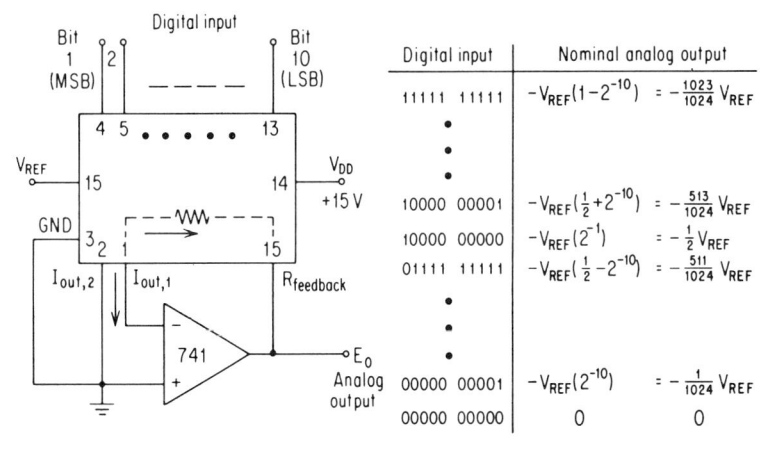

Digital input	Nominal analog output
11111 11111	$-V_{REF}(1-2^{-10}) = -\frac{1023}{1024}V_{REF}$
⋮	
10000 00001	$-V_{REF}(\frac{1}{2}+2^{-10}) = -\frac{513}{1024}V_{REF}$
10000 00000	$-V_{REF}(2^{-1}) = -\frac{1}{2}V_{REF}$
01111 11111	$-V_{REF}(\frac{1}{2}-2^{-10}) = -\frac{511}{1024}V_{REF}$
⋮	
00000 00001	$-V_{REF}(2^{-10}) = -\frac{1}{1024}V_{REF}$
00000 00000	0 = 0

Fig. 3-8 The AD7520 as a Unipolar Binary Digital-to-Voltage Converter (two-quadrant multiplier). *(Courtesy Analog Devices.)*

to +10 V. Equally important, the conversion linearity error (0.05 percent) is independent of the sign or magnitude of V_{REF}.

The extremely low analog linearity error at constant digital input results in excellent fidelity to the input waveform, which suggests some interesting possibilities for the AD7520 in the calibration and control of gain in signal generators, high-fidelity amplifiers, and response-testing systems.

The two most common forms of application are in unipolar D/A conversion (two-quadrant multiplication) and bipolar offset binary conversion (four-quadrant multiplication) shown in Figs. 3-8 and 3-9.

The response equation for unipolar conversion for Fig. 3-8 is nominally

$$V_0 = \frac{N_{binary}}{1024}V_{REF} \tag{3-1}$$

Response to typical codes are tabulated. Since V_{REF} may be positive or negative, two-quadrant multiplication is inherent. Circuit gain is easily trimmed by adjusting V_{REF}, inserting adjustable resistance in series with V_{REF} or $R_{feedback}$, or by tweaking scale factors elsewhere in the system. As noted previously, once set, using low TC (temperature coefficient) trim resistors, gain stability with temperature is excellent.

The offset binary response equation for Fig. 3-9 is nominally

$$V_0 = -\left[\frac{N_{binary}}{512} - 1\right]V_{REF} \tag{3-2}$$

74 Data Conversion and Telecommunications Circuits

Digital input		Nominal analog output	
11111	11111	$-V_{REF}(1-2^{-9})$	$= -\frac{511}{512} V_{REF}$
•			
•			
•			
10000	00001	$-V_{REF}(2^{-9})$	$= -\frac{1}{512} V_{REF}$
10000	00000	0	0
01111	11111	$V_{REF}(2^{-9})$	$= +\frac{1}{512} V_{REF}$
•			
•			
•			
00000	00001	$V_{REF}(1-2^{-9})$	$= +\frac{511}{512} V_{REF}$
00000	00000	V_{REF}	$= +1 \times V_{REF}$

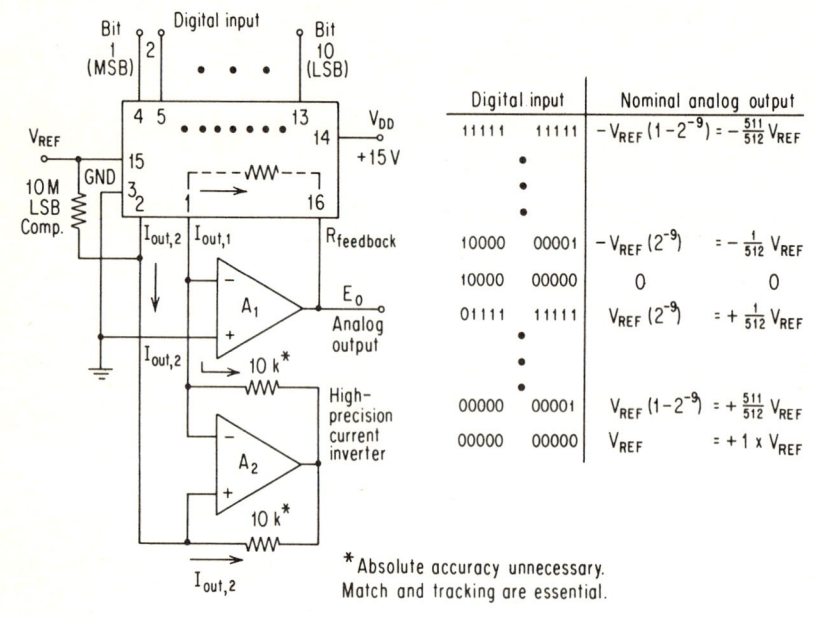

*Absolute accuracy unnecessary. Match and tracking are essential.

Fig. 3-9 The AD7520 as a Bipolar Offset Binary Digital-to-Voltage Converter (four-quadrant multiplier). *(Courtesy Analog Devices.)*

Responses to typical codes are again tabulated. If the MSB is complemented, the conversion relationship will be recognized as appropriate for a 2's complement input, but with a negative scale factor. The MSB determines the sign, and the last 9 b determine the magnitude in 2's complement notation. Since V_{REF} may be either positive or negative, four-quadrant multiplication is inherent.

In this configuration, $I_{out,2}$, which is the complement of $I_{out,1}$, is inverted and added to $I_{out,1}$, halving the resolution (of each polarity) and doubling the gain. The 10-MΩ resistor corrects for a 1/1024 of the difference (inherent in this technique) between $I_{out,1}$ and $I_{out,2}$ at zero (10000 00000). A_2 is shown as a current inverter.

If sign-magnitude coding is desired, to obtain bipolar conversion with the full 10-b plus-sign resolution, the output of the unipolar conversion circuit may be fed into a sign-magnitude converter, such as that shown in Fig. 3-10.

AD7520 Applications: Power Series Generation Several practical applications using the AD7520 DAC are now presented.

Figure 3-11 illustrates a method of obtaining squared, cubed, and so

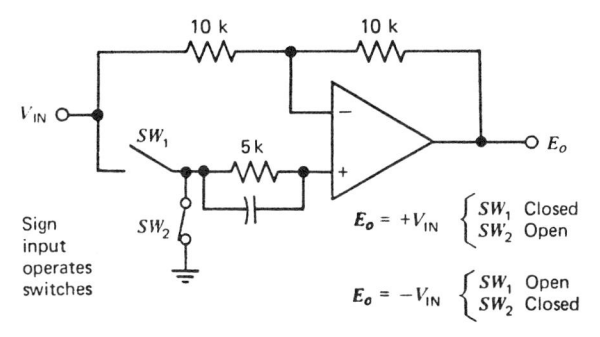

Fig. 3-10 Sign-Magnitude to Bipolar Converter. *(Courtesy Analog Devices.)*

on functions by driving all digital inputs from common bus lines, and by feeding the V_{REF} terminals of each DAC with the output of the previous DAC. Figure 3-12 goes a step further, showing the transfer functions for up to the fifth power.

AD7520 Applications: Time Delay Figure 3-13 shows a method of generating time delays using the AD7520 as the control element. The amount of current fed to the active integrator is determined by the DAC's digital input. When a trigger input to the 555 timer goes low, the timer's output goes high until the integrator's output charges to $\frac{2}{3} V_{DD}$—at which time the timer's output goes low, and the integrator is reset to zero.

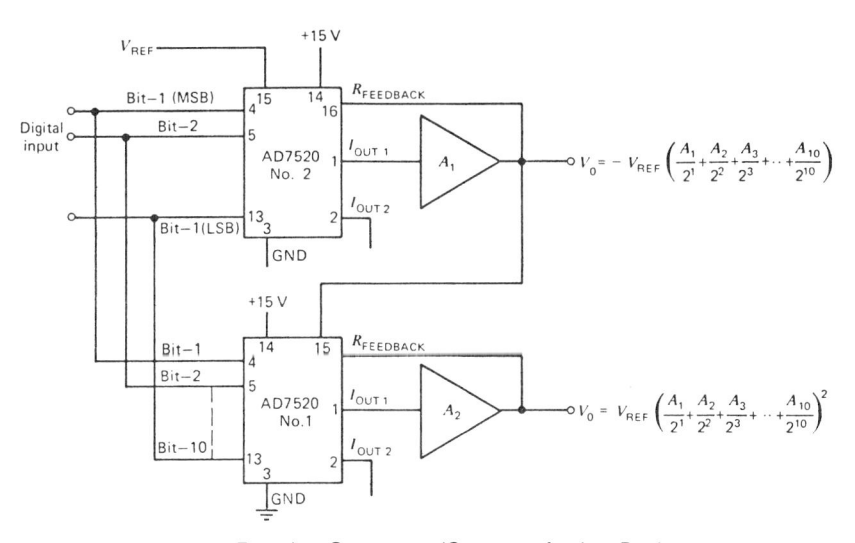

Fig. 3-11 Nonlinear Function Generator. *(Courtesy Analog Devices.)*

$$V_0 = -V_{REF}\left(\frac{A_1}{2^1}+\frac{A_2}{2^2}+\frac{A_3}{2^3}+\cdots+\frac{A_{10}}{2^{10}}\right)$$

$$V_0 = V_{REF}\left(\frac{A_1}{2^1}+\frac{A_2}{2^2}+\frac{A_3}{2^3}+\cdots+\frac{A_{10}}{2^{10}}\right)^2$$

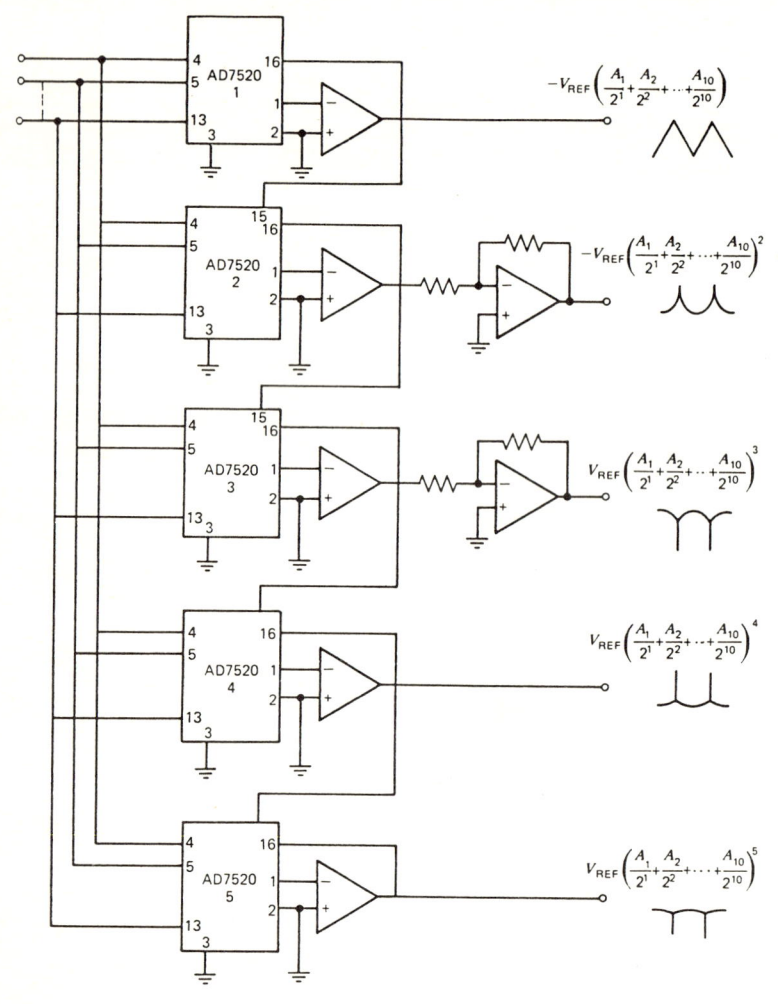

Fig. 3-12 Power Series Generation. *(Courtesy Analog Devices.)*

The AD7522 10-b Buffered Multiplying D/A Converter

The AD7522 (Fig. 3-14) is a monolithic CMOS 10-b multiplying D/A converter, with an input buffer and a holding register, allowing direct interface with microprocessors. Most applications require the addition of only an operational amplifier and a reference voltage.

The AD7522 is packaged in a 28-pin DIP (dual in-line package), as shown in Fig. 3-15, and operates with a +15-V main supply at 2 mA max, and a logic supply of +5 V for TTL interface or +10 to +15 V for CMOS interface.

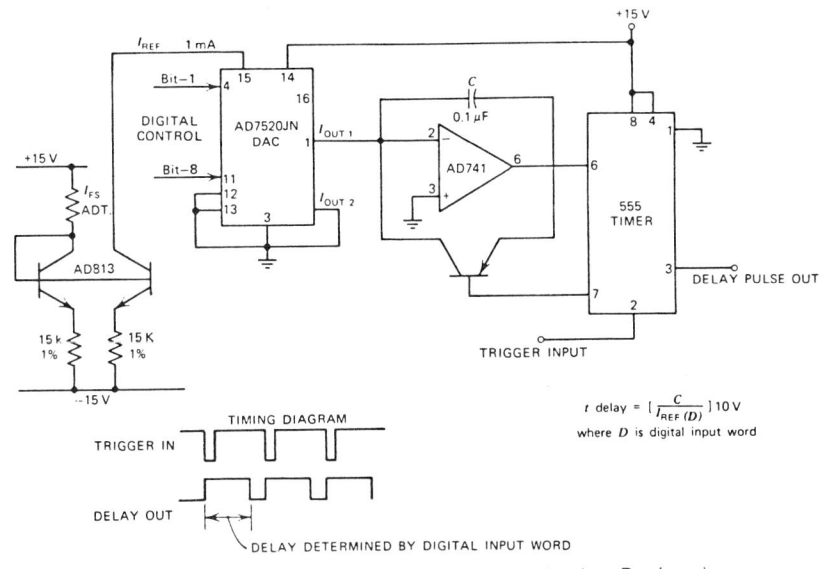

Fig. 3-13 Digitally Controlled Time Delay. *(Courtesy Analog Devices.)*

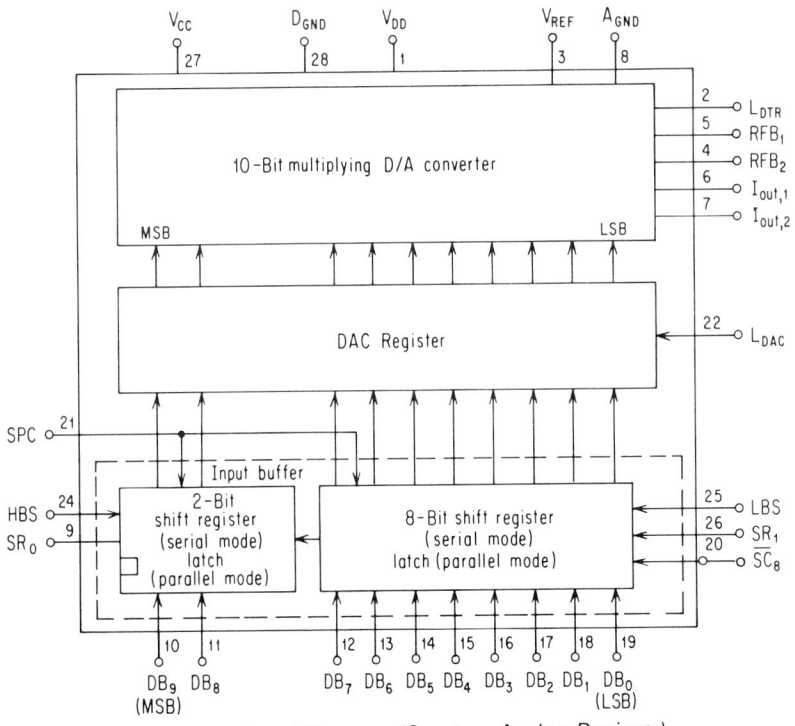

Fig. 3-14 AD7522 Functional Diagram. *(Courtesy Analog Devices.)*

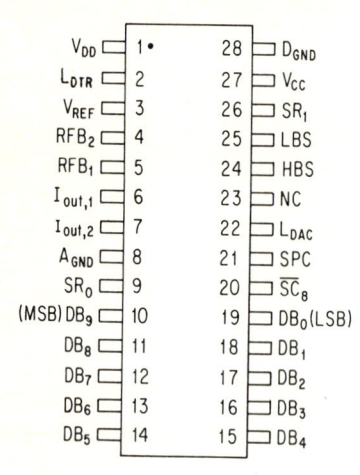

78 Data Conversion and Telecommunications Circuits

```
         ┌──────┐
   V_DD ─┤1•   28├─ D_GND
   L_DTR ─┤2    27├─ V_CC
   V_REF ─┤3    26├─ SR_1
   RFB_2 ─┤4    25├─ LBS
   RFB_1 ─┤5    24├─ HBS
  I_out,1─┤6    23├─ NC
  I_out,2─┤7    22├─ L_DAC
   A_GND ─┤8    21├─ SPC
    SR_0 ─┤9    20├─ SC_8
(MSB)DB_9─┤10   19├─ DB_0(LSB)
    DB_8 ─┤11   18├─ DB_1
    DB_7 ─┤12   17├─ DB_2
    DB_6 ─┤13   16├─ DB_3
    DB_5 ─┤14   15├─ DB_4
         └──────┘
```

Fig. 3-15 AD7522 Pin Connection Diagram. *(Courtesy Analog Devices.)*

A thin-film–on–high-density-CMOS process, using silicon nitride passivation, ensures high reliability and excellent stability.

The AD7522 has 10 SPDT N-channel current-steering switches and a thin-film–on–CMOS R/2R ladder attenuator for current weighting. In addition, it has a dual-rank input storage system consisting of 10 D-type level-triggered holding latches and a 10-b edge-triggered serial/parallel input-loading register (which in turn consists of 2 controllable bytes of 8- and 2-b capacity).

Basic unipolar operation (either fixed-reference or two-quadrant multiplication) requires only the addition of an external positive or negative, constant or variable reference voltage or current, and an operational amplifier, as is shown in Fig. 3-16. For bipolar conversion (four-quadrant multiplication) with offset binary or 2's complement coding, one additional operational amplifier is needed.

The main (V_{DD}) supply requires a nominal +15 V at 2 mA max; 1 μA is typical since most of the current is required only during switching. The choice of the logic (V_{CC}) supply depends on the logic-interface requirement. For example, if $V_{CC} = +5$ V, the digital inputs are TTL-compatible. If $V_{CC} = +10$ to $+15$ V, the digital inputs and outputs are CMOS-compatible.

Three grades of conversion linearity are offered—8, 9, and 10 b. Typical current settling time following a full-scale code change on the digital inputs is 500 ns.

The most interesting aspect of the AD7522 to the system designer is the DAC's double-buffered input structure, which offers tremendous

Data Conversion Circuits 79

versatility yet is seldom found even in discrete module D/A converters. Salient features include the following:

1. Logic-controlled choice of serial or parallel loading.
2. A "load-display" choice, which either allows new data to update the DAC or locks out unwanted data appearing at the digital inputs. If the AD7522 is used with a CPU (central processing unit) data bus, this lockout function allows the CPU or other I/O peripheral to place data on the bus without altering data that was previously loaded into the AD7522.
3. Byte-serial (or-parallel) loading allows a 10-b word to be loaded into the DAC from either an 8-b microcomputer data bus or from a 10-bit (or more) paralleled line.
4. A serial output allows recovery of data from the input register.
5. A short-cycle feature allows 8- b to the MSB to be loaded serially.

The AD7522 can operate in both a unipolar and a bipolar mode. Figure 3-17 shows the analog circuit connections required for unipolar operation. The input-code/output-voltage relationship is shown in Table 3-1.

Figure 3-18 shows the analog circuit connections required for bipolar operation. Table 3-2 depicts the input-code/output-voltage relationship.

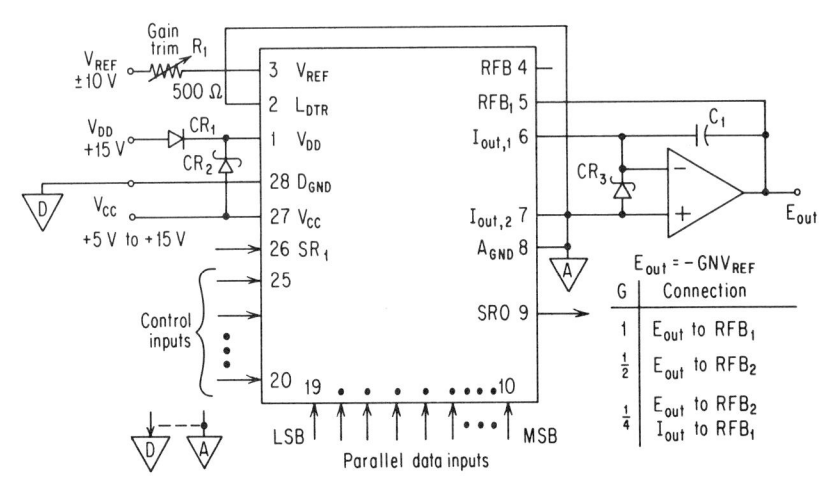

Fig. 3-16 Connecting the AD7522 for Unipolar D/A Conversion. *N* Is a Fractional 10-b Binary Number from 0 to $(1 - 2^{-10})$.

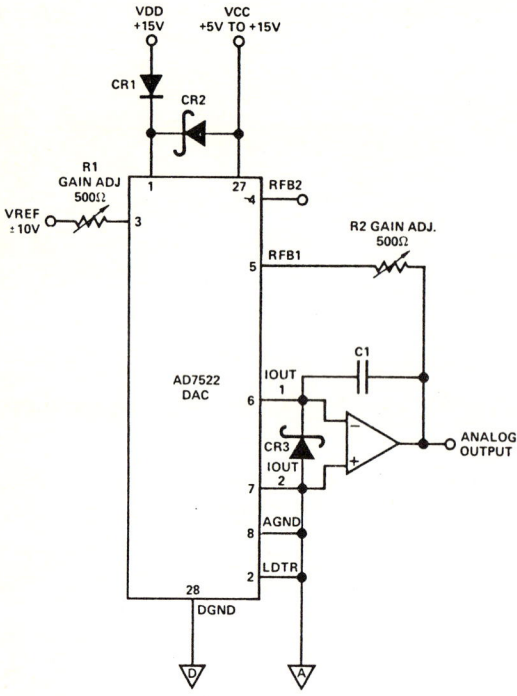

Fig. 3-17 Unipolar Binary Operation (two-quadrant multiplication). *Courtesy Analog Devices.)*

Figure 3-19 illustrates the logic connections for loading single-byte parallel data into the input buffer. DB_0 should be grounded on K and T versions and DB_0 and DB_1 should be grounded on J and S versions for monotonic operation of the DAC. DB_9 is always the MSB, whether 8-, 9-, or 10-b linear AD7522s are used.

TABLE 3-1 Unipolar Code Table

DIGITAL INPUT	ANALOG OUTPUT
1 1 1 1 1 1 1 1 1 1	$-VREF(1 - 2^{-10})$
1 0 0 0 0 0 0 0 0 1	$-VREF(1/2 + 2^{-10})$
1 0 0 0 0 0 0 0 0 0	$-VREF/2$
0 1 1 1 1 1 1 1 1 1	$-VREF(1/2 - 2^{-10})$
0 0 0 0 0 0 0 0 0 1	$-VREF(2^{-10})$
0 0 0 0 0 0 0 0 0 0	0

SOURCE: Courtesy Analog Devices.

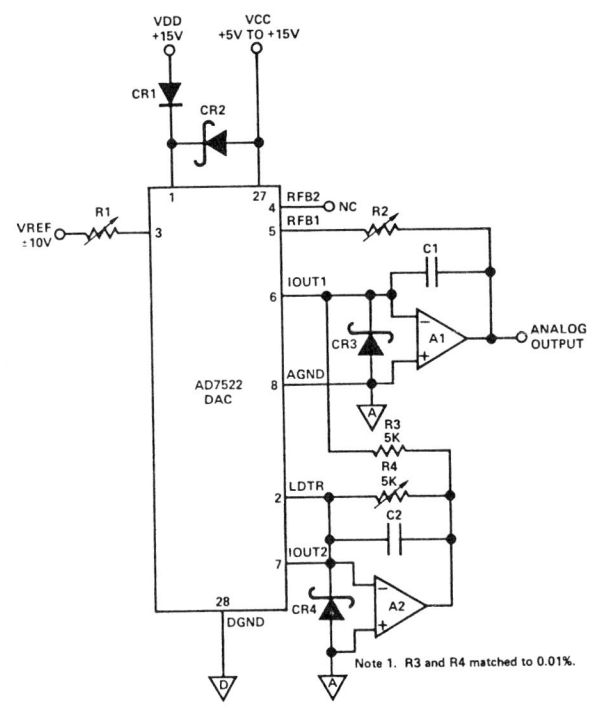

Fig. 3-18 Bipolar Operation. *(Courtesy Analog Devices.)*

When data is stable on the parallel inputs (DB$_0$ through DB$_9$), it can be transferred into the input buffer on the positive edge of the strobe pulse.

Data is transferred from the input buffer to the DAC register when LDAC (load digital-to-analog converter) is a logic 1. LDAC is a level-actuated (versus edge-triggered) function, and must be held high at least 3 μs for data transfer to occur.

TABLE 3-2 Bipolar Code Table

DIGITAL INPUT	ANALOG OUTPUT
1 1 1 1 1 1 1 1 1 1	$-$VREF $(1 - 2^{-9})$
1 0 0 0 0 0 0 0 0 1	$-$VREF (2^{-9})
1 0 0 0 0 0 0 0 0 0	0
0 1 1 1 1 1 1 1 1 1	VREF (2^{-9})
0 0 0 0 0 0 0 0 0 1	VREF $(1 - 2^{-9})$
0 0 0 0 0 0 0 0 0 0	VREF

SOURCE: Courtesy Analog Devices.

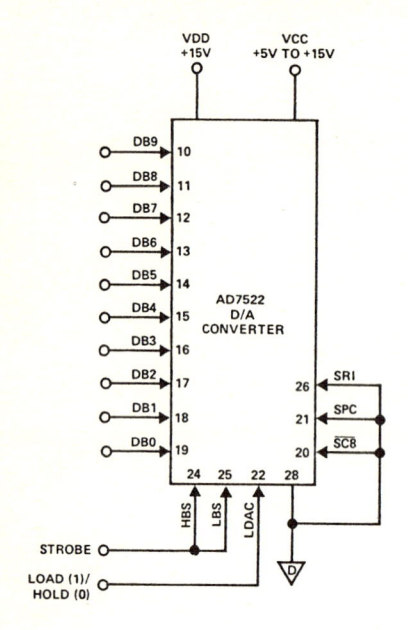

82 Data Conversion and Telecommunications Circuits

Fig. 3-19 Single-Byte Parallel Loading. *(Courtesy Analog Devices.)*

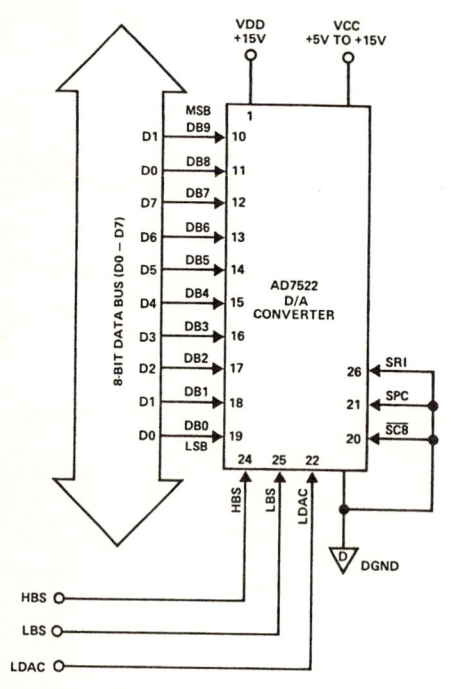

Fig. 3-20 Two-Byte Parallel Loading. *(Courtesy Analog Devices.)*

Data Conversion Circuits   83

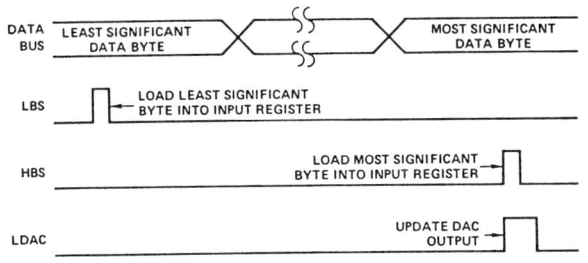

Fig. 3-21 Timing Diagram for Two-Byte Parallel Loading. *(Courtesy Analog Devices.)*

Figures 3-20 and 3-21 show the logic connections and timing requirements for interfacing the AD7522 to an 8-b data bus for 2-byte parallel loading of a 10-b word.

First, the least-significant data byte (DB_0 through DB_7) is loaded into the input buffer on the positive edge of LBS (low byte strobe). Subsequently, the data bus is used for status indication and instruction fetching by the CPU. When the most-significant data byte (DB_8 and DB_9) is available on the bus, the input buffer is loaded on the positive edge of HBS (high byte strobe). The DAC register updates to the new 10-b word when LDAC is high. LDAC may be exercised coincident with, or at any time after, HBS loads the second byte of data into the input buffer.

Figures 3-22 and 3-23 show the connections and timing diagram for serial loading. To load a 10-b word ($\overline{SC_8} = 1$), HBS and LBS must be strobed simultaneously with exactly 10 positive edges to clock the serial

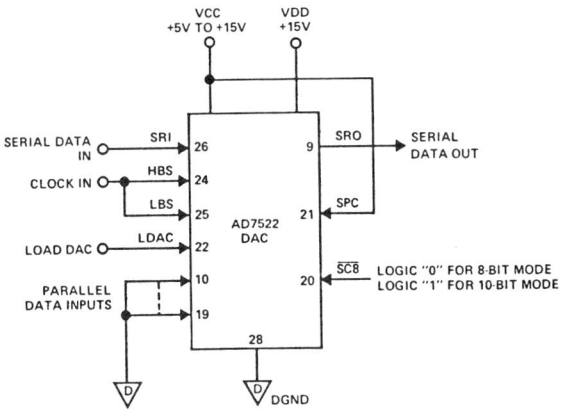

Fig. 3-22 Serial 8- and 10-b Loading. (Analog Outputs Not Shown for Clarity). *(Courtesy Analog Devices.)*

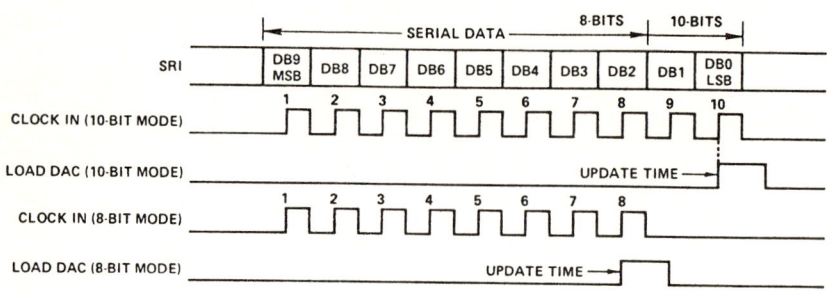

84 Data Conversion and Telecommunications Circuits

Fig. 3-23 Timing Diagram for Serial 8- and 10-b Loading. *(Courtesy Analog Devices.)*

data into the input buffer. For 8-b words ($\overline{SC}_8 = 0$), only eight positive edges are required.

The DAC register can now be loaded by holding LDAC high.

The Teledyne 8640/8641 12-b Multiplying D/A Converters

The 8640 and 8641 are 12-b monolithic CMOS D/A converters featuring double-layer metal interconnections for improved high-speed operation and lower cost. The use of precision thin-film deposition resistors provides 12-b linearity without laser trimming, thus eliminating any long-term instabilities laser trimming might introduce. The use of compensating FET switches in the feedback resistor and at the end of the ladder chain reduces the gain error temperature coefficient to a maximum of 2 ppm/°C.

The 8640/8641 12-b multiplying D/A converters consist of a highly stable thin-film R/2R ladder and 12 CMOS current switches on a monolithic chip. Most applications require the addition of only an output operational amplifier and a voltage or current reference.

The 8641 is guaranteed to have a linearity of $\pm\frac{1}{2}$ LSB (0.012 percent), while the 8640 is identical except for ± 1 LSB (0.024 percent) linearity error. Both devices are encased in an 18-pin DIP.

The simplified D/A circuit is shown in Fig. 3-24. An inverted R/2R ladder structure is used—that is, the binarily weighted currents are switched between the I_{out1} and the I_{out2} bus lines, thus maintaining a constant current in each ladder leg independent of the switch state. The CMOS current switches are similar to that shown in Fig. 3-6 for Analog Devices' AD7520. The geometries of devices 1, 2, and 3 are optimized to make the digital control inputs DTL-, TTL-, and CMOS-compatible over the full military temperature range. The input stage drives two inverters (devices 4, 5, 6, and 7) which in turn drive the two output N channels. The ON resistances of the switches are binarily scaled so the voltage drop

Data Conversion Circuits   85

across each switch is the same. For example, switch 1 of Fig. 3-24 was designed for an ON resistance of 10Ω, switch 2 for 20 Ω, and so on. For a 10-V reference input, the current through switch 1 is 0.5 mA, the current through switch 2 is 0.25 mA, and so on, thus maintaining a constant 5-m V drop across each switch. It is essential that each switch voltage drop be equal if the binarily weighted current division property of the ladder is to be maintained.

The primary electrical characteristics of the 8640/8641 are summarized in Table 3-3. Such DACs find use in CRT (cathode-ray tube) graphics generation, synchro-to-digital converters, digitally controlled power supplies, function generators, digital filters, and programmable amplifiers, to name a few applications.

Commercial D/A Converter Summary

Table 3-4 summarizes the key electrical characteristics of the commercially available monolithic CMOS DACs.

A/D Converters*,†

As with D/A converters, A/D converters (ADCs) can be fabricated using standard CMOS logic circuits, as shown in Fig. 3-25 on p. 90.

The CD4040A binary counter, used in conjunction with an R/2R resistor ladder network, generates a staircase ramp at the negative input to the comparator, as shown in Fig. 3-25. When the ladder voltage matches the analog input, the comparator output goes low. This signal is inverted by the CD4007A and becomes a logic 1 latched into the flip-flop. This

---

* See E. R. Hnatek, *A User's Handbook of D/A and A/D Converters,* John Wiley & Sons, Inc., New York, 1976, for detailed discussions on A/D converters.

† Portions of this section excerpted from D. Block, "A Typical Data Gathering and Processing System Using CD4000 Series COS/MOS Parts," RCA Application Note ICAN6210, RCA, Somerville, NJ, 1974. Used with permission.

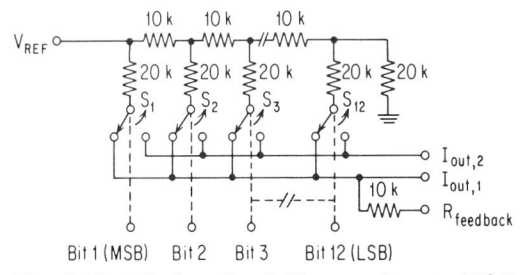

Fig. 3-24 D/A Functional Diagram (Inputs HIGH). Digital Inputs are DTL-, TTL-, and CMOS-Compatible.

TABLE 3-3 Electrical Characteristics of 8640/8641 Multiplying DACs

Parameter	$T_A = 25°C$ Min	Typ	Max	T_A = Full-Temp. Range Min	Typ	Max	Units	Conditions
Static Accuracy								
Resolution	12			12			bits	
Nonlinearity 8640			±1			±1	LSB	$V_{out.1} = V_{out.2} = 0$ V
8641			±0.5			±0.5	LSB	$V_{out.1} = V_{out.2} = 0$ V
Nonlinearity Tempco			2			2	ppm/°C	
Gain Error[a]		±0.3					%FSR[b]	
Gain Error Tempco			2			2	ppm/°C	
Power Supply Sensitivity			0.003			0.005	%per%	$V_{DD} = 15$ V ± 0.5 V
Output Leakage Current			±50			±200	nA	$V_{REF} = ±10$ V
Dynamic Performance								
Output Current Settling Time			500			500	ns	To 2 LSB (0.05%)
			1			1	µs	To ¼ LSB (0.01%)
Feedthrough Error			1			1	VPP	$V_{REF} = 20$ VPP @ 10 kHz
Reference Input								
Input Resistance	5	10	20	5		20	kΩ	
Digital Inputs								
V_{INH}		2.4			2.4		V	
V_{INL}	0.8			0.8			V	
Input Leakage Current			±1			±1	µA	$V_{in} = 0$ or 15 V
Input Capacitance			8			8	pF	
Analog Output Capacitance								
$C_{out.1}$			200			200	pF	Digital Inputs = V_{INH}
$C_{out.2}$			60			60	pF	Digital Inputs = V_{INH}
$C_{out.1}$			60			60	pF	Digital Inputs = V_{INL}
$C_{out.2}$			200			200	pF	Digital Inputs = V_{INL}
Power Requirements								
V_{DD} Range	5		16	5		16	V	Accuracy is not Guaranteed over This Range
I_{DD}			2			2	mA	Digital Inputs = V_{INH} or V_{INL}

NOTE: Unless otherwise specified, $V_{DD} = +15$ V, $V_{REF} = +10$ V.
[a] Using internal feedback resistor.
[b] FSR = full-scale range.

action inhibits additional clocks to the counter and indicates that the conversion is complete. The output of the counter, which is buffered by high-current CD4041As to minimize switch impedance effects on the resistor ladder, is the digital equivalent of the analog input voltage. A RESET clears the flip-flop and resets the counter so that the next conversion can begin.

To generate a staircase from -5 to $+10$ V, one end of the resistor ladder is connected to -5 V and the counter is connected with a V_{DD} of $+10$ V and a V_{SS} of -5 V. Since the CLOCK and RESET signals to the counter must then swing from -5 to 10 V, a CD4054A is used as a level translator.

A micropower op amp, the CA3080A, is used as a voltage comparator for the D/A converter. The op amp is gated off after the conversion is completed by turning off a P device of the CD4007A supplying bias current to the unit. In this way, power dissipation is reduced to a few microwatts in the standby state during those times in which a conversion is not actually being performed. With the active bias current (I_{ABC}) set at 15 μA, typical power dissipation for the CA3080 is about 500 μW with a V^+ of 10 V and a V^- of -5 V. Dissipation during a conversion would be approximately 20 mW at a clock rate of 100 kHz for the circuit shown in Fig. 3-25.

COMMERCIALLY AVAILABLE A/D CONVERTERS

High-Speed 6-b A/D Converter

A high-speed, 15–18 MHz (video), low-cost, low-power (50 mW) 6-b A/D converter has been developed by RCA Solid State Technology Center Laboratories* (the CA3300) using CMOS/SOS technology for radar signature analysis, transient analysis, distortion analysis, and high-speed data transmission.

8-b Differential Input A/D Microprocessor Converter

National Semiconductor's low-cost, 8-b A/D converter, called the "Naked 8" and designated as ADC 0800, is compatible with most microprocessors and in most applications does not require added logic.

Conversion time is 100 μs, and the device operates from a single 5-V supply. Four accuracies are available: $\pm\frac{1}{4}$ LSB full scale adjusted, $\pm\frac{1}{2}$ LSB unadjusted, $\pm\frac{1}{2}$ LSB adjusted, and ± 1 LSB unadjusted.

The devices are available in standard 0.3-in-wide, 20-pin packages, hermetic or plastic, and in three temperature ranges.

In addition, National is working on 10-b and 12-b versions of this ADC to provide a complete family of devices.

Analog Device's AD7574 A/D Converter

Analog Devices' AD7574 is a low-cost, 8-b CMOS successive-approximation A/D converter that not only has three-state outputs, but also interfaces with a microprocessor as if it were a memory.

* See A.G.F. Dingwall, "Monolithic Expandable 6b 15MHz CMOS/SOS A/D Converter," 1979 IEEE ISSCC Digest of Technical Papers, Philadelphia, PA, February 1979, for a detailed discussion.

TABLE 3-4 Summary of Commercially Available Single-Chip CMOS DACs

Manufacturer	Model	Resolution (Bits)	Linearity (±%FS)	Settling Time to ½ LSB (μs)	Gain TC (ppm/°C)	Power Required V(mA)	Current (mA)	Package, DIP (Pins)	Input Registers	Extended Temperature Range Available	Multiplying (Quadrants)	Special Features
Analog Devices	AD7523	8	0.2	150	10	5–16(0.1)	1.0	16			4	Second-Sourced by Intersil.
	AD7520	10	0.05	500	10	5–15(2)	1.5	16			4	Leakage I = 200 nA.
	AD7522	10	0.05	500	10	5–15(2)	1.5	16		×	4	Serial or Parallel Input.
	AD7530	10	0.05	500	10	5–15(2)	1.0	16	×	×	4	Leakage I = 300 nA.
	AD7533	10	0.05	600	10	5–15(2)	2.0	16		×	4	Provides 8-, 9-, or 1C-b Accuracy. Pin-Compatible with AD7520. Low Cost.
Intersil	AD7520	10	0.05	500	10	5–15(2)	2.0	16		×	4	Reference Input ±25 V. Short-Circuit Diodes Built In.
National	DAC1020	10	0.05	500	10	5–15(1.6)	1.6	16		×	4	DAC1030 = μP-compatible DAC1020
	DAC1022	10	0.2	500	10	5–15(1.6)	1.6	16		×	4	
Teledyne	8640/8641	12	0.2/0.1	1	2	5–16(2)	2.0	18		×	4	Pin-Compatible with AD7521 and AD7541.

Manufacturer	Device	Bits								Notes
Analog Devices	AD7541	12	0.01	1	10	5–16(2)	1.5	18	×	4	True 12-b monolithic. Second-Sourced by Intersil.
	AD7521	12	0.05	500	10	5–15(2)	1.5	18	×	4	Leakage $I = 200$ nA. Second-Sourced by Intersil.
	AD7531	12	0.05	500	10	5–15(2)	2.0	18	×	4	Leakage $I = 300$ nA.
National	DAC1220	12	0.05	500	10	5–15(1.6)	1.6	18	×	4	DAC1230 = μP-compatible DAC1220
Intersil	ICL7112	12	0.01	500	5	5–15(2)	2.0	18	×	4	Pin-Compatible with AD7521/AD7541.
	ICL7113	3	0.05	500	10	5–15(2)	2.0	18	×	4	3-Digit BCD.
Micropower	MP7621	12	0.02/0.01	1	2	5–15(2)	2.0	18	×	4	Pin Compatible with AD7521/AD7541.
Beckman	7581C	12	±½ LSB	6[a] / 12[d]	30FS[b] / 5FS[e]	±15	13 (6 Typ)[c] / 9 (4 Typ)[f]	36	×	4	Microprocessor-Compatible; Serial or parallel input format.

NOTE: AD7520/7530 are identical except for output leakage current and feedthrough, while AD7521/7531 are also identical except for output leakage current and feedthrough

[a] 0 to 10 V.
[b] With Ref.
[c] Positive output.
[d] −10 to +10 V.
[e] Without Ref.
[f] Negative output.

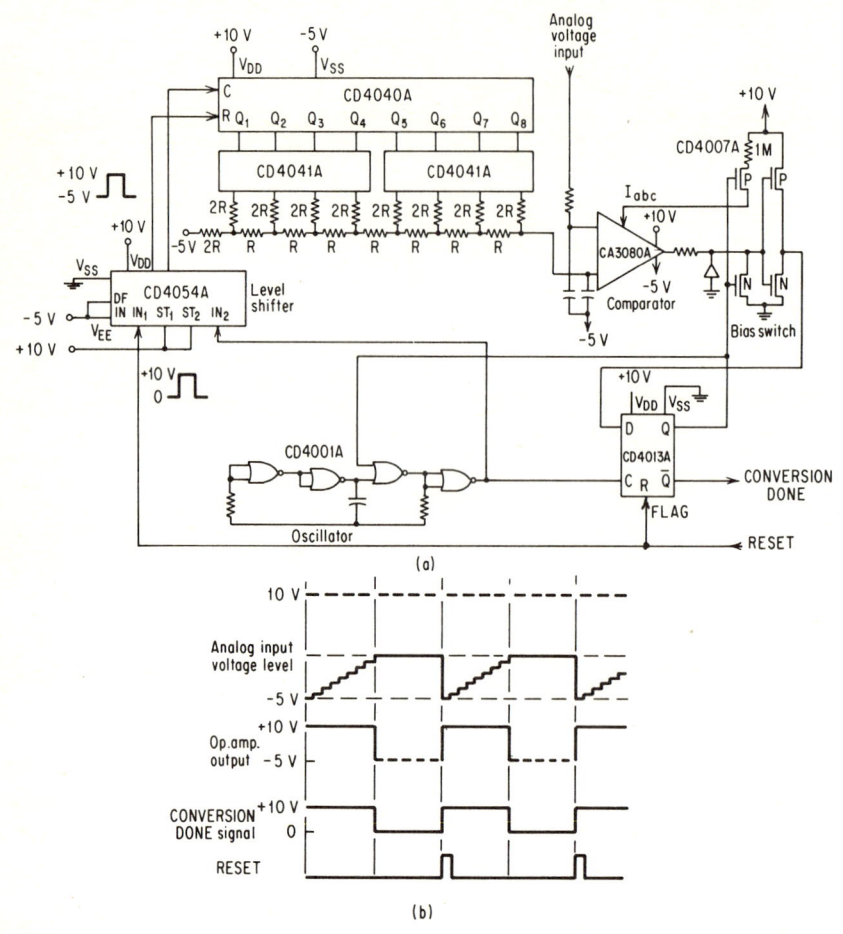

Fig. 3-25 (a) The A/D Converter Assembled with CMOS Standard Parts; (b) Staircase Ramp Generated by the CD4040A Binary Counter. *(Courtesy RCA Solid State Division.)*

At the user's option, the AD7574 may be connected and operated as if it were an RAM, ROM, or bulk memory. Therefore, both the data conversion and the data readout can be directly controlled by the processor.

The AD7574 is fully monotonic over its operating temperature ranges. It is possible to obtain a relative accuracy of $\pm\frac{1}{2}$ LSB, a differential nonlinearity of $\pm\frac{3}{4}$ LSB, a gain error of 3 LSB, and a ± 30-mV offset error in any of the three temperature ranges (from consumer through commercial to military).

Conversion time is rated at 15 μs and is controlled by an outboard RC network. Somewhat faster or slower conversions are possible, though performance is not guaranteed at speeds below 15 μs.

Working like a memory-mapped input device, the converter has three modes of operation with a microprocessor. In the static-RAM mode, conversion begins when the microprocessor sends a MEMORY-WRITE command to the unit. A DATA-READ occurs when the microprocessor fires a MEMORY-READ toward the converter's address.

In the ROM mode, a MEMORY-READ instruction triggers data dump as well as automatically triggering the beginning of a new conversion.

In the "slow-" or bulk-memory mode, the AD7574's status output is used to control the READY input of a microprocessor. Conversion starts when a MEMORY-READ command is generated for the 7574. $\overline{\text{BUSY}}$ immediately goes low, indicating that conversion is underway and forcing the microprocessor into a wait state. It waits until $\overline{\text{BUSY}}$ goes high, then finishes executing the DATA-READ instruction.

The AD7574 is compatible with most widely used microprocessors such as the Z80 and the 8080. It is almost self-contained, requiring a resistor and a capacitor for clocking, plus a 5-V power supply and a -10-V outboard reference.

The AD7574 uses little power, consuming only 25 mW during standby. Also, the converter's clock oscillator is run only during conversions. Finally, the unit can be operated as a ratiometric device, but is not tested or guaranteed for this purpose; its rated transfer accuracy applies only with a -10-V reference.

Analog Devices' AD7570 10-b CMOS A/D Converter

The AD7570 is a monolithic CMOS 10-b successive-approximation A/D converter, requiring only an external comparator and reference and passive clocking components. Ratiometric operation is inherent, since an extremely accurate multiplying DAC is used in the feedback loop. A block diagram of the AD7570 is shown in Fig. 3-26.

The AD7570 parallel output DATA lines and BUSY line utilize three-state logic to permit bussing with other A/D output and control lines or with other I/O interface circuitry. Two enables are available, one of which controls the 2 MSBs and the second of which controls the remaining 8 LSBs. This feature provides the control interface for most microprocessors, which can accept only an 8-b byte.

The AD7570 also provides a serial DATA OUTPUT line to be used in conjunction with the serial SYNCHRONIZATION line. The clock can be driven externally or, with the addition of a resistor and a capacitor, can run internally as high as 0.6 MHz, allowing a total conversion time (8 b) of typically 20 μs. An 8-b short-cycle CONTROL pin stops the clock after exercising 8 b, normally used for the J version (8-b resolution).

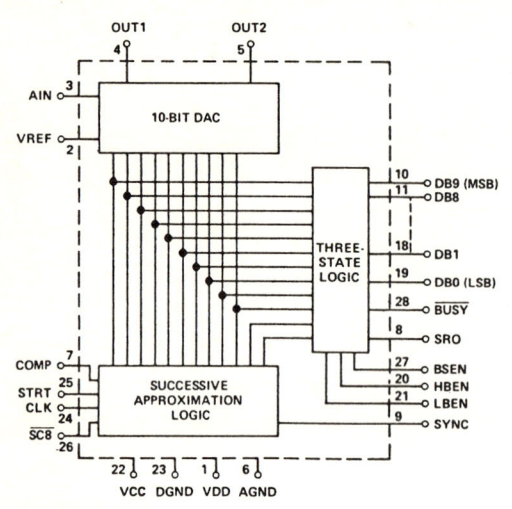

92 Data Conversion and Telecommunications Circuits

Fig. 3-26 AD7570 Functional Diagram. *(Courtesy Analog Devices.)*

The AD7570 requires two power supplies, a +5-V main supply and a +5-V (for TTL and DTL logic) to +15-V (for CMOS logic) supply for digital circuitry. The AD7570 is encased in a 28-pin DIP, as shown in Fig. 3-27.

In the successive-approximations technique, the output of a D/A converter is compared against the analog input for a succession of combinations of digits. Figure 3-28 is the AD7570 timing diagram showing the

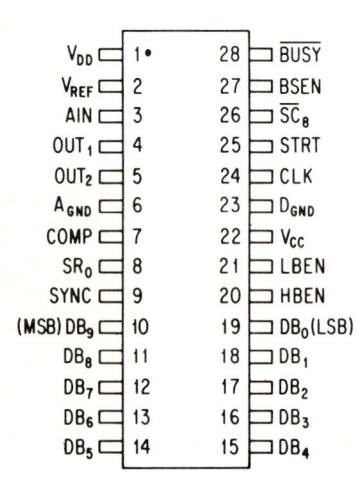

Fig. 3-27 AD7570 Pin Connection Diagram. *(Courtesy Analog Devices.)*

Data Conversion Circuits 93

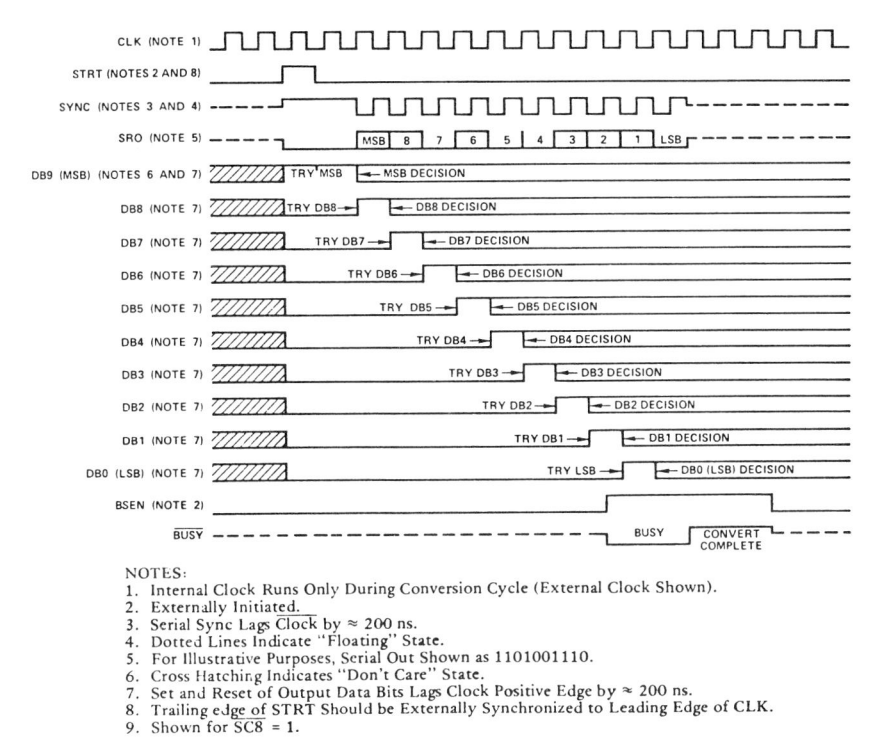

Fig. 3-28 AD7570 Conversion Timing Sequence. *(Courtesy Analog Devices.)*

NOTES:
1. Internal Clock Runs Only During Conversion Cycle (External Clock Shown).
2. Externally Initiated.
3. Serial Sync Lags Clock by ≈ 200 ns.
4. Dotted Lines Indicate "Floating" State.
5. For Illustrative Purposes, Serial Out Shown as 1101001110.
6. Cross Hatching Indicates "Don't Care" State.
7. Set and Reset of Output Data Bits Lags Clock Positive Edge by ≈ 200 ns.
8. Trailing edge of STRT Should be Externally Synchronized to Leading Edge of CLK.
9. Shown for $\overline{SC8}$ = 1.

successive trials and decisions for each data bit. When the START signal is given, the MSB latch output (appearing at DB_9, if enabled) goes high and causes the DAC to apply a current equal to one-half of full scale to the input network, where it is compared with the current developed by the input voltage. If the input is less, the comparator output causes the MSB latch to go low at the second clock pulse plus 200 ns; if the input is greater, the MSB stays high, retaining the DAC output at one-half full scale. In either case, the decision initiates the trial of the second bit (one-quarter full scale); it is compared and accepted (input one-quarter or three-quarters) or rejected (input one-quarter or three-quarters). The comparison proceeds until the LSB has been tried and accepted or rejected. The outputs DB_9 through DB_0, if all bits are enabled, will indicate a valid binary representation of the magnitude of the analog input, relative to the reference. This result will remain latched until another conversion is initiated.

From the timing diagram, it can be seen that when STRT (CONVERT START) goes high, DB_9 is set while DB_0 through DB_8 are reset. Two

94 Data Conversion and Telecommunications Circuits

clock pulses plus 200 ns after the STRT pulse returns to low, the MSB (DB$_9$) decision is made. Each succeeding trial and decision is made at T_{CLK} + 200 ns (a fixed delay time designed into the AD7570 to ensure that data from the comparator is available at the DATA input of the output latch before clocking the latch). The output data lines (DB$_0$ through DB$_9$) are buffered from the output data latches by three-state drivers (similar to transmission gates in series with the outputs). The transmission gates are controlled by HBEN (HIGH BYTE ENABLE), which controls DB$_9$ and DB$_8$, the 2 MSBs, and LBEN (LOW BYTE ENABLE), which controls DB$_7$ through DB$_0$, the 8 LSBs.

Serial NRZ (nonreturn-to-zero) data is available during conversion at the SRO (SHIFT REGISTER OUTPUT) terminal. SYNC (SYNCHRONIZATION) provides 10 positive edges which occur in the middle of each serial output bit. SYNC out must be used in conjunction with SRO to avoid misinterpretation of data. Both SYNC and SRO "float" when conversion is not taking place.

The AD7570 can be operated in both UNIPOLAR BINARY and BIPOLAR (BINARY OFFSET) modes.

Figure 3-29 shows how the AD7570 might be employed in an 8-channel DAS (data-acquisition system) with 8-b resolution. A single converter is used with an 8-channel multiplexer to perform time-division multiplexing of the eight 0 to +10 V analog signals in a sequential scan

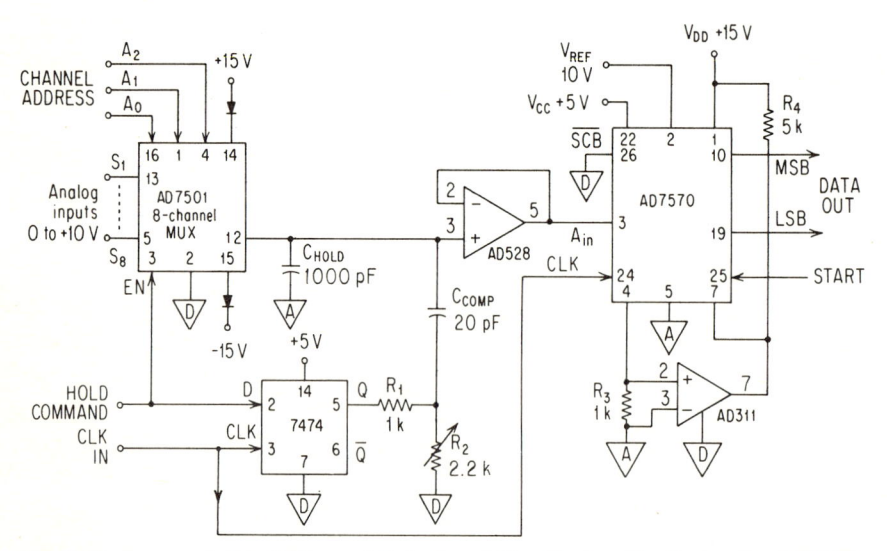

Fig. 3-29 Low-cost, 8-b, 8-Channel Data Acquisition System Employing the AD7570. *(Courtesy Analog Devices.)*

mode. It can provide 8 b of data at a per-channel throughput rate of 3.8 kbytes/s, or a total system throughput of 30.7 kbytes/s.

The AD7501 multiplexer's ENABLE line controls the SAMPLE HOLD function. In low, the capacitor holds the previous charge; in high it samples the input that is connected. The follower-connected amplifier unloads the hold capacitor, and provides a low-impedance input signal to the A/D converter. The flip-flop applies a delayed negative spike to C_{COMP} to cancel the multiplexer's charge injection; R_2 helps optimize the compensation.

The AD7570 is specifically designed for ease of use in data bus systems, where its three-state outputs are under external control.

Since most 8-b microprocessors utilize a bidirectional data bus, each input peripheral (such as the AD7570) must be capable of isolating itself from the data bus when other I/O devices, the memory, or the CPU take(s) control of the bus. The AD7570 output DATA and STATUS (BUSY) lines all utilize three-state logic to provide this requirement.

Figure 3-30 illustrates a method of interfacing a TTY keyboard and printer to the AD7570 using an 8080 microprocessor as the interface controller.

The program (stored in ROM) waits for a keystroke on the TTY keyboard. When a keystroke is detected, an A/D conversion is started. When conversion is complete, the 8080 reads in the binary data from the AD7570, converts it to ASCII (American Standard Code for Information Interchange), and prints out the decimal number (preceded by a carriage return and linefeed) on the teletype printer.

The Intersil ICL7109 12-b A/D Converter

The ICL7109 dual-slope integration 12-b single-chip A/D converter is specifically oriented toward a wide variety of microprocessor datalogging applications.

Its byte-organized, TTL-compatible, three-state outputs allow it to be interfaced directly with microprocessor data buses that are 8 b wide or wider. And for remote data-transmission applications, the 7109's handshaking capability means it can be directly interfaced with UARTs.

The ICL7109 will sequence through two 8-b bytes either synchronously or on demand from the microprocessor to the UART, and no additional components are needed because the device has on-board logic to control the UART. Therefore, if a designer is logging temperature, pressure, humidity, light intensity, or any other real analog variable, the ICL7109 provides a one-chip solution, straight to the data bus.

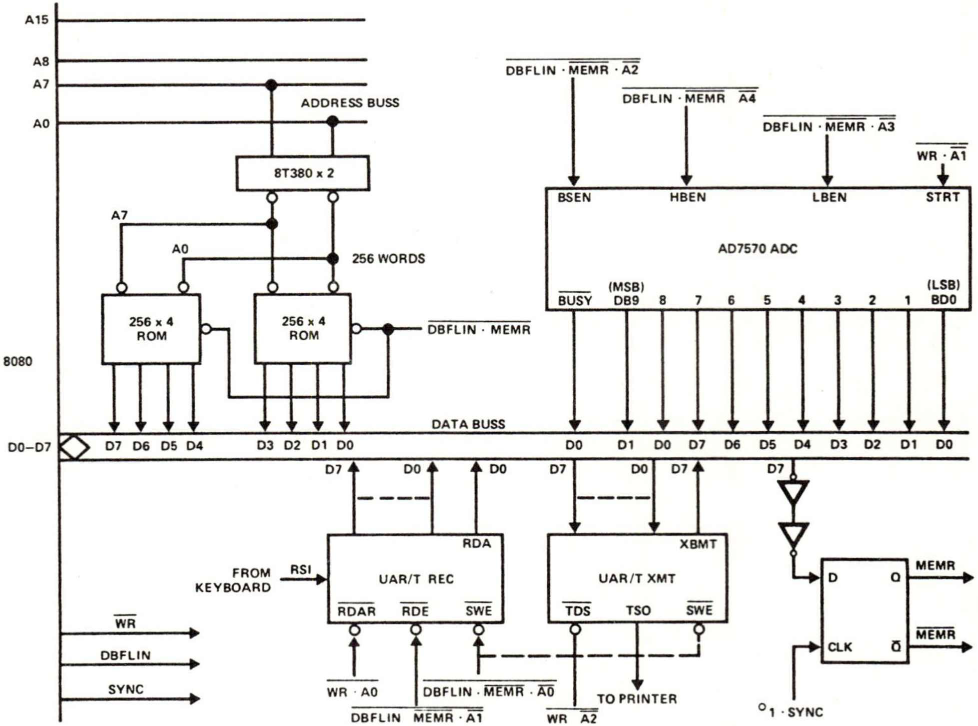

Fig. 3-30 Microprocessor-Controlled TTY/ADC Interface. *(Courtesy Analog Devices.)*

Specific features of the ICL7109 include true differential input for noise rejection, a zero drift of less than 1 μV/°C, nonlinearity of less than 0.01 percent, an input impedance of 1 TΩ, a conversion rate of 0.1 to 15 conv/s, and power consumption of less than 20 mW.

When in the byte-organized parallel mode, the ICL7109 can interface directly with the data buses of such popular microprocessors as Intel's 8080 and 8048, Motorola's MC6800, and Intersil's own 6100. There are 14 data output lines, providing 12 b of magnitude plus polarity and out-of-range bits. These output lines can be grouped in two 8-b bytes, each of which is activated by its own BYTE ENABLE signal. In the handshaking mode, the ICL7109 has two inputs so that it can sequence through 2 bytes either synchronously or on demand without the use of external components.

For the analog section, Intersil was able to use the experience gained from its ICL7106/7107 $3\frac{1}{2}$-digit A/D converter to produce tight specifications. Among them are a true polarity at zero for precise null detection and a typical input current of 1 pA. The true differential input helps keep the noise level below 15 μV peak to peak and is useful when it comes to measuring the output of load cells, strain gages, and other bridge-type transducers.

Although the 7109, like the 7106, has its own voltage reference, it is recommended that an external reference be used. The ICL7109 is encased in a 40-pin DIP package, as shown in Fig. 3-31.

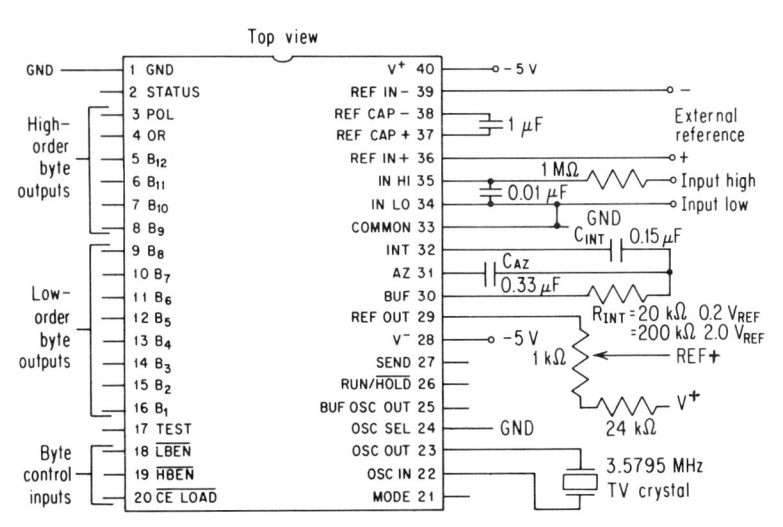

Fig. 3-31 ICL7109 Pin Connection Diagram and Test Circuit. *(Courtesy Intersil.)*

98 Data Conversion and Telecommunications Circuits

Some practical circuits utilizing the parallel three-state output capabilities of the ICL7109 are shown in Figs. 3-32 through 3-35. Figure 3-32 shows a straightforward application to the Intel MCS-48, -80, and -85 systems via an 8255PPI, where the ICL7109 data outputs are active at all times. The I/O ports of an 8155 may be used in the same way. This interface can be used in a READ-ANYTIME mode, although a read performed while the data latches are being updated will lead to scrambled data. This will occur very rarely, in the proportion of set-up-skew times to conversion time. One way to overcome this is to read the STATUS output as well, and, if it is high, read the data again after a delay of more than half a converter clock period. If STATUS is now low, the second reading is correct, and if it is still high, the first reading is correct. Alternatively, this timing problem is completely avoided by using a READ-AFTER-UPDATE sequence, as shown in Fig. 3-33. Here the high-to-low transition of the STATUS output drives an interrupt to the microprocessor, causing it to access the data. This application also shows the RUN/$\overline{\text{HOLD}}$ input being used to initiate conversions under software control.

A similar interface to Motorola MC6800 or MOS Technology MCS650X systems is shown in Fig. 3-34. The high-to-low transition of the STATUS output generates an interrupt via the control register B CB1 line. Note that CB2 controls the RUN/$\overline{\text{HOLD}}$ pin through control register B, allowing software-controlled initiation of conversions in this system also.

Figure 3-35 shows an interface to the Intersil IM6100 CMOS microprocessor family using the IM6101 PIE (parallel interface element) to control the data transfers. Here the data is read by the microprocessor in an 8- and a 6-b word, directly from the ICL7109 to the microprocessor data bus. Again, the high-to-low transition of the STATUS output generates an interrupt, leading to a software routine controlling the two READ operations. As before, the RUN/$\overline{\text{HOLD}}$ input to the ICL7109 is shown as being under software control.

A Microprocessor-compatible A/D Converter*

The rapid advancement of the microprocessor for use on analog data acquisition systems has increased the demand for low-cost, high-performance, microprocessor-compatible A/D converters. To meet the

---

* E. Masuda et al., "A Single-Chip C^{2}MOS A/D Converter for Microprocessor Systems," *1978 IEEE ISSCC Digest of Technical Papers*, San Francisco, CA, 1978. Copyright © 1978 by the Institute of Electrical and Electronics Engineers, Inc. Reprinted with permission.

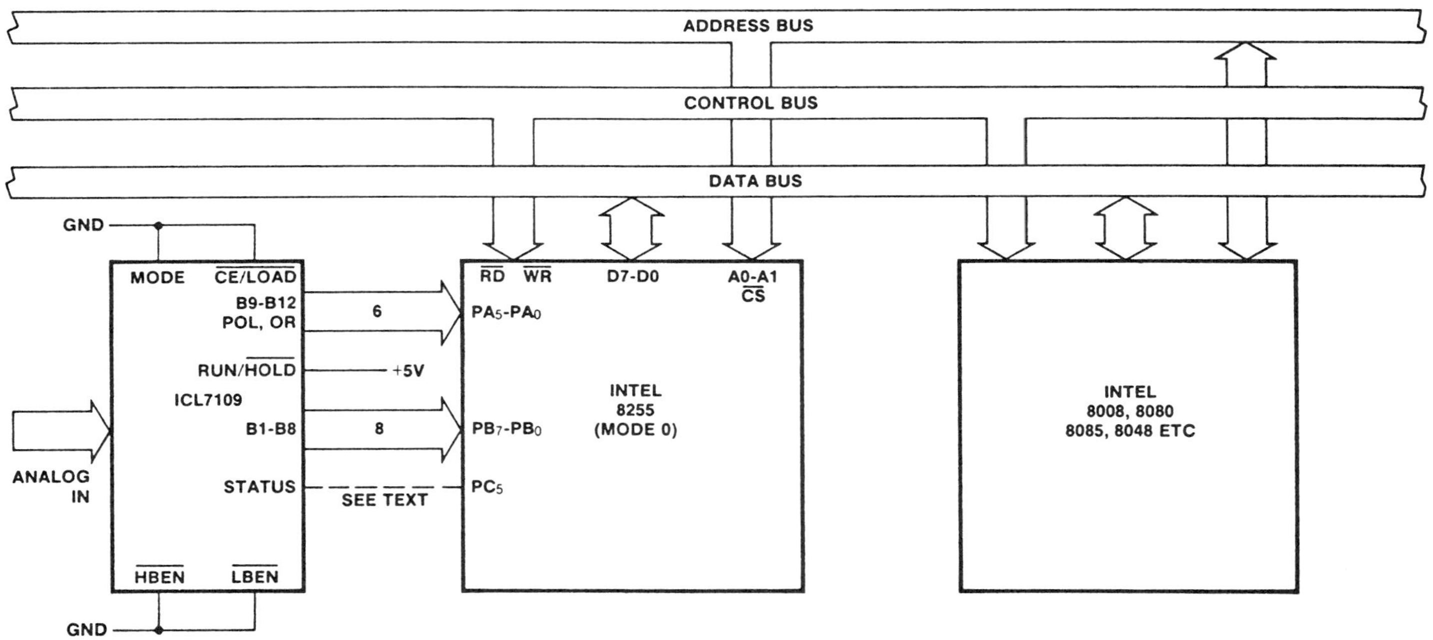

Fig. 3-32 Full-Time Parallel Interface to Intel Microcomputer Systems. (Courtesy Intersil.)

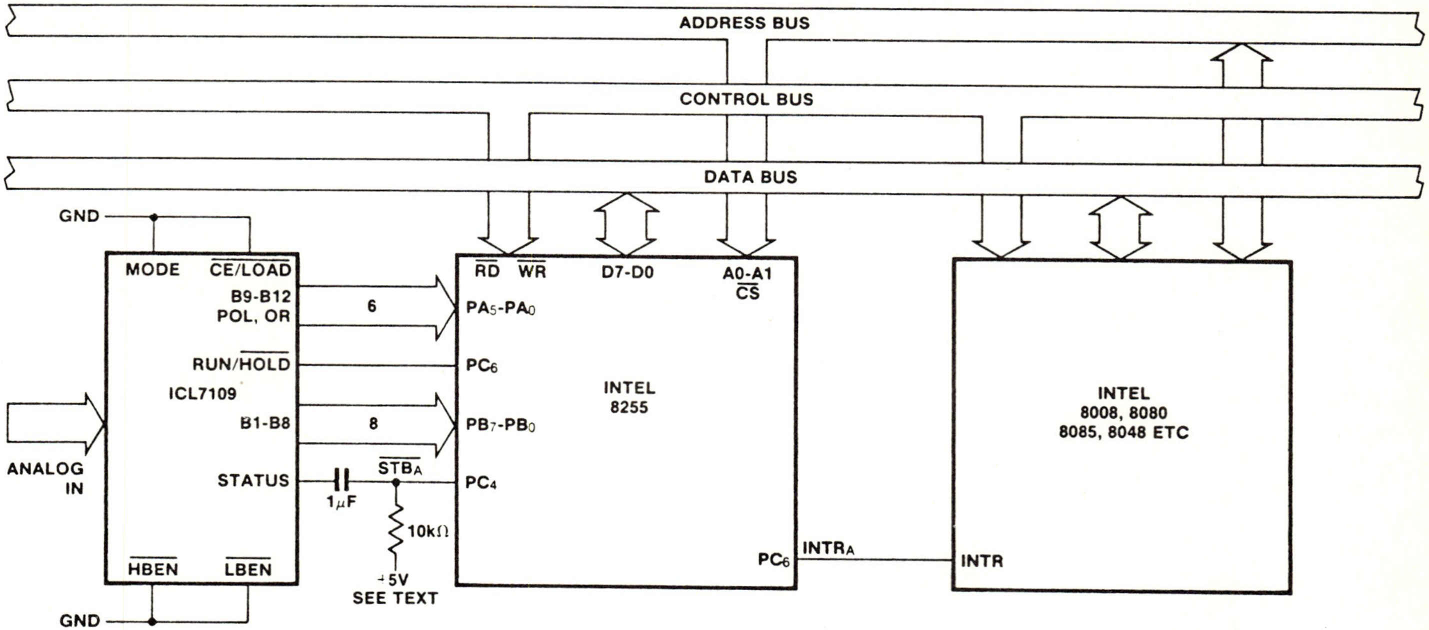

Fig. 3-33 Full-Time Parallel Interface To Intel Microcomputers with Interrupt. (Courtesy Intersil.)

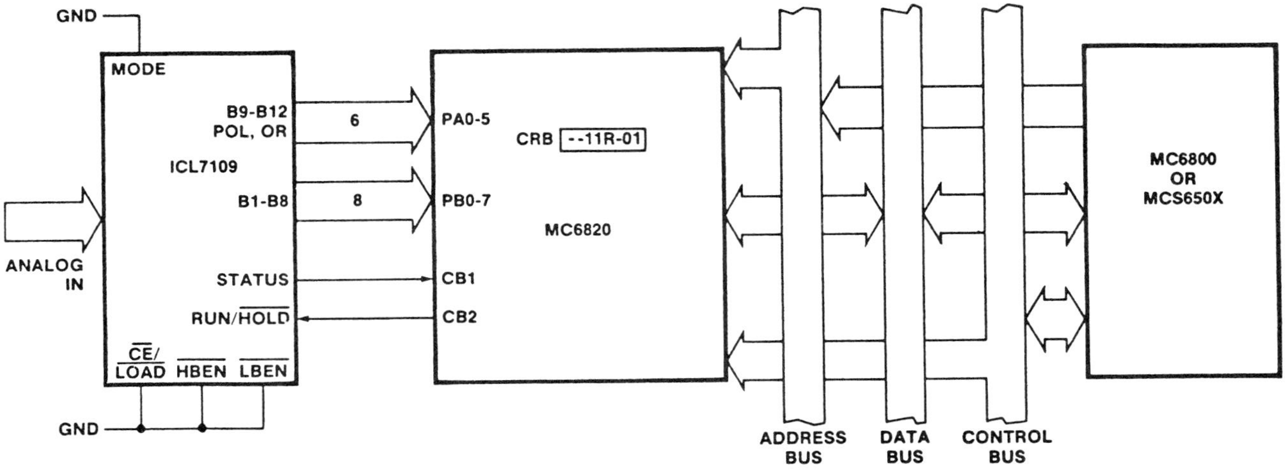

Fig. 3-34 Full-time Parallel Interface to MC6800 or MCS650X Microprocessors. *(Courtesy Intersil.)*

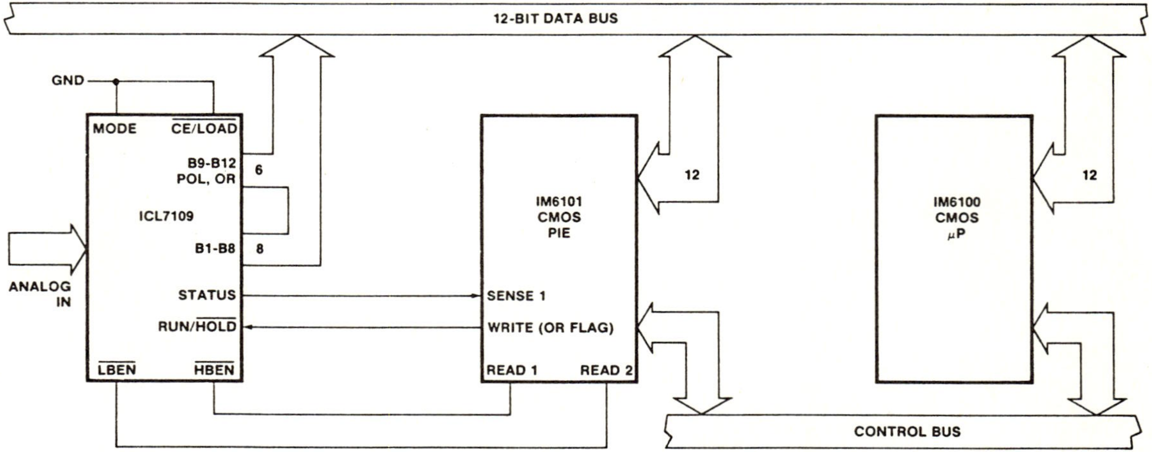

Fig. 3-35 ICL7109-IM6100 Interface Using IM6101 PIE. *(Courtesy Intersil.)*

demand, a single-chip C²MOS 12-b A/D converter has been developed using a standard metal-gated CMOS process.

The block diagram of the A/D converter with full external components is shown in Fig. 3-36. To interface with a microprocessor and to reduce the total system cost, the device contains an 8-channel analog multiplexer, a channel-address decoding logic, a read-write interface logic, an integrating A/D conversion circuit, and TriState output buffers on a single chip powered by a single 5-V supply. The device is controlled directly by the microprocessor through software. At the beginning of WRITE operations, one channel is addressed out of the eight inputs through bidirectional bus lines. An END signal is received as an interrupt by the microprocessor when the conversion is complete. Data access time from the start of the READ operation to the DATA VALID on bus lines measures 200 μs typically, with a load of 100 pF.

The use of relatively high clock frequency (3 MHz typical inner clock, which is half of the external input clock) results in a conversion time of 3.6 ms.

The system employs an improved dual-slope technique which provides a unipolar and ratiometric conversion. A conversion cycle, as shown in Fig. 3-37, consists of five phases: 0—the integrator runs as a voltage follower with subreference $\frac{1}{2}V_{REF}$ input and charges its offset voltage to an external capacitor C_C; I—the analog ground voltage A_{GND} is integrated after the operation of offset compensation which inserts the charged C_C

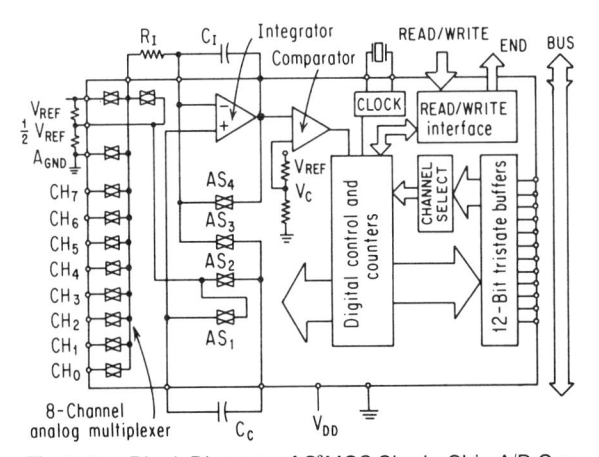

Fig. 3-36 Block Diagram of C²MOS Single-Chip A/D Converter. *(Reprinted from 1978 IEEE ISSCC Digest of Technical Papers, San Francisco, CA, February 1978. Copyright © 1978 by the Institute of Electrical and Electronics Engineers, Inc. Used with permission.)*

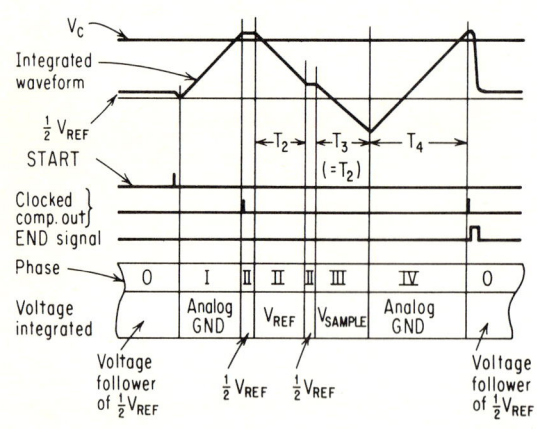

Fig. 3-37 Timing Diagram of Conversion Cycle. *(Reprinted from 1978 IEEE ISSCC Digest of Technical Papers, San Francisco, CA, February 1978. Copyright © 1978 by the Institute of Electrical and Electronics Engineers, Inc. Used with permission.)*

between $\frac{1}{2}V_{REF}$ and the noninverting input of the integrator in reverse polarity; II—reference voltage V_{REF} is integrated for a fixed interval T_2 (2048 clock counts); III—selected sample voltage V_{sample} is integrated for the same interval as T_2; and IV—analog ground voltage A_{GND} is integrated for an interval T_4 related to the digital output.

The threshold voltage V_C of a comparator is supplied by on-chip dividing resistors. Note that either the absolute accuracy of the threshold voltage V_C or the offset voltage of the comparator has no influence on data conversion as long as the ratio of the dividing resistors remains stable within a conversion cycle.

The main factors which dominate the conversion accuracy stem from the signal loop between the digital and analog circuits (Fig. 3-38). High accuracy in the system using a relatively high voltage of the integrator is held precisely at a constant level. These phases contribute to reduce the error factors generated by the stray capacitance at the output of the analog multiplexer in transient states.

The accuracy achieved is 0.05 percent of full scale over the −30 to +85°C temperature range.

The relationships between conversion time, conversion accuracy, and current consumption to input clock frequency are plotted in Fig. 3-39. Table 3-5 summarizes the key features of this 12-b A/D Converter.

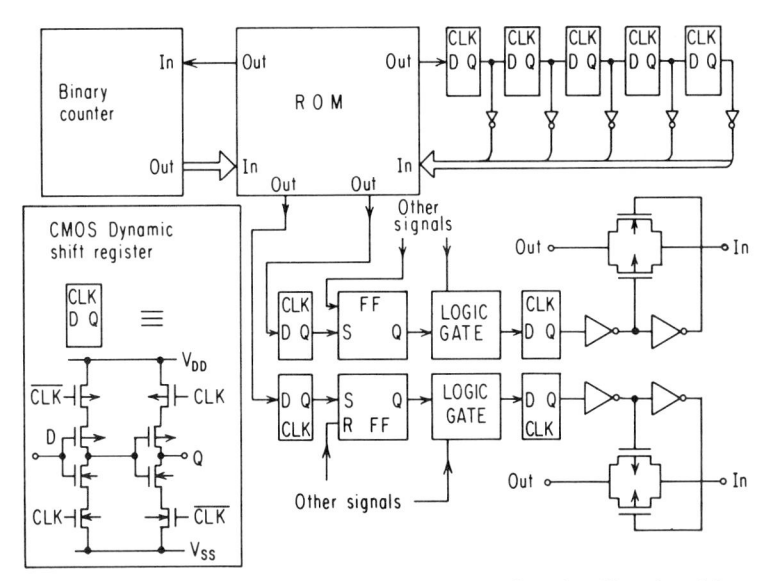

Fig. 3-38 Schematic of Signal Synchronization Circuits. *(Reprinted from 1978 IEEE ISSCC Digest of Technical Papers, San Francisco, CA, February 1978. Copyright © 1978 by the Institute of Electrical and Electronics Engineers, Inc. Used with permission.)*

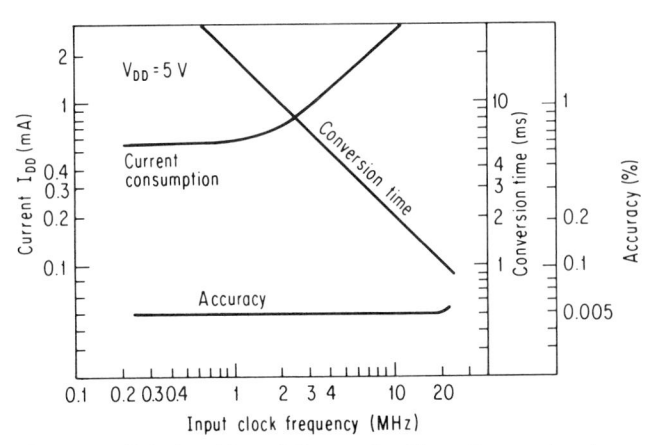

Fig. 3-39 Relationships of Conversion Time, Conversion Accuracy, and Current Consumption to Input Clock Frequency; $T_A = 25°C$. *(Reprinted from 1978 IEEE ISSCC Digest of Technical Papers, San Francisco, CA, February 1978. Copyright © 1978 by the Institute of Electrical and Electronics Engineers, Inc. Used with permission.)*

TABLE 3-5 Summary of C²MOS A/D Converter Features

Resolution	12 b
Accuracy	0.05% FS
Conversion Time	3.6 ms (6-MHz Input Clock)
Analog Input	Analog Ground to Reference V_{REF} with 8-Channel Analog Multiplexer
Digital Output	Binary Parallel Code with Tri-State Buffers
Input Clock	6-MHz (Typical) or Equivalent Crystal
Reference	V_{REF} (5 V Typical), $\frac{1}{2}V_{REF}$ Ratiometric Conversion Ranging to V_{DO}
Power Supply	V_{DO} = 4.5 to 8.0 V
Power Dissipation	10 mW (V_{DO} = 5 V, Conversion = 100 conv/s)
Chip Size	3.9 × 4.3 mm² with 2000 FETs
Package	42-Pin Plastic Dip

SOURCE: *1978 IEEE ISSCC Digest of Technical Papers,* San Francisco, CA, 1978. Copyright © 1978 by the Institute of Electrical and Electronics Engineers, Inc. Reprinted with permission.

Simultaneous-Integration A/D Converter*

Nippon Electric Company has developed a 12-b current mode A/D converter for use with microcomputer data handling systems that includes a single 5-V A/D converter, current mirror circuits, and a high gain amplifier.

The operational concept of this A/D converter is shown in the block diagram of Fig. 3-40. With a WRITE command, A/D conversion is initiated and an analog input port is selected. A floating-point 12-b conversion data is left-justified within a 16-b field composed of high and low bytes.

Instead of the conventional dual-slope technique, a simultaneous-integration technique (Fig. 3-41) has been employed. The output of the comparator (CMP₂) at the end of each interval $(t_1, t_2, \ldots, t_7)$ determines either the simultaneous integration of I_X and I_{REF} or the simple integration of I_X to be performed for each of the subsequent integrating intervals $(T_2, T_3, \ldots, T_8)$. This technique has made it possible to obtain a much larger equivalent output voltage range of the integrator, a reduction of the external integrating capacitor value, and an improvement of the S/N (signal-to-noise) ratio by an order of magnitude, as well as a relatively high conversion speed.

* K. Hareyama et al., "A Monolithic CMOS 12b Simultaneous Integration ADC," *1979 IEEE ISSCC Digest of Technical Papers,* Philadelphia, PA, February 1979.

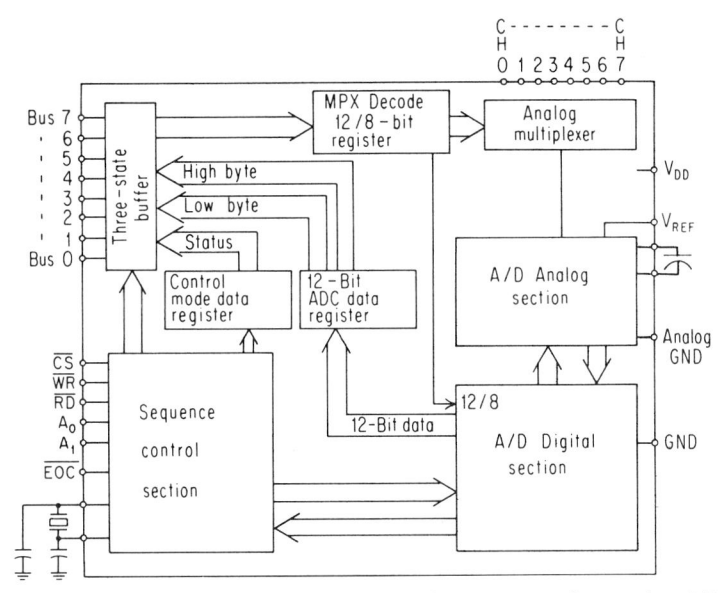

Fig. 3-40 Block Diagram of CMOS 12-b Simultaneous Integration A/D Converter. *(Reprinted from 1979 IEEE ISSCC Digest of Technical Papers, Philadelphia, PA. February 1979. Copyright © 1979 by the Institute of Electrical and Electronics Engineers, Inc. Used with permission.)*

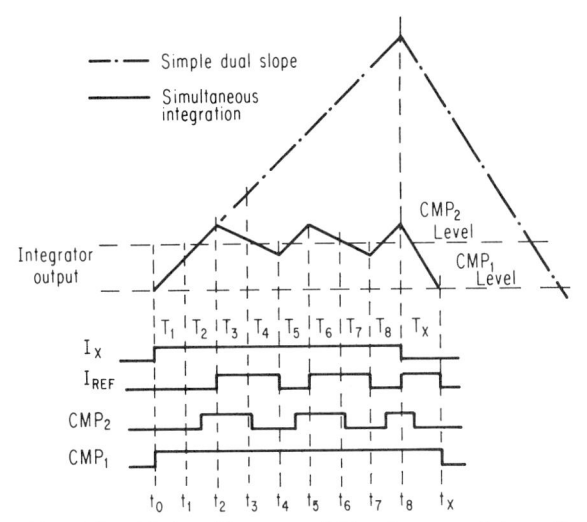

Fig. 3-41 Timing Diagram of Simultaneous Integration Technique. *(Reprinted from 1979 IEEE ISSCC Digest of Technical Papers, Philadelphia, PA. February 1979. Copyright © 1979 by the Institute of Electrical and Electronics Engineers, Inc. Used with permission.)*

Typical device characteristics include TTL-compatible logic inputs, 1000 MΩ of analog input impedance, 15 mW power consumption, and 20 ppm/°C of temperature drift at zero scale and at full scale.

13-b Integrating CMOS A/D Converter

Analog Devices' AD7550 CMOS 13-b A/D converter is directly compatible with microprocessors, holds gain and offset drift to less than 1 ppm/°C, and automatically compensates itself for temperature variations. Moreover, the AD7550 has its own amplifier, comparator, clock, and digital logic, so that it requires only an external resistor, capacitor, and reference voltage for operation.

The AD7550 employs a conversion technique called *quad slope,* which involves a four-phase integration period in addition to a reset phase, as shown in Fig. 3-42. In contrast, most existing integrating A/D converters operate with only a two-phase (dual-slope) integration period.

The additional two phases of the AD7550's integration period are put to good use, and the integrating time includes a digitally corrected autozero cycle. This means that the converter automatically compensates for offsets, making its temperature stability excellent with gain and offset drifts that are practically unmeasurable.

Furthermore, because of quad slope, the device can accept bipolar inputs, and its output is inherently monotonic, as well as free of the problem of missing codes.

The AD7550 parallel OUTPUT DATA lines have three-state logic and are microprocessor-compatible through the use of two ENABLE

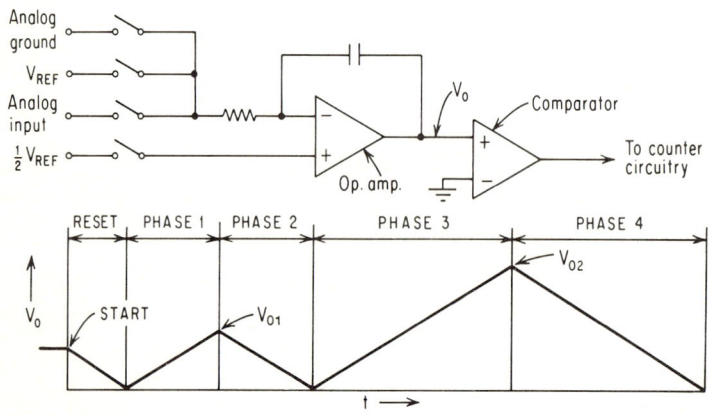

Fig. 3-42 Quad Slope Conversion *(Courtesy Analog Devices.)*

lines which control the lower 8 LSBs (LOW BYTE ENABLE) and the 5 MSBs (HIGH BYTE ENABLE).

An overrange output warns of input voltages that are beyond the bipolar input range, which extends over about plus and minus half the reference voltage level. An external positive pulse can be applied to start a single conversion cycle, or the AD7550 can be wired in its self-starting mode for continuous conversion by connecting a capacitor to the START pin of the device.

The AD7550 conversion time is about 40 ms with a 1-MHz clock, which can be externally controlled or internally generated by simply connecting a capacitor to the CLOCK pin. A positive START pulse can be self-generated by having a capacitor on the START pin, or can be externally applied.

The unit furnishes 13 b of parallel binary data in microprocessor-compatible 2's complement code, which can be interfaced directly with the 8-b microprocessor data bus lines through its three-state buffers. Its digital outputs are also compatible with both CMOS and TTL.

Additionally, the AD7550 provides a serial linear output pulse stream, which contains 8192 (or 2^{13}) pulses for a full-scale input. Since it develops both parallel and serial outputs, the AD7550 is equipped to provide the simultaneous digital data needed for microcomputer data acquisition, as well as the serial data needed for a digital display involving, for example, a counter and an LED decoder-driver.

The AD7550 converter can accurately digitize signals as large as one-half the value of the applied reference voltage. And it can convert very-low-level signals, too. This low-level capability is limited only by the ability of the unit's internal amplifier to integrate small microvolt signals accurately, without errors due to noise.

What is more, the AD7550 can operate over a wide range of supply voltages. It requires a drain (V_{DD}) supply of 5 to 15 V, a source (V_{SS}) supply of −5 to −15 V, and a logic (V_{CC}) supply of 5 V to the V_{DD} level, depending on whether TTL or CMOS logic level interfacing is needed. The device can also be wired for operation from a single supply.

Potential applications for the AD7550 are vast, including those in which previous monolithic converters could not be used. They range from battery-operated single-supply A/D conversion systems to sophisticated microprocessor-controlled analog interface peripherals. And because the converter's response is very stable, it can be used in automotive and industrial applications, where harsh environments formerly precluded the use of digital techniques.

The AD7550 is housed in a 40-pin ceramic DIP. Its operating temperature range is −25 to +85°C.

Figure 3-43 shows the basic circuit connection for binary operation. With all the data output commands held high as shown, parallel data will be present on all the outputs. By selectively exercising the various commands HBEN (4 MSBs plus polarity bit), LBEN (8 LSBs), and STEN (overrange, $\overline{\text{BUSY}}$, BUSY), the desired data can be placed out on an 8-b data bus.

To ensure linearity of the AD7550's amplifier, a potential of +4 V minimum should exist between A_{GND} and V_{SS}. However, because A_{GND} is not a supply path, it can be biased upward to allow single-supply operation, as shown in Fig. 3-44. This configuration would be especially useful for battery pack operation, but is not without its limitations. The value of the MSB in this circuit is directly dependent upon the TC of the resistor used, so to assure accuracy over a wide temperature range, a metal film or wire-wound resistor should be selected.

Figure 3-45 shows a block diagram of how the serial count output of the AD7550 can be modified before being clocked back into its three-state buffers, providing both a linearized BCD value of a thermocouple output for digital panel meter display plus the same value represented in binary format which can be outputted on a microprocessor data bus line.

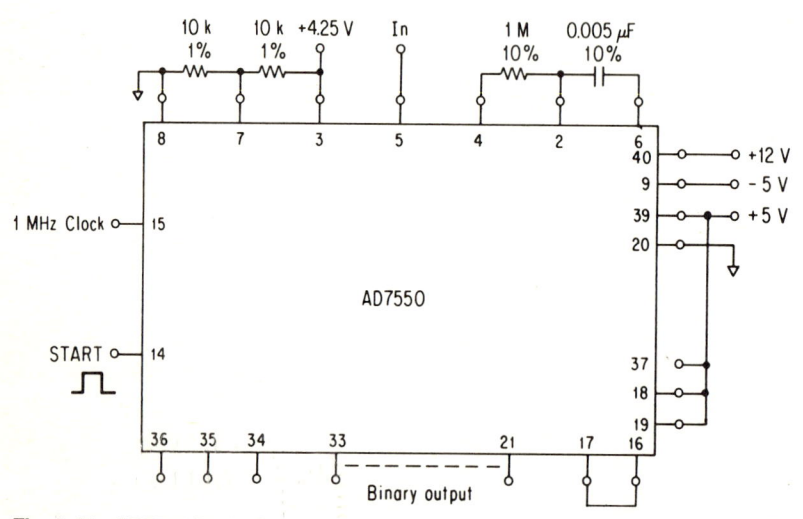

Fig. 3-43 AD7550 Basic Circuit Connection Diagram. Notes: (1) The integrating capacitor has to be of the polystyrene or polypropylene type; (2) conversion time is about 40 ms; (3) all digital inputs/outputs are TTL-compatible (V_{CC} = +5 V). *(Courtesy Analog Devices.)*

Data Conversion Circuits   111

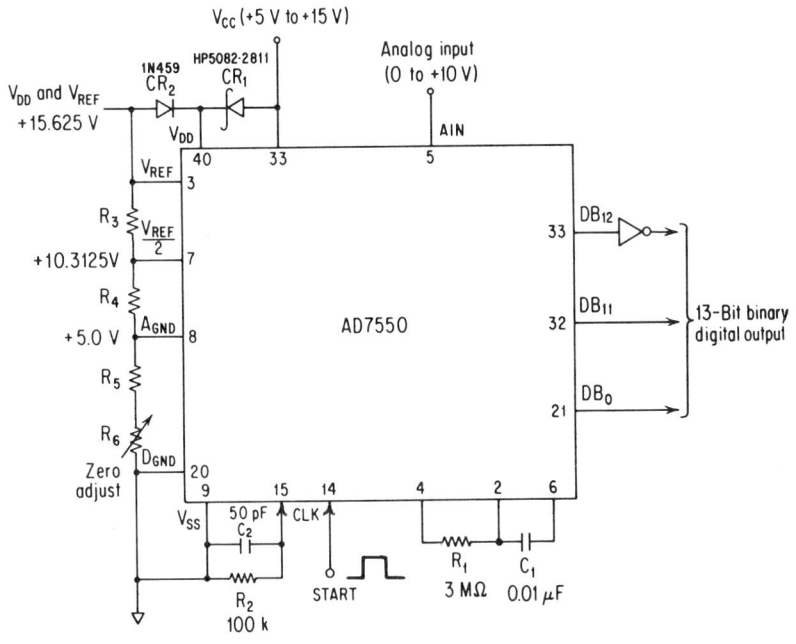

Fig. 3-44 Single-Supply Operation. *(Courtesy Analog Devices.)*

For a data acquisition system, addition of a line driver-receiver will allow the AD7550 output to be supplied over a two-wire line to the linearizing circuitry, allowing digital transmission of the analog value for visual display.

In this configuration, the analog input to the AD7550 is derived from an instrumentation-type amplifier, such as an AD521, to which the thermocouple and cold-junction inputs are applied. A conversion is

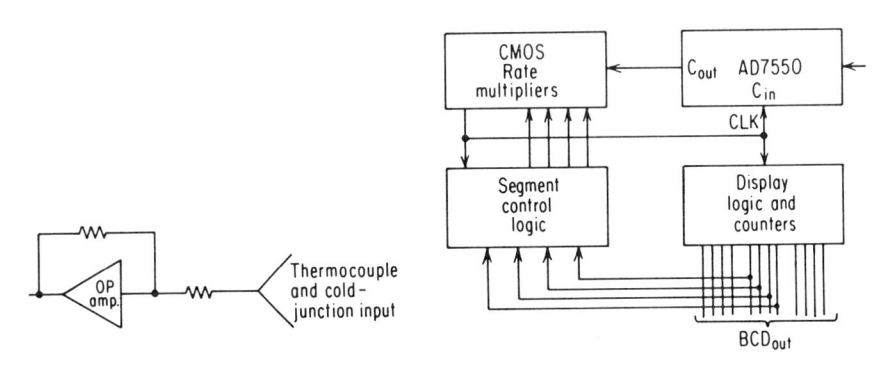

Fig. 3-45 Digital Linearization. *(Courtesy Analog Devices.)*

started and an output pulse stream is generated by the AD7550 which is determined by the thermocouple voltage and the scale factor of the amplifier compared to the reference voltage. This pulse stream is clocked through a set of CMOS rate multipliers connected in cascade, which perform the linearizing function by modifying the pulse count output as determined by the binary value applied to its inputs. This binary value is selected by determining the scale factor used, the temperature range over which the readings will be taken, and the accuracy required. By dividing the temperature range into several segments and calculating the BCD value needed to linearize each segment, a digital approximation of the piecewise linear correction technique can be realized. Clocking the pulses out of the CMOS rate multipliers into a set of BCD counters, then monitoring the outputs of the counters to see that they successively change ranges as the pulse count continues, allows the BCD value to the rate multiplier to be changed for each segment.

In this circuit, accuracy is primarily a function of the number of segments used and the range desired. The digital linearization technique ensures drift-free performance over temperature without the usual expense and problems of using programmed ROMs. By applying an analog voltage to the input of the op amp which corresponds to a desired temperature center point, very accurate readings can be obtained for any particular temperature range.

The Intersil ICL7106/7107 A/D Converter

The Intersil ICL7106/7107 dual-slope A/D converter demonstrates that CMOS can ideally combine both analog and digital signal processing on a single chip. This device contains all the circuitry to drive either a $3\frac{1}{2}$-digit LCD or LED display directly, plus all the analog circuitry, including the voltage reference.

Requiring about 1 mA of current from a 9-V battery or dc power supply (LED display drive not included), it will operate over the supply voltage range of about 4 to 15 V.

Accuracy is better than one count or 1 least-significant digit at full scale, and noise and offset error is about half a count around zero. The input noise, about $12\mu V$, is an improvement over most previous integrated designs, including some of the better bipolar designs.

Figure 3-46 is a block schematic of the A/D's analog section, which contains three op amps, one comparator, and 11 analog gates. While the analog section operates at the full supply voltage as high as 15 V, the digital section operates from an internal supply of 5 V, which allows the

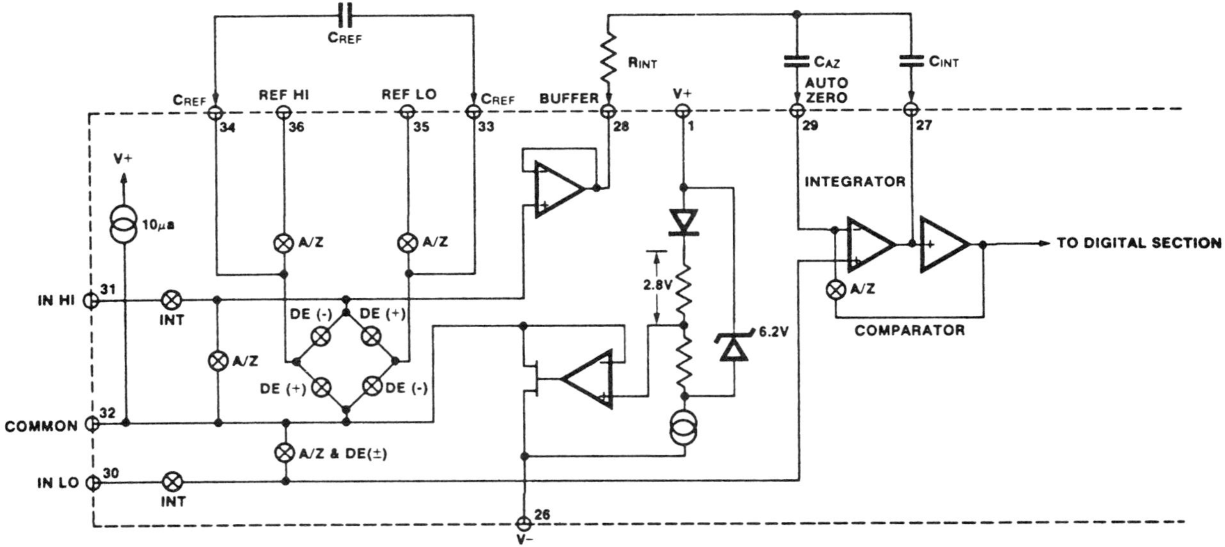

Fig. 3-46 The Analog Section of the Intersil ICL7106/7 Dual-Slope CMOS A/D Converter Uses Many Analog Gates. While Consuming Little Power, It Achieves High Precision—1 Part in 5000. *(Courtesy Intersil.)*

large digital section to be designed with compact, low-voltage design rules. The result is a chip savings of about 25 percent compared to operating both the analog and digital sections from the same supply.

Various panel meter circuits can be built using the ICL7106/7107 ICs plus a few external components. The 7106 drives LCDs and the 7107 drives LEDs.

The Siliconix LD130 ±3-Digit A/D Converter

The LD130 combines both the analog and digital subsystems of a 3-digit A/D system in a single monolithic CMOS IC (Fig. 3-47). Figure 3-48 shows the pin connection diagram. The "quantized feedback" conversion scheme, introduced by Siliconix, provides the LD130 with an autozero, Autopolarity A/D system requiring only a single reference voltage. External parts are minimized by the on-chip resistors and buffer amplifiers. These high-impedance input and reference buffer amplifiers eliminate source-loading errors, providing the outstanding temperature coefficient and ratio operation inherent in this system. Break-before-make switch action ensures that neither the analog input nor the reference voltage will be shorted to ground at any time.

The LD130 3-digit A/D is made functionally complete by the following additions (refer to Fig. 3-48):

1. C_{AZ} (0.10 μF) between AZ (autozero) and Σ (summation) pins
2. C_{INT} (0.033 μF) between INT (integrator) and Σ pins
3. C_{OSC} (0.001 μF) between OSC (oscillator) and digital ground
4. $V_{REF} \simeq 2.000$ V
5. ± 5-V supplies (at 3 mA)

The operation of the LD130 is as follows (refer to Fig. 3-47).

Time-Base Counter The internal oscillator circuit becomes fully functional with an external capacitor to ground. The OSC input can be driven by an external oscillator (0 to V_1 logic levels) if desired. A squaring circuit divides the oscillator frequency by two before it drives the BCD counter and time-base counter.

The two fundamental intervals of the sampling period, the AZ and measure intervals, are established by the time-base counter as 1024 and 2048 clock periods, respectively. The total sampling interval is then 3072 clock periods long. Since the internal clock is one-half of the oscillator frequency, the sampling period is then 6144 (2 × 3072) oscillator periods. The time-base counter also divides the internal clock by 8. This

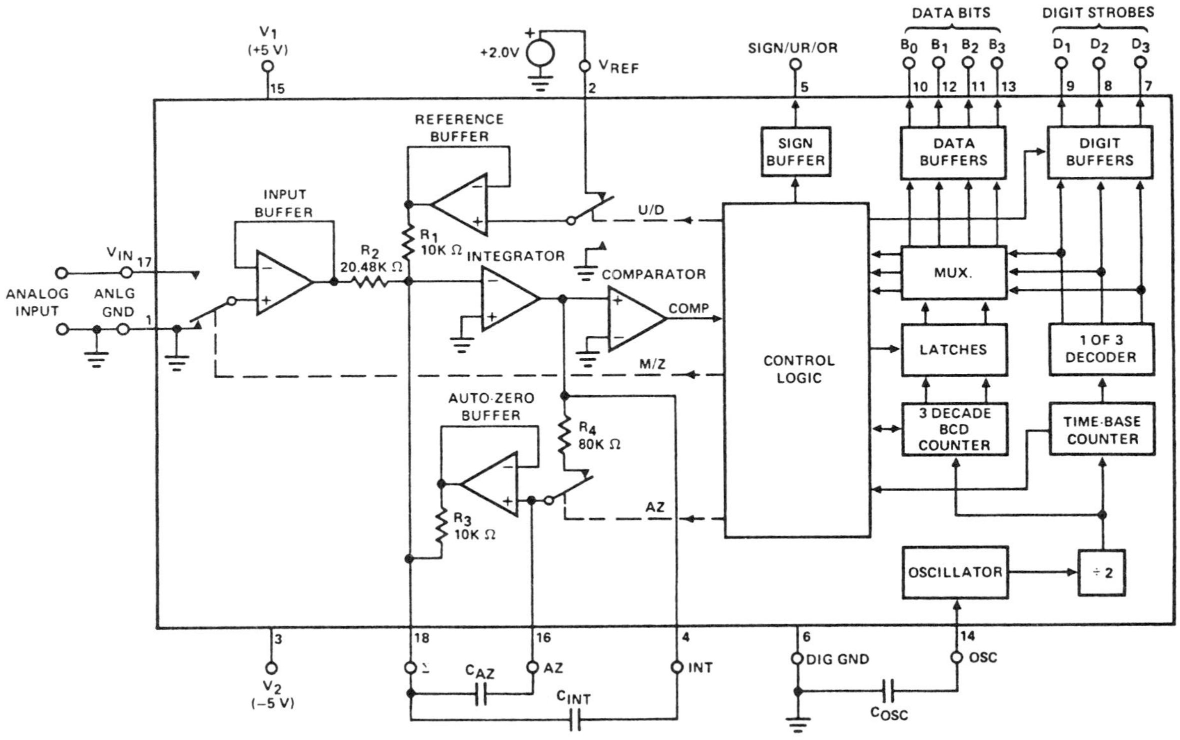

Fig. 3-47 LD130 Functional Connection Diagram. *(Courtesy Siliconix, Inc.)*

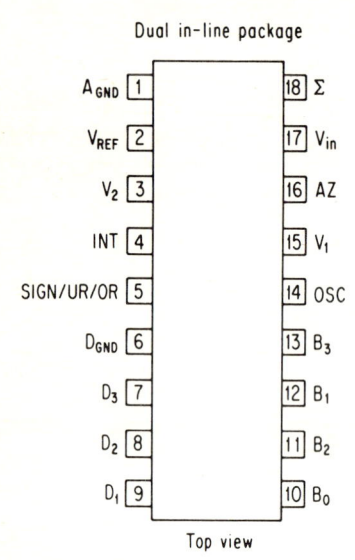

116 Data Conversion and Telecommunications Circuits

Dual in-line package

A$_{GND}$	1	18	Σ
V$_{REF}$	2	17	V$_{in}$
V$_2$	3	16	AZ
INT	4	15	V$_1$
SIGN/UR/OR	5	14	OSC
D$_{GND}$	6	13	B$_3$
D$_3$	7	12	B$_1$
D$_2$	8	11	B$_2$
D$_1$	9	10	B$_0$

Top view

Fig. 3-48 LD130 Pin Connection Diagram. *(Courtesy Siliconix, Inc.)*

division provides sets of eight clock periods (octets), which are used by both the data multiplexer as digit on times and the control logic as U/D (UP/DOWN logic) duty cycle periods.

Autozero Interval The autozero interval provides a means to null out the offset voltages of the amplifiers used in the LD130. In addition, it automatically establishes a second tracking reference voltage necessary for bipolar A/D conversion.

The autozero sequence is initiated when the M/Z (MEASURE/ZERO) signal switches the input buffer amp to analog ground. After a brief count-correcting override period, the AZ switch is closed, connecting the AZ amplifier and integrator together in a closed-loop second-order system. During this time, the control logic ignores the comparator output and pulses the U/D switch at a 50 percent duty cycle of four clock periods up and four down (see Fig. 3-49). Equilibrium of this closed-loop system is attained when the average currents through R_1 and R_3 are equal and opposite. This is achieved when V_{AZ}, the autozero voltage, is equal to $-\frac{1}{2}V_{REF}$ since $R_1 = R_3$. Establishing V_{AZ} and storing it on C_{AZ} gives the U/D logic the capability of switching either a plus or minus reference current to the integrator during conversion. Thus when U/D is up, $I_1 + I_3 = -V_{REF}/2R_1$. The autozero interval is of sufficient duration to insure that V_{AZ} will be well established.

Prior to the start of the measure interval, the integrator output (which had been cycling around −1 V) is brought back to analog ground, the comparator threshold. The system is now ready for a conversion.

Measurement Interval The so-called "quantized feedback" conversion system is characterized by a single-phase digitization interval in which a digital control system feeds back quantized units of charge in response to the sampled state of an analog comparator. These quanta of charge balance the charge being supplied to the integrator by the analog voltage. The magnitude ($V_{REF}/2R_1 \times 6/f_{clock}$) of the quantized charge being fed back and its sign (+ or −) arise from the fact that the control logic has two U/D duty cycles available during the measure interval, as shown in Fig. 3-50.

The U/D logic is up 1 clock cycle and down 7 cycles for a high comparator output in the clock cycle preceding a set of 8 cycles. This will be designated duty cycle A. With a low comparator output in clock cycle 7, the U/D logic will be up for 7 cycles and down for 1 cycle in the following 8 clock cycles. This is duty cycle B. The effect of these two reference current duty cycles on the integrator output is shown in Fig. 3-50. It can be seen that the up state of the U/D logic drives the integrator output voltage up. The U/D BCD counter increments by each clock pulse when the U/D logic is up and decrements by each clock pulse when the U/D logic is down. Consequently the net count goes up 6 counts for a B duty cycle and down 6 for an A duty cycle.

Input polarity is determined by the first appearance of two consecutive

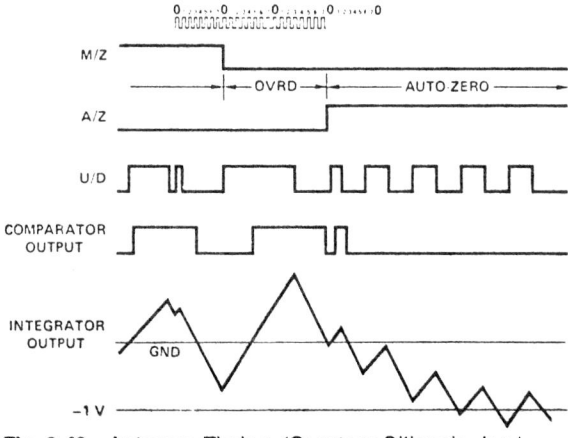

Fig. 3-49 Autozero Timing. *(Courtesy Siliconix, Inc.)*

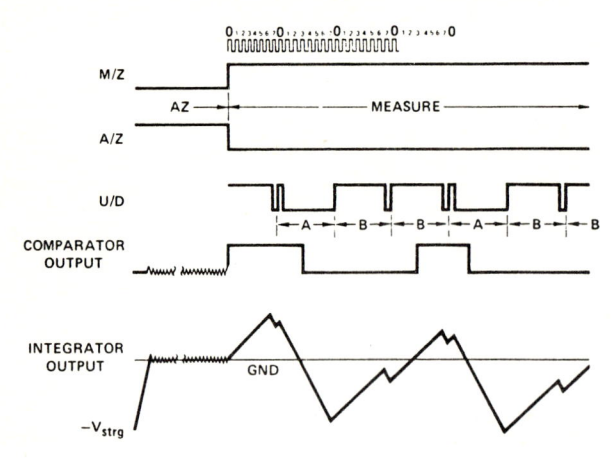

118 Data Conversion and Telecommunications Circuits

Fig. 3-50 Measure Interval Timing. *(Courtesy Siliconix, Inc.)*

duty cycles of the same type. The control logic would determine the analog input to be negative if two A duty cycles occur in succession and positive if two B duty cycles occur in succession.

Since the counting process is done by increments (or decrements) of 6 during the measure interval, a short override interval is required at the end of the measurement to "fine-tune" the count to the nearest LSB. This occurs within the first 32 clock periods of the AZ interval.

Data Flow Following the count-correcting override sequence, the contents of the BCD counters and sign flip-flop are loaded into the internal latches. Counter states of less than 80 or greater than 999 are decoded as underrange or overrange conditions, respectively. The presence of an out-of-range signal gates a single pulse (one clock period) to the SIGN/UR/OR (SIGN/UNDERRANGE/OVERRANGE) output during either D_1 or D_2 digit time (D_2 identifies overrange, D_1 identifies underrange). The overrange condition also provides a visual signal by holding the digit strobe outputs low during the measure interval. This holds the display off for two-thirds of the sampling interval, giving a blinking effect.

The BCD data stored in the latches is continuously scanned every 24 clock periods (eight clock times per digit). Sign information appears at the SIGN/UR/OR pin coincident with the D_3 strobe. Interdigit blanking of the digit strobes is achieved by taking one full clock period from both the leading and trailing edges of the strobes. Thus the digit is on for six clock periods while the BCD data for that digit appears for the full eight clock periods. Figure 3-51 shows the data output timing.

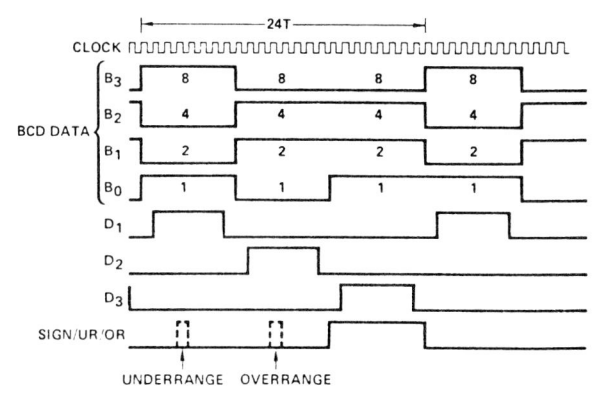

Fig. 3-51 Data Output Format (Output = 769). *(Courtesy Siliconix, Inc.)*

A/D Converter Display Interface Figure 3-52 shows a LED interface using a ±3-digit LD130 A/D converter in a ±3-digit DPM (digital panel meter) application. This circuit features a 1000-V scale and a sampling rate of 5 samples/s. High-efficiency LEDs and a CMOS decoder-driver keep power consumption to a minimum (I_1 is typically 60 mA).

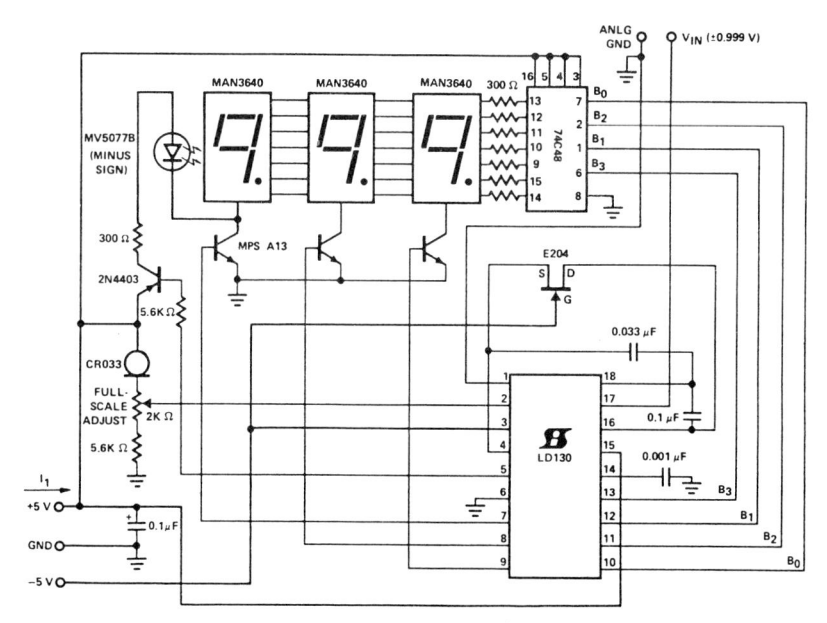

Fig. 3-52 Three-digit DVM (±999 mV, 5 samples/s). *(Courtesy Siliconix, Inc.)*

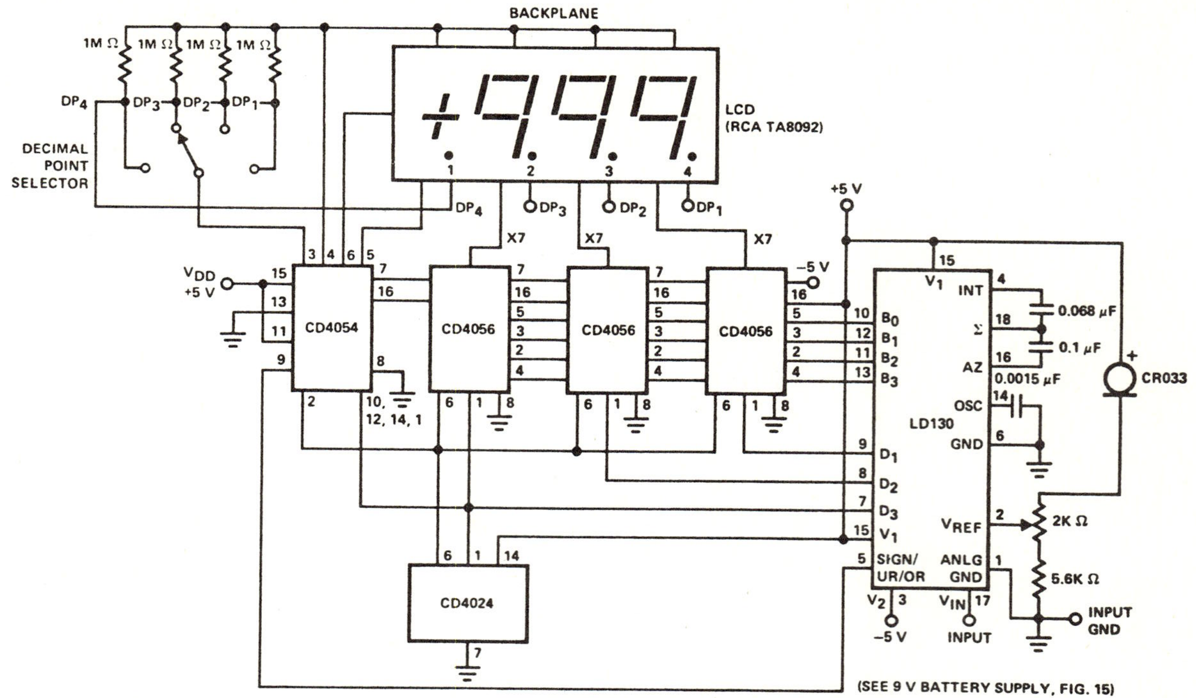

Fig. 3-53 LCD Interface. *(Courtesy Siliconix, Inc.)*

Greater power savings can be realized by interfacing the LD130 with an LCD, as shown in Fig. 3-53. This circuit uses four LCD decoder-driver-latches to demultiplex the BCD output of the LD130 and drive the display. The interdigit blanking assures that the latched code will always be valid. Alternating current drive ($\approx$30 Hz) to the backplane of the LCD is derived from a divided down digit strobe (D_3). A single LCD drive signal, out of phase with the backplane, is switched to turn on the appropriate decimal point.

The monolithic LED displays are attractive in terms of small size and power drain. The fact that all segments are multiplexed, however, can make them difficult to interface with when a polarity sign is required. What is needed is a means to turn on the g segment of an unused digit on the left of the display. This segment would give a minus sign when the input voltage was negative. Such a function is provided by the circuit shown in Fig. 3-54. The presence of a zero SIGN/UR/OR output during

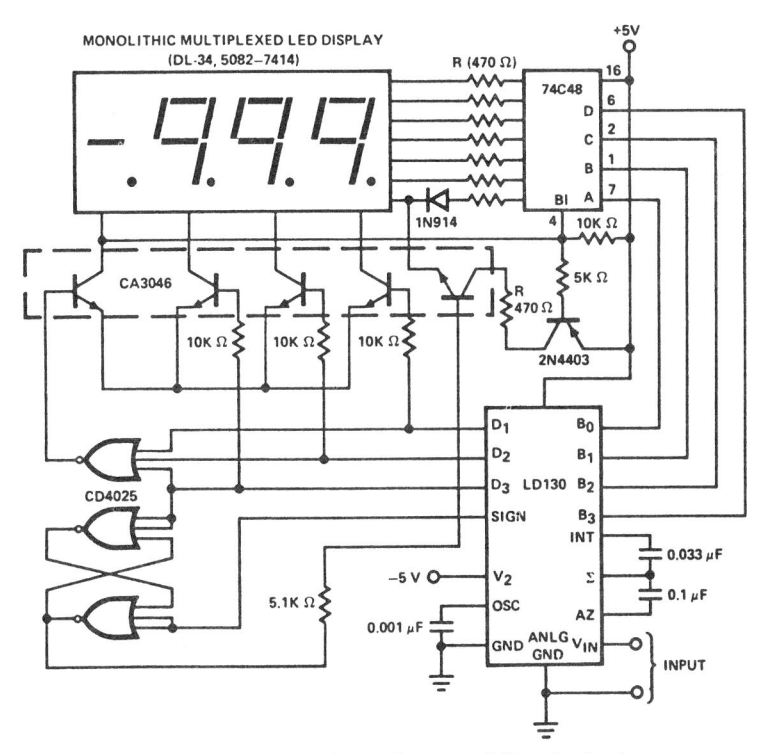

Fig. 3-54 Monolithic LED Interface. *(Courtesy Siliconix, Inc.)*

D_3 time (negative input) sets the RS flip-flop (CD4025), which in turn lights segment g of the 4-digit display during each interdigit blanking interval. Since the cumulative on time of the interdigit blanking periods equals a digit on time, this minus sign is of equal intensity to the rest of the display. A positive voltage at the LD130 will result in a 1 at the SIGN/UR/OR output during D_3 time, resetting the RS flip-flop and turning off the minus sign.

Digital Thermometer A digital thermometer, reading in either degrees Fahrenheit or degrees Celsius, can be constructed, as shown in Fig. 3-55, with a basic LD130 DPM. This circuit converts temperature to a voltage by using the temperature-dependent forward voltage of a PN junction as the sensing element. The change in this voltage is typically -2.3 mV/°C at room temperature and can be suitably linear if the PN junction is biased with a constant current much greater then the reverse saturation current.

Since the diode will have a finite voltage at either a Celsius or Fahrenheit zero, this voltage component must be subtracted out. Figure 3-55 shows the temperature-sensing diode (base emitter of a 2N2222) biased with a zero TC current regulator diode (CR033). Zeroing is achieved by summing the currents from the CR033 and E506 diodes at the wiper of a potentiometer.

The scaling for either Celsius or Fahrenheit is achieved as follows:

$$\text{Count} = 2000 \, \frac{V_{in}}{V_{REF}}$$

$$\Delta\text{Count} = 2000 \, \frac{\Delta V_{in}}{V_{REF}}$$

for $\Delta T = 100°C, \Delta V_{in} \cong -230\text{mV}; \Delta T = 1000°F, \Delta V_{in} \cong -1.278$ V. Thus

$$\begin{aligned} V_{REF} &= 0.46 \text{ V} \quad \text{for °C} \\ &= 2.5 \text{ V} \quad \text{for °F} \end{aligned}$$

The fact that the forward voltage decreases with temperature requires that the sense of the LD130 sign bit be inverted (0 for +, 1 for −).

Digital Dwell Tachometer The dwell tachometer can be a very useful automotive analysis and adjustment tool. The circuit shown in Fig. 3-56 will turn the LD130 into a digital dwell tachometer usable on 4-, 6-, or 8-cylinder four-stroke engines (with distributor ignition). The dwell measurement is made simply by taking the average value of the distributor duty cycle and scaling to degrees of rotation. Engine r/min is

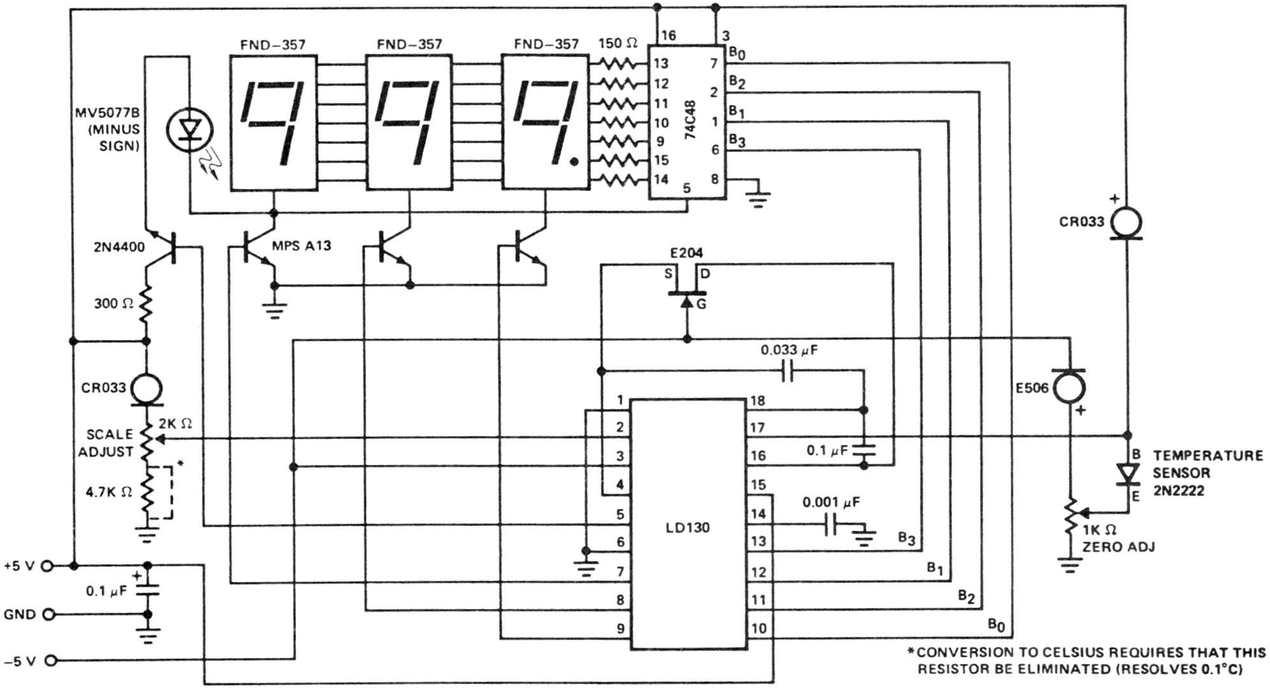

Fig. 3-55 Three-Digit Thermometer (Fahrenheit). *(Courtesy Siliconix, Inc.)*

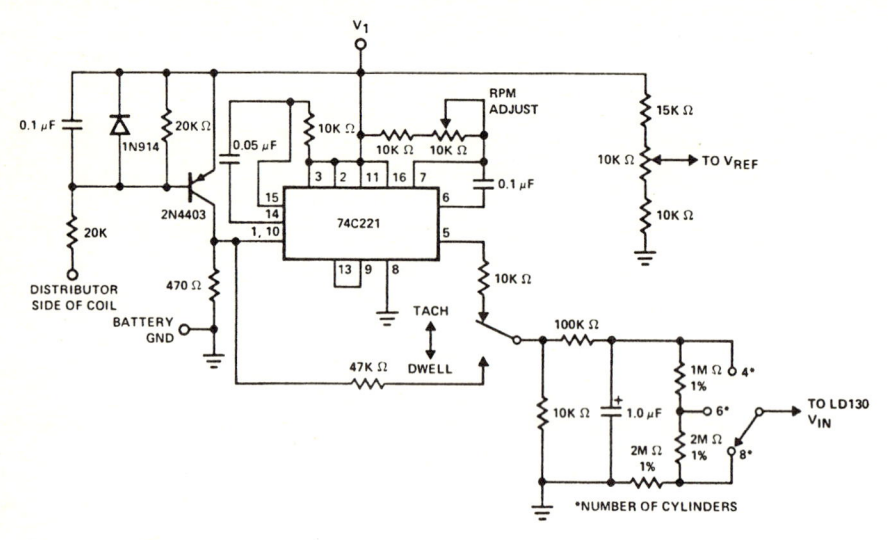

124 Data Conversion and Telecommunications Circuits

Fig. 3-56 Dwell-Tachometer Function. *(Courtesy Siliconix, Inc.)*

measured by triggering a one-shot multivibrator with each point closure. The average dc value of the pulse train produced is scaled to read r/min × 10.

Extraneous triggering from ignition system ringing is avoided by using the second one-shot to disable the frequency-to-voltage converter (first one-shot) when the points open. A CMOS one-shot is used because the MOSFET output does not have the inherent $V_{CE}(sat)$ offset that a bipolar one-shot would have. This allows the system to achieve a proper zero.

The dwell tachometer features full-scale readings of 10,000 r/min as a tachometer and 100.0° as a dwell meter. The reference for the LD130 is divided from the supply voltage since the outputs are all proportional to the positive supply. Selection of the 4-, 6-, or 8-cylinder output is with a SP3T switch.

The National Semiconductor ADD3501 3½-Digit Digital Voltmeter

An A/D conversion method based on pulse duration modulation does the same digitizing job as the dual-slope method usually used in digital panel meters—but it takes better advantage of the capabilities inherent in CMOS technology.

The dual-slope approach requires linear circuits—an integrator and a comparator. But the pulse-width approach, developed by National Semiconductor for use in its ADD3501 CMOS DVM (digital voltmeter) chip, eliminates virtually all linear circuits. It uses a comparator made up of CMOS inverters that are similar to those used in CMOS digital logic, and integrates with digital circuits by counting pulses (Fig. 3-57).

Like the dual-slope technique, the new method integrates the input signal to reduce measurement errors due to noise and 60-Hz contamination of the signal. But while the dual-slope method uses an opposite-polarity reference, the new method uses a reference voltage of the same polarity as the signal to be measured. Consequently, the pulse-width method is better for applications that can use a single power supply for both transducer excitation and the reference source (one 5-V TTL supply is all that is required).

Operating with an isolated supply allows the conversion of positive as well as negative voltages. The sign of the input voltage is automatically

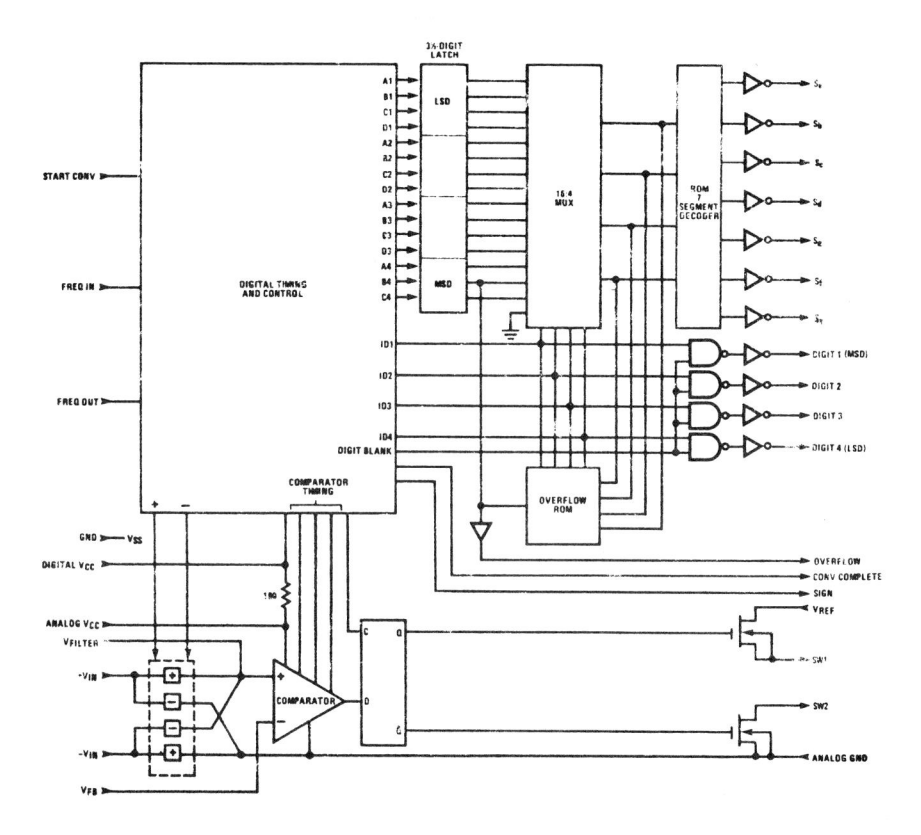

Fig. 3-57 ADD3501 3½-Digit Block Diagram. *(Courtesy National Semiconductor Corp.)*

determined, and is outputed or made available on the SIGN pin. If the power supply is not isolated, only one polarity of voltage may be converted.

The conversion rate is set by an internal oscillator. The frequency of the oscillator can be set by an external RC network or the oscillator can be driven from an external frequency source. When using the external RC network, a square wave output is available. It is important to note that great care has been taken to synchronize digit multiplexing with the A/D conversion timing to eliminate noise due to power supply transients.

The ADD3501 has been designed to drive 7-segment multiplexed LED displays directly with the aid of external digit buffers and segment resistors. Under conditions of overrange, the overflow output will go high and the display will read +OFL or −OFL, depending on whether the input voltage is positive or negative. In addition to this, the most significant digit is blanked when zero.

The ADD3501 converts 0 V to ±1.999 V, operates at medium speed −200 ns/conversion, and is encased in a 28-pin DIP, as shown in Fig. 3-58.

Figure 3-59 illustrates how the ADD3501 measures a voltage. Counter 2 sets the duration of each A/D conversion by accumulating 2000 consecutive pulses of the clock, f_{in}. Counter 1, which feeds information to the LEDs via a ROM, obtains the numeric value of the measured voltage by accumulating the clock pulses passed by the gate. The key input to the gate is Q, from the D flip-flop. Its transitions are at clock times, but its average duty cycle is proportional to V_{in}.

Slaving the duty cycle of Q to the magnitude of V_{in} is the responsibility of the oscillating analog loop made up of the comparator, the D flip-flop, and the pair of switch transistors, SW_1 and SW_2. During the conversion, the analog loop makes the feedback voltage, V_{FB}, closely approach V_{in}.

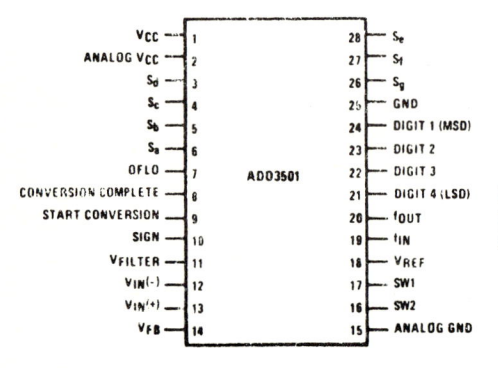

Fig. 3-58 ADD3501 Pin Connection Diagram. *(Courtesy National Semiconductor Corp.)*

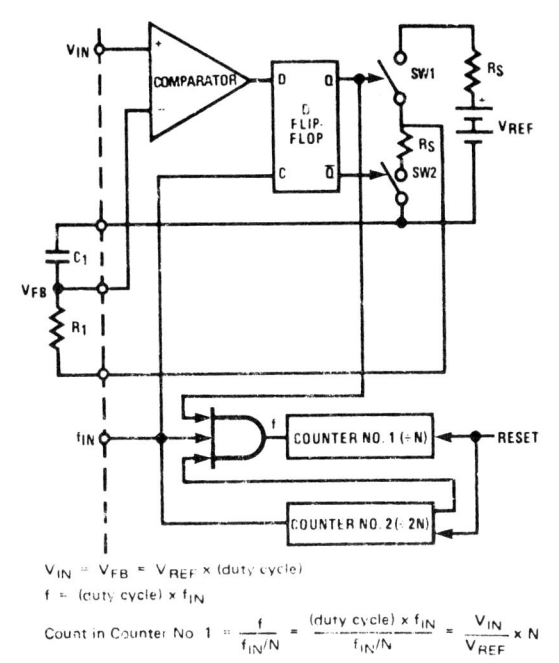

Fig. 3-59 Block Diagram of PDM A/D Conversion Technique. *(Courtesy National Semiconductor Corp.)*

$V_{IN} = V_{FB} = V_{REF} \times \text{(duty cycle)}$

$f = \text{(duty cycle)} \times f_{IN}$

$$\text{Count in Counter No. 1} = \frac{f}{f_{IN}/N} = \frac{\text{(duty cycle)} \times f_{IN}}{f_{IN}/N} = \frac{V_{IN}}{V_{REF}} \times N$$

The loop keeps flipping between opposite states—makes V_{FB} oscillate around V_{in}—while remaining within a fraction of a millivolt of V_{in}.

The SPDT switch pair connects R first to the reference voltage V_{REF}, then to ground. Since feedback voltage V_{FB} is the RC filtered output of the switch, it is controlled by the duty cycle of D. Each clock time, the comparator decides whether D should be on or off to keep V_{FB} tracking V_{in} with the smallest error.

In this way, the comparator forces the average duty cycle of D to approach the ratio of the clock pulses, V_{in}/V_{REF}, which is gated into counter 2.

The complete measurement takes 100 ms when the clock frequency is 20 kHz. The actual master clock, which may be generated on the chip or externally supplied, is 640 kHz, from which the 20-kHz rate is derived. To develop the various timing signals for the comparator, the 640 kHz is divided by 8. One of the eight 80 kHz outputs is further divided by 4 by prescalers at the inputs of counters 1 and 2.

The successive-approximation method of A/D conversion is faster than either dual-slope or pulse-width methods, but it is rarely used for DPM

128 Data Conversion and Telecommunications Circuits

applications because it lacks the noise reduction advantage of an integrating approach.

The timing diagram, shown in Fig. 3-60, gives operation for the free-running mode. Free-running operation is obtained by connecting the START CONVERSION input to logic 1 (V_{CC}). In this mode the analog input is continuously converted and the display is updated at a rate equal to $64,512 \times 1/f_{in}$.

The rising edge of the CONVERSION COMPLETE output indicates that new information has been transferred from the internal counter to the display latch. This information will remain in the display latch until the next low-to-high transition of the CONVERSION COMPLETE output. A logic 1 will be maintained on the CONVERSION COMPLETE output for a time equal to $64 \times 1/f_{in}$.

Figure 3-61 gives the operation using the START CONVERSION input. It is important to note that the START CONVERSION input and CONVERSION COMPLETE output do not influence the actual A/D conversion in any way.

Internally the ADD3501 is always continuously converting the analog voltage present at its inputs. The START CONVERSION input is used to control the transfer of information from the internal counter to the display latch.

An RS (RESET/SET) latch on the START CONVERSION input allows a broad range of input pulse widths to be used on this signal. As shown in Fig. 3-61, the CONVERSION COMPLETE output goes to a logic 0 on the rising edge of the START CONVERSION pulse and goes

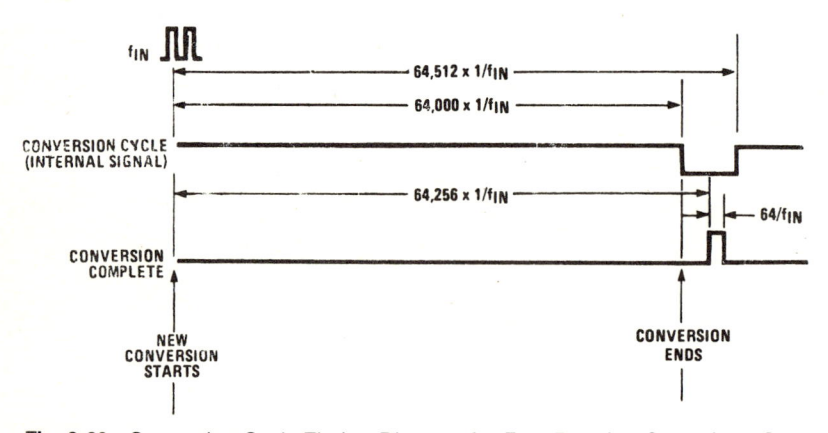

Fig. 3-60 Conversion Cycle Timing Diagram for Free Running Operation. *(Courtesy National Semiconductor Corp.)*

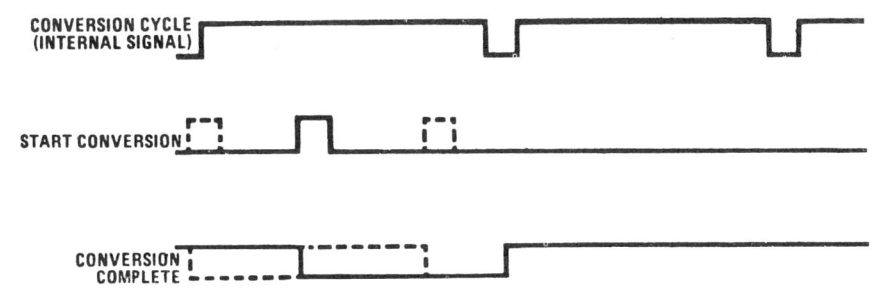

Fig. 3-61 Conversion Cycle Timing Diagram Operating with START CONVERSION Input. *(Courtesy National Semiconductor Corp.)*

to a logic 1 some time later when the new conversion is transferred from the internal counter to the display latch. Since the START CONVERSION pulse can occur at any time during the conversion cycle, the amount of time from START CONVERSION to CONVERSION COMPLETE will vary. The maximum time is $64,512 \times 1/f_{in}$ and the minimum time is $256 \times 1/f_{in}$.

Perhaps the most important thing to consider when designing a system using the ADD3501 is power supply noise on the V_{CC} and ground lines. Because a single power supply is used and currents in the 300-mA range are being switched, good circuit layout techniques cannot be overemphasized. Great care has been exercised in the design of the ADD3501 to minimize such problems, but poor printed circuit layout can negate these features.

Figures 3-62 and 3-63 show schematics of several DVM systems. An attempt has been made to show, on these schematics, the proper distribution for ground and V_{CC}. To help isolate digital and analog portions of the circuit, the analog V_{CC} and ground have been separated from the digital V_{CC} and ground. Care must be taken to eliminate high current from flowing in the analog V_{CC} and ground wires. The most effective method of accomplishing this is to use a single ground point and a single V_{CC} point where all wires are brought together. In addition to this, the conductors must be of sufficient size to prevent significant voltage drops.

To prevent switching noise from causing jitter problems, a voltage regulator with good high-frequency response is necessary. The LM309 and the LM340-5 voltage regulators, which both function well, are shown in Figs. 3-62 and 3-63. Adding more filtering than is shown will in general increase the jitter rather than decrease it. The most important characteristic of transients on the V_{CC} line is their duration and not their amplitude.

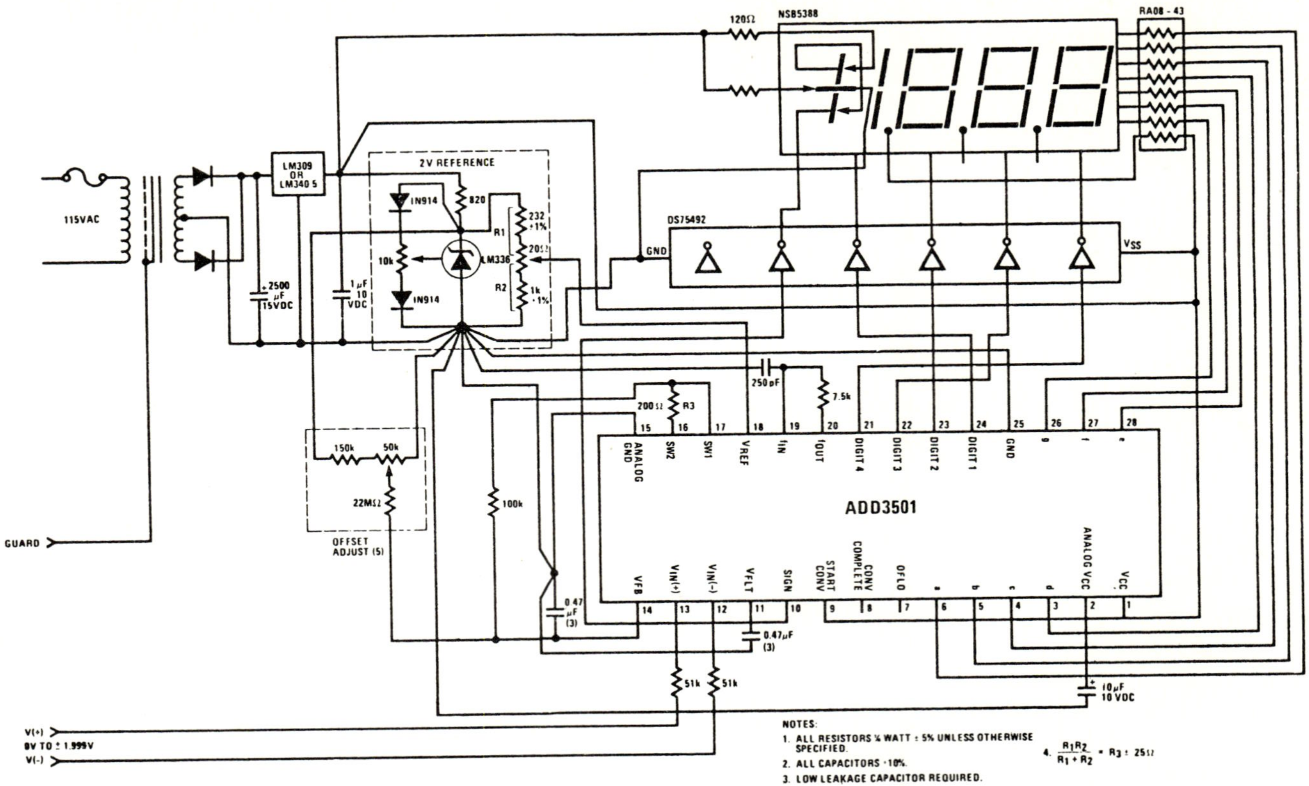

Fig. 3-62 A 3½-Digit DPM, ±1.999 V Full Scale. *(Courtesy National Semiconductor Corp.)*

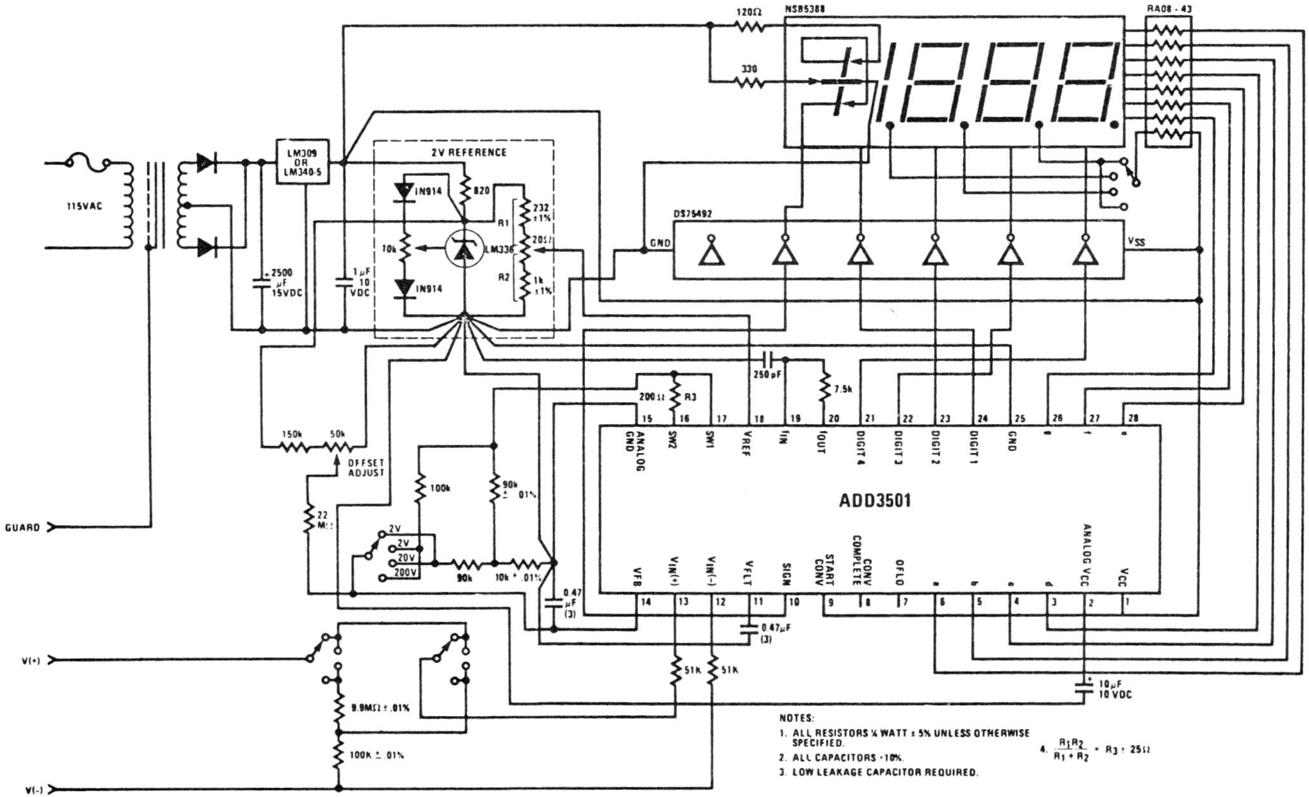

Fig. 3-63 A 3½-Digit DVM, Four Decade, ±0.2 V, ±2 V, ±20 V, and ±200 V Full Scale. *(Courtesy National Semiconductor Corp.)*

Figures 3-62 and 3-63 show systems operating with an isolated supply that will convert positive and negative inputs. 60-Hz common-mode input becomes a problem in this configuration, and a transformer with an electrostatic shield between primary and secondary windings is shown. The necessity for using a shielded transformer depends on the performance requirements and the actual application.

The filter capacitors connected to V_{FB} (pin 14) and V_{FLT} (pin 11) should be low leakage. In the application examples shown, every 1.0 nA of leakage current will cause 0.1 mV of error (1.0×10^{-9} A $\times$ 100 kΩ = 0.1 mV). If the leakage current in both capacitors is exactly the same, no error will result since the source impedances driving them are matched.

The National Semiconductor ADC0816 Data Acquisition System

The ADC0816 is a CMOS DAS (data acquisition system) chip that contains a true 8-b A/D converter with bus-oriented (microcomputer) outputs, a 16-channel expandable multiplexer with address input latches, latched Tri-State outputs, provision for handling external signal conditioning, and all the logic control needed for interfacing the chip to all the standard microcomputers.

The 16-channel multiplexer can directly access any one of 16 single-ended analog signals and provides the logic for additional channel expansion. Signal conditioning of any analog input signal is eased by direct access to the input of the 8-b A/D converter.

The ADC0816 operates from a single 5-V supply, consumes only 15 mW of power, and has a conversion speed specified at 100 μs. The typical speed is 50 μs, which is useful for applications with slow-changing inputs from, say, pressure, temperature, and velocity sensors. Also, by using a chopper-stabilized comparator, the design reduces both long-term drift and temperature coefficient errors to hybrid-version levels, yielding a linearity error of less than $\pm\frac{1}{2}$ LSB over the commercial temperature range of -40 to $+85°$C for the ADC0816. The ADC0817 has a ± 1 LSB linearity error.

The device eliminates the need for external zero and full-scale adjustments and features an absolute accuracy of ≤ 1 LSB, including quantizing error. Easy interfacing to microprocessors is provided by the latched and decoded address inputs and latched TTL Tri-State outputs.

The ADC0816/0817 is encased in a 40-pin DIP, as shown in Fig. 3-64.

One especially nice feature of the ADC0816 is its ability to perform without external components in systems using ratiometric transducers, such as potentiometer strain gauges, thermistor bridges, pressure transducers, and so on.

Data Conversion Circuits   133

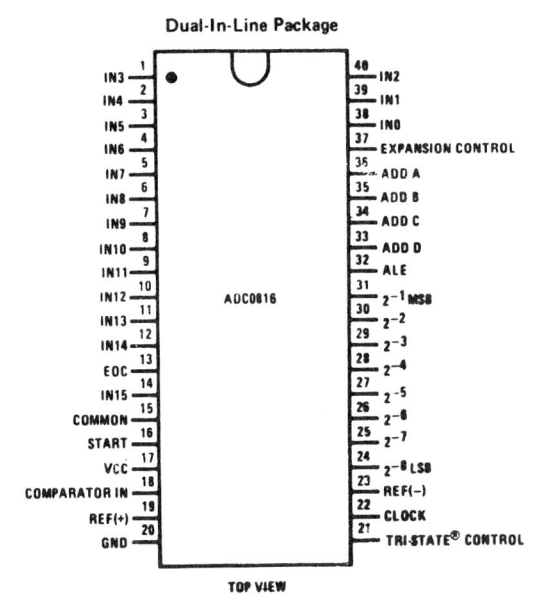

```
              Dual-In-Line Package
                  1              40
          IN3 ─┤              ├─ IN2
               2              39
          IN4 ─┤              ├─ IN1
               3              38
          IN5 ─┤              ├─ IN0
               4              37
          IN6 ─┤              ├─ EXPANSION CONTROL
               5              36
          IN7 ─┤              ├─ ADD A
               6              35
          IN8 ─┤              ├─ ADD B
               7              34
          IN9 ─┤              ├─ ADD C
               8              33
         IN10 ─┤              ├─ ADD D
               9              32
         IN11 ─┤              ├─ ALE
              10              31
         IN12 ─┤   ADC0816    ├─ 2⁻¹ MSB
              11              30
         IN13 ─┤              ├─ 2⁻²
              12              29
         IN14 ─┤              ├─ 2⁻³
              13              28
          EOC ─┤              ├─ 2⁻⁴
              14              27
         IN15 ─┤              ├─ 2⁻⁵
              15              26
       COMMON ─┤              ├─ 2⁻⁶
              16              25
        START ─┤              ├─ 2⁻⁷
              17              24
          VCC ─┤              ├─ 2⁻⁸ LSB
              18              23
 COMPARATOR IN ─┤              ├─ REF(−)
              19              22
        REF(+) ─┤              ├─ CLOCK
              20              21
          GND ─┤              ├─ TRI-STATE® CONTROL

                    TOP VIEW
```

Fig. 3-64 ADC0816/0817 Pin Connection Diagram.
(Courtesy National Semiconductor Corp.)

Since in these systems only the change in the parameter is measured, the device can operate without an external voltage reference. Here the transducer is connected directly across the supply voltage and the outputs are connected directly into the multiplexer inputs. On the other hand, for systems that require an absolute value measurement, a standard, commercially available voltage reference will be required in addition to the ADC0816.

Figure 3-65 illustrates the block diagram of the ADC0816. As shown, the device contains a 16-channel single-ended analog signal multiplexer. A particular input channel is selected by using the address decoder. Table 3-6 shows the input states for the address line and the expansion control line to select any channel. The address is latched into the decoder on the low-to-high transition of the ADDRESS LATCH ENABLE signal.

Additional single-ended analog signals can be multiplexed to the A/D converter by disabling all the multiplexer inputs. These additional external signals are connected to the comparator input and the device ground. Additional signal conditioning (i.e., prescaling, sample and hold, instrumentation amplification, etc.) may also be added between the analog input signal and the comparator input.

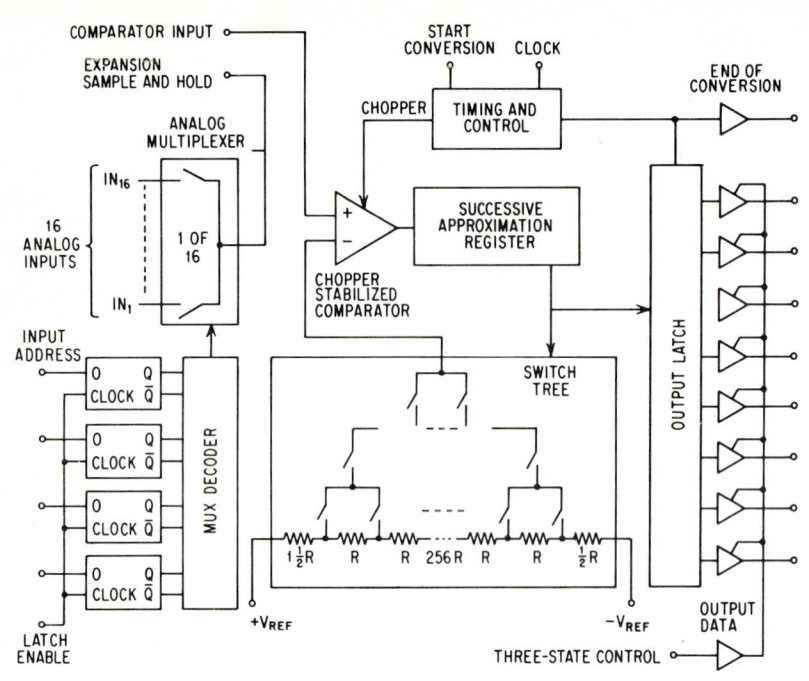

Fig. 3-65 ACD0816 Block Diagram. *(Courtesy National Semiconductor Corp.)*

TABLE 3-6 Address Line Input State Table

Selected Analog Channel	\multicolumn{4}{c}{Address Line}	Expansion Control
	D	C	B	A	
IN_0	L	L	L	L	H
IN_1	L	L	L	H	H
IN_2	L	L	H	L	H
IN_3	L	L	H	H	H
IN_4	L	H	L	L	H
IN_5	L	H	L	H	H
IN_6	L	H	H	L	H
IN_7	L	H	H	H	H
IN_8	H	L	L	L	H
IN_9	H	L	L	H	H
IN_{10}	H	L	H	L	H
IN_{11}	H	L	H	H	H
IN_{12}	H	H	L	L	H
IN_{13}	H	H	L	H	H
IN_{14}	H	H	H	L	H
IN_{15}	H	H	H	H	H
All Channels Off	×	×	×	×	L

NOTE: L = low; H = high; × = don't care.
SOURCE: Courtesy National Semiconductor Corp.

The heart of this single-chip data acquisition system is its 8-b A/D converter. The converter is designed to give fast, accurate, and repeatable conversions over a wide range of temperatures. The converter is partitioned into three major sections: the 256R ladder network, the SAR (successive-approximation register), and the comparator. The converter's digital outputs are positive true.

The 256R ladder network approach (Fig. 3-65) was chosen over the conventional R/2R ladder because of its inherent monotonicity, which guarantees no missing digital codes. Monotonicity is particularly important in closed-loop feedback control systems. A nonmonotonic relationship can cause oscillations that will be catastrophic for the system. Additionally, the 256R network does not cause load variations on the reference voltage.

The bottom resistor and the top resistor of the ladder network in Fig. 3-65 do not have the same value as the remainder of the network. The difference in these resistors causes the output characteristic to be symmetrical with the zero and full-scale points of the transfer curve. The first output transition occurs when the analog signal has reached $+\frac{1}{2}$ LSB, and succeeding output transitions occur every 1 LSB later up to full scale.

The SAR performs eight iterations to approximate the input voltage. For any SAR-type converter, n iterations are required for an n-b converter. Figure 3-66 shows a typical example of a 3-b converter. In the ADC0816/ADC0817, the approximation technique is extended to 8 b using the 256R network.

The A/D converter's SAR is reset on the positive edge of the SC (START CONVERSION) pulse. The conversion is begun on the falling edge of the START CONVERSION pulse. A conversion in process will be interrupted by receipt of a new START CONVERSION pulse. Continuous conversion may be accomplished by tying the EOC (END OF

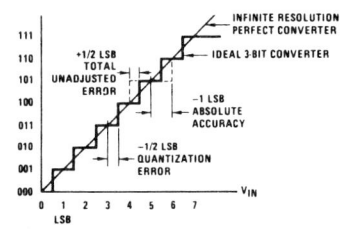

Fig. 3-66 A 3-b A/D Transfer Curve. *(Courtesy National Semiconductor Corp.)*

CONVERSION) output to the SC input. If used in this mode, an external START CONVERSION pulse should be applied after powerup. EOC will go low between one and eight clock pulses after the rising edge of START CONVERSION.

The most important section of the A/D converter is the comparator. It is this section which is responsible for the ultimate accuracy of the entire converter. It is also the comparator drift which has the greatest influence on the repeatability of the device. A chopper-stabilized comparator provides the most effective method of satisfying all the converter requirements.

The chopper-stabilized comparator converts the dc input signal into an ac signal. This signal is then fed through a high-gain ac amplifier and has the dc level restored. This technique limits the drift component of the amplifier since the drift is a dc component which is not passed by the ac amplifier. This makes the entire A/D converter extremely insensitive to temperature, long-term drift, and input offset errors.

The ADC0816/ADC0817 is designed as a complete DAS for ratiometric conversion systems. In ratiometric systems, the physical variable being measured is expressed as a percentage of full scale which is not necessarily related to an absolute standard.

A good example of a ratiometric transducer is a potentiometer used as a position sensor. The position of the wiper is directly proportional to the output voltage, which is a ratio of the full-scale voltage across it. Since the data is represented as a proportion of full scale, reference requirements are greatly reduced, eliminating a large source of error and cost for many applications. A major advantage of the ADC0816/ADC0817 is that the input voltage range is equal to the supply range so that the transducers can be connected directly across the supply and their outputs connected directly into the multiplexer inputs (Fig. 3-67).

Ratiometric transducers such as potentiometers, strain gauges, thermistor bridges, pressure transducers, etc., are suitable for measuring proportional relationships; however, many types of measurements must be referred to an absolute standard such as voltage or current. This means a system reference must be used which relates the full-scale voltage to the standard volt. For example, if $V_{CC} = V_{REF} = 5.12$ V, then the full-scale range is divided into 256 standard steps. The smallest standard step is 1 LSB, which is then 20 mV.

Figure 3-68 shows the connection for interfacing the ADC0816/ADC0817 to a microprocessor.

Mostek Corp.'s MK50816 CMOS 16-channel input 8-b A/D converter has been designed as a pin-compatible alternative to the ADC0816.

$$Q_{OUT} = \frac{V_{IN}}{V_{REF}} = \frac{V_{IN}}{V_{CC}}$$

$$4.75V \leq V_{CC} = V_{REF} \leq 5.25V$$

*Ratiometric transducers

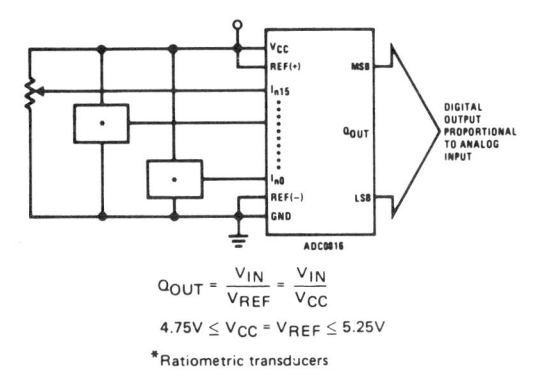

Fig. 3-67 Ratiometric Conversion System. *(Courtesy National Semiconductor Corp.)*

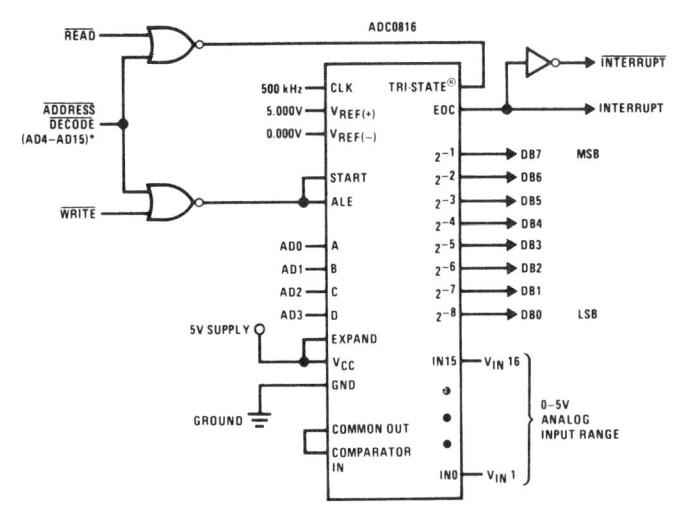

*Address latches needed for 8085 and SC/MP interfacing the ADC0816 to a microprocessor

MICROPROCESSOR INTERFACE TABLE

PROCESSOR	READ	WRITE	INTERRUPT (COMMENT)
8080	MEMR	MEMW	INTR (Thru RST Circuit)
8085	RD	WR	INTR (Thru RST Circuit)
Z-80	RD	WR	INT (Thru RST Circuit, Mode 0)
SC/MP	NRDS	NWDS	SA (Thru Sense A)
6800	VMA · φ2 · R/W	VMA · φ2 · R/W	IRQA or IRQB (Thru PIA)

Fig. 3-68 ADC0816 Microprocessor Interface. *(Courtesy National Semiconductor Corp.)*

TABLE 3-7 Summary of Commercially Available Single-Chip CMOS Analog-to-Digital Converters

Supplier	Model	Resolution (Bits)	Linearity (±% LSB)	Conversion Time [To ½ LSB (µs)]	Gain TC (ppm/°C)	Power Required V (mA)	Package, DIP
RCA	CA3300	6	0.50	0.07	N/A	4–12 (10), 50 mW	18
Analog Devices	AD7570	8	0.39	20	10	5–15 (2)	28
	AD7574	8	0.75	15	a	5 (5)	18
	AD7583	8	±1 count	400	N/A	5–15 (20)	40
National Semicond.	ADC0808	8	0.2	100	17	5 (1)	28
	ADC0816	8	0.2	100	17	5 (1)	40
Mostek	MK50808	8	0.25	110ᶜ		6.8 mW	28
	MK50816	8	0.25	110ᶜ		6.8 mW	40
Teledyne Semicond.	8700	8	0.2	1.8 ms	75	+5V, 20 mW	24
	8703	8	0.2	1.8 ms	80	+5V, 20 mW	24
Analog Devices	AD7570	10	0.2	40	10	5–15 (2)	28
Teledyne Semicond.	8701	10	0.05	6 ms	75	20 mW	24
	8704	10	0.05	6 ms	80	20 mW	24
Intersil	ICL7109	12	0.01	2.79	5	20 mW	40
Teledyne Semicond.	8702	12	0.012	24 ms	75	20 mW	24
	8705	12	0.012	24 ms	75	20 mW	24
NEC	UPD7002	12	0.025	125 ms	20	25 mW	28
Analog Devices	AD7550	13	0.5 LSB	40 ms	1	5–15 (2)	40

		Conversn Technique			Output Coding						Minimum External Components Required		
PDM	Dual Slope	Successive Approximation	Charge Balancing	Parallel (Flash)	Straight Binary	Decimal	ComplimentBinary	Three-State Outputs	Internal Reference	Input Buffer AMP	Actives	Passives	Special Features
				x				x		x	N/A	N/A	
		×			×			×			2	0	Parallel and Serial Outputs.
		×			×			×			2	0	μP-compatible ADC that Operates like a Memory. Complete ADC.
	x[d]				x			x			4	7	9-channel MUX; μP
		×			×			×			1	0	NMC; 8-channel MUX.[b]
		×			×			×			1	0	NMC; 16-channel MUX.
		×			×			×	×		0	0	8-channel MUX.
		×			×			×	×		0	0	16-channel MUX. Pin-compatible with ADC0808 and ADC0816, respectively.
			×		×						1	6	10-μA FS input—100 mV to 1 kV; NMC; Internal Clock; Strobed or Free-Running Conversion.
			×		×			×			1	6	Full temperature Operation. NMC.
		×			×			×			2	0	Parallel and Serial Outputs.
			×		×						1	6	See 8-b Versions.
			×		×			×			1	6	See 8-b Versions.
	×				×			×	×		0	3	Microprocessor Interface/UART Handshake. Programmable Latched Parallel Tri-State Binary Outputs.
			×		×						1	6	See 8-b Versions.
			×		×			×			1	6	See 8-b Versions.
			×		×			×	×		0	2	
	x[d]						×	×	×		0	4	μP compatible Serial and Parallel Outputs.

139

TABLE 3-7 *(Continued)*

Supplier	Model	Resolution (Bits)	Linearity (±% LSB)	Conversion Time [To ½ LSB (μs)]	Gain TC (ppm/°C)	Power Required V (mA)	Package, DIP
Intersil	ICL7106/ 7107	3½	0.025	333 ms	5	9(1.8)/5(1.8)	40
National Semicond.	ADC3511	3½	0.05	200 ms	...	5 (0.5)	24
	ADC3711	3¾	0.05	400 ms	...	5 (0.5)	24
	ADD3501	3½	0.05	200 ms	...	5 (0.5)	24
	ADD3701	3¾	0.05	400 ms	...	5 (0.5)	24
Teledyne Semicond.	8750	3½	0.025	12 ms	80	20 mW	24
Siliconix	LD130	3	0.1	60 conv/s	...	30 mW	18
	LD131	3½	0.1	60 conv/s	...	30 mW	18
Intersil	ICL7116/ 7117	3½	0.025	333 ms	...	5–9 V(1.8)/5(1.8)	40

[a] TC = 0.6%/°C.
[b] NMC = no missing codes; MUX = multiplexer.
[c] Using a 640-kHz clock; 150–350 μs using internal clock.
[d] Quad-slope—four-phase integration period.
N/A = not available.

Data Conversion Circuits   141

PDM	Dual Slope	Conversn Technique Successive Approximation	Charge Balancing	Parallel (Flash)	Straight Binary	Output Coding Decimal	Complement Binary	Three State Outputs	Internal Reference	Input Buffer AMP	Minimum External Components Required Actives	Passives	Special Features
		×			×			×	×	×	0	9	7-Segment Parallel Outputs; True Differential Inputs; Directly Drives LCD Displays; ±5-V Model Available. 7107 Drives LEDs. On-Chip Clock.
			×		×		×		×	1	5	Microprocessor-compat.
			×		×		×		×	1	5	Microprocessor-compat.
			×				×		×	1	5	7-segment multiplexed LED driver.
			×				×		×	1	5	
			×		×	×				0	10	See 8700 comments. Strobed or free-running conversion. NMC
	×				×			×		0	3	
	×				×			×		0	3	
	×				×			×	×	0	9	7116 Drives LCD, 7117 LEDS. Have All Features of 7106/7107 with addition of HOLD READING inputs. True Differential Input and Ref. On-Chip Clock and Ref.

142 Data Conversion and Telecommunications Circuits

Commercial A/D Converter Summary

Table 3-7 summarizes the key characteristics of the commercially available single-chip CMOS ADCs.

MICROPROCESSOR-INTERFACE APPLICATIONS

A/D Converter–Microprocessor Interface*

The growth of microcomputers has included an expansion into process monitoring and control systems, as well as other applications requiring interaction with "real world" physical variables via monolithic data converters.

For example, the Intel 8080A microprocessor and Teledyne Semiconductor's 8700 series CMOS A/D converters are well suited to such a combination. This section describes the basic techniques for interfacing the two.

The 8700 A/D Converter Description   Teledyne Semiconductor's 8700 series is a family of monolithic CMOS A/D converter ICs. All versions—the 8700 8-b, 8701 10-b, and 8702 12-b—are integrating converters which can accept an unlimited input voltage range (changed to a current input by external scaling resistor), can provide a latched parallel binary digital output, and are available in 24-pin ceramic DIPs. The 8-b version is shown in the block diagram of Fig. 3-69. Each device contains all of the essential elements for a complete A/D converter; only minimal support components are needed.

---

* Portions of this section taken from "Interfacing the 8700 A/D Converter with the 8080 µP System," by D. Guzeman, Teledyne Semiconductor Applications Note 8, Teledyne Semiconductor, Mountain View, CA, August 1976. Used with permission.

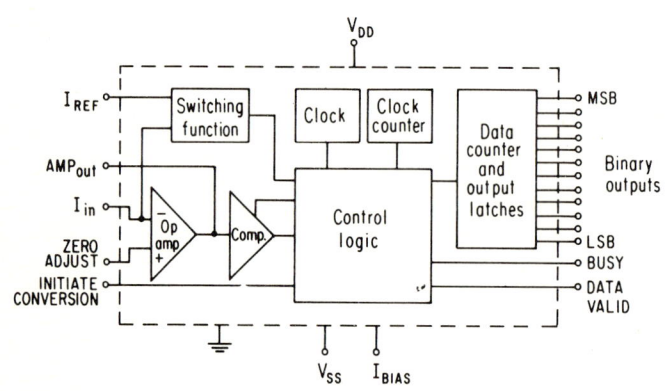

Fig. 3-69   8700 Series ADCs Internal Elements. *(Courtesy Teledyne Semiconductor.)*

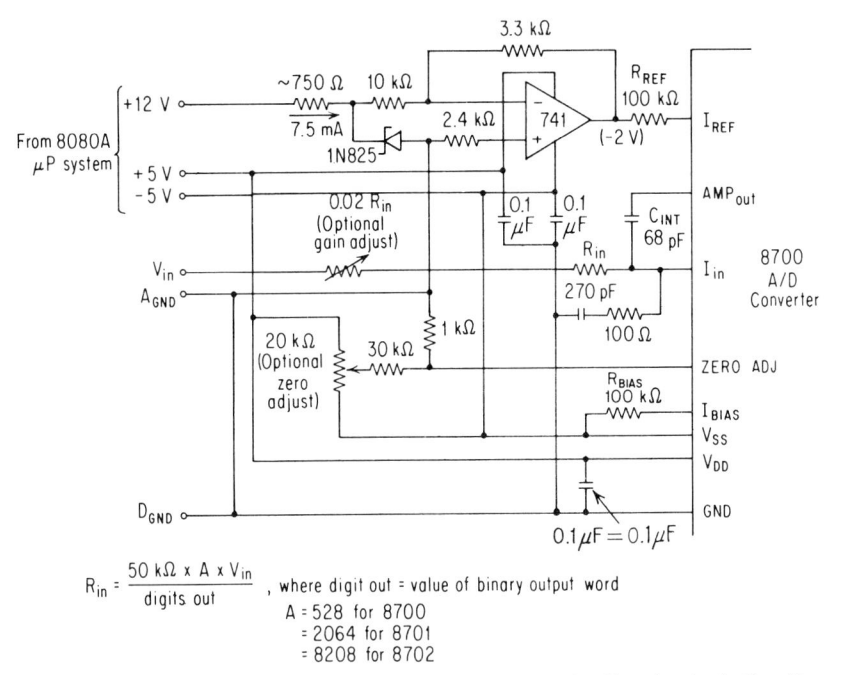

Fig. 3-70 8700 Analog Input and Power Supply Hook-Up Circuits, Including V_{REF} Derived from 8080A System. *(Courtesy Teledyne Semiconductor.)*

$$R_{in} = \frac{50 \text{ k}\Omega \times A \times V_{in}}{\text{digits out}}$$

, where digit out = value of binary output word
A = 528 for 8700
= 2064 for 8701
= 8208 for 8702

In addition to the 8, 10, or 12 buffered data output lines, three handshaking signals are provided to ease the interface to the host system. All outputs are CMOS- and LPTTL-compatible (LPTTL = low-power TTL). The DATA VALID output signal is normally high, indicating that the data in the output latches is valid for the entire cycle, except for approximately 5 μs before the end of the conversion when the data is being updated. Notice that the latches maintain the data from the previous conversion even while the next conversion is being performed. A second output, BUSY, is high whenever a conversion is being performed. Finally, an input to the device, INITIATE CONVERSION, allows the function to be operated under system control. A positive-going pulse of at least 500-ns duration causes the conversion to begin. If this input is tied high, the conversion will occur in a free-running mode at approximately 800 conv/s for the 8700 (200 conv/s for the 8701 and 50 conv/s for the 8702).

Since the 8700 series devices operate from +5- and −5-V supplies, they are particularly easy to interface with the 8080A microprocessor system. Figure 3-70 shows a possible connection for the 8700's analog inputs and power supply; also incorporated is a circuit to supply the necessary negative reference, using a temperature-compensated zener diode and an inverting op amp. Note that the ±5-V supplies needed for the 8700, as well as the additional +12 V used in the reference circuit, are all available from the 8080A system.

The 8080A Microprocessor Description The 8080A, an 8-b microprocessor, communicates within the microcomputer system over two buses, a 16-b address bus and an 8-b data bus. During each machine cycle, the current contents of the program counter are sent out over the address bus; the memory receives the address and returns the contents of the selected memory location to the 8080A via the data bus. During an INSTRUCTION-FETCH cycle, the returning data is interpreted as an instruction.

Communications between the microcomputer and the outside world are via I/O ports addressed by the address bus. I/O instructions utilize 8-b addresses; the port address is duplicated on both the low-order address lines and the high-order address lines of the address bus.

In addition to the address and data buses, the 8080A communicates with the memory and I/O ports via a set of control signals. In particular two control lines $\overline{\text{IN}}$ and $\overline{\text{OUT}}$ are used to enable the I/O ports. A logic 0 on the $\overline{\text{IN}}$ line will enable the input port that corresponds to the address on the address bus at that time. The $\overline{\text{OUT}}$ line functions in a similar fashion.

The 8700–8080A Interface

Figure 3-71 shows the basic approach to interfacing the 8080A and the 8700 8-b A/D converter. The conversion is started on command of the 8080A, using the INITIATE CONVERSION input of the 8700. When the conversion is complete, the DATA VALID output of the A/D requests an interrupt; the interrupt service routine transfers the current data from the working registers to the stack memory, and the A/D input port is read. A control signal then is sent to the INITIATE CONVERSION input to restart the conversion and the main program activity is resumed.

It is assumed that the data bus will be shared by many devices, both in the ports and in memory, and that inverting drivers-receivers (such as 8228) will be included in the 8080A system to service this bus. Therefore, 80L98 buffers have been provided at the 8700 to drive an inverted input over the data bus, as well as to provide a three-state function, electrically

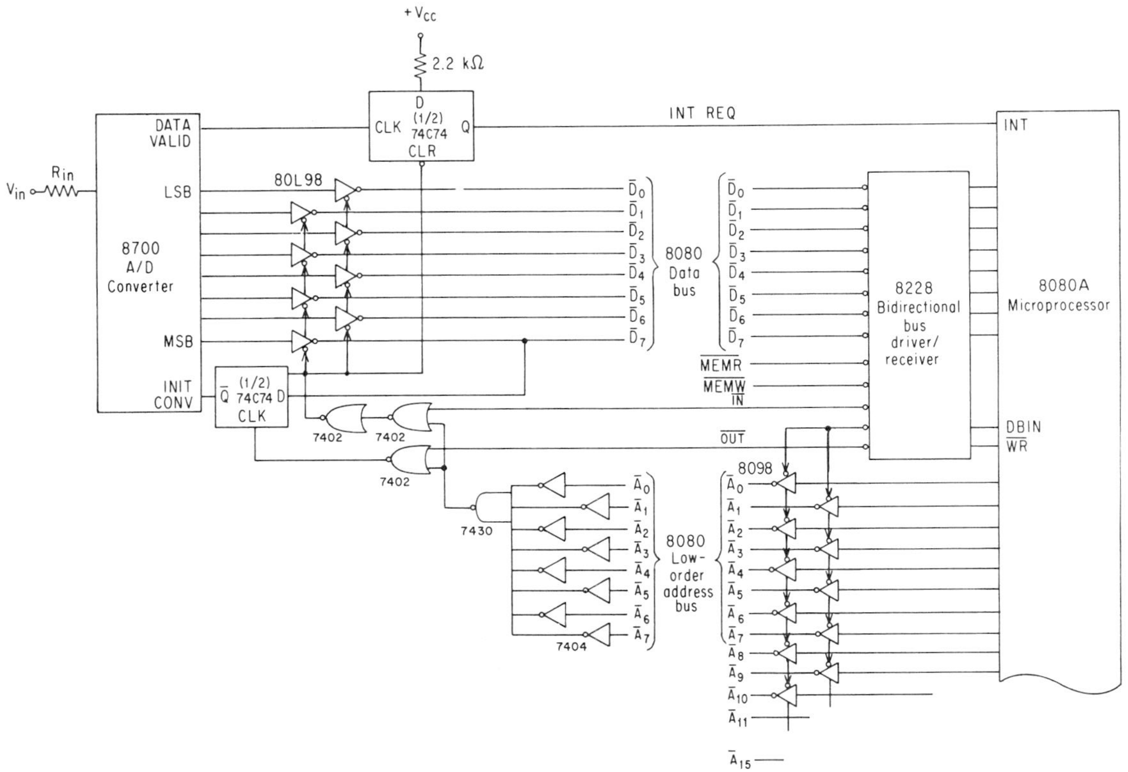

Fig. 3-71 Basic 8700/8080A Interface. *(Courtesy Teledyne Semiconductor.)*

removing the A/D from the bus when its input port has not been selected.

Each port of the system is assigned an address by virtue of the way the address bus is decoded to select the port. In the basic input port shown in Fig. 3-71, the output of the 7430 gate is low only when all of its inputs are high. This corresponds to address FF_H.

To initiate a conversion in the A/D, an output port, also address FF_H, is used. By defining both the input and output ports as address FF_H, the same address decoder, the 7430, may be used for both functions. In this case the output of the 7430 and the $\overline{OUT}$ signal are gated by the 7402 to clock half of a 74C74 flip-flop. The D input of the flip-flop is tied to the D_7 line of the data bus. The flip-flop is, in effect, a 1-b output port. Sending the data word 80_H to port FF_H with an output ($\overline{OUT}$) instruction will cause the flip-flop to be set, thus supplying an INITIATE CONVERSION signal to the 8700. A second output instruction, sending 00_H to the same port, will reset the flip-flop and remove the INITIATE CONVERSION signal. Since an output instruction requires ten 0.5-μs clock cycles to execute, the INITIATE conversion pulse will be approximately 5 μs long. After beginning the conversion process by the double output instructions, the 8080A is free to perform other processing operations.

When the 8700 completes its conversion cycle and latches the result onto its internal output latches, the DATA VALID output goes high. This triggers the other half of the 74C74 flip-flop, clocking a logic 1 from the D input (tied high) onto the INTERRUPT REQUEST line. The result is that the microprocessor is interrupted when the conversion is complete. The interrupt service routine (see Table 3-8) saves the CPU's working register contents by pushing them onto the stack and then reads the output of the 8700.

To read the 8700 input port, it is necessary to supply the address of port FF_H on the address bus while simultaneously sending out a logic 0 on the $\overline{IN}$ control line. The combination of the 7430 and 7402 gates supplies a logic 0 to the enabling input of the 80L98 three-state buffers on the outputs of the 8700 and to the clear input of the 74C74 flip-flop on the INTERRUPT line; this puts the 8700 data on the data bus and removes the interrupt request.

After reading the converter data and saving it in one of the registers, the system again pulses the INITIATE CONVERSION input to start the next conversion, restores the stack with a series of POP instructions, and resets the internal interrupt-enable flip-flop. Thus the 8080A only

TABLE 3-8 Interrupt-Service Routine

;INITIATION			
	MVI	A,BØH	;THE CONVERSION
	OUT	ØFFH	;IS INITIATED
	MVI	A,Ø	;BY SENDING A
	OUT	ØFFH	;BRIEF PULSE
			;TO PORT FF
;INTERRUPT			
	PUSH	B	;THE PROCESSOR
	PUSH	D	;REGISTERS AND
	PUSH	H	;STATUS ARE SAVED
	PUSH	PSW	;IN THE STACK, AND THE
	IN	ØFFH	;DATA IS READ AND
	MOV	B,A	;STORED IN REG B.
	MVI	A,BØH	;THE CONVERSION IS
	OUT	ØFFH	;INITIATED AND
	MVI	A,Ø	;THE DATA IS
	OUT	ØFFH	;PROCESSED.
	.		
	.		
	.		
	POP	PSW	;WHEN COMPLETE.
	POP	H	;THE REGISTERS
	POP	D	;ARE RESTORED, THE
	POP	B	;INTERRUPTS ENABLED
	RET		;AND PROGRAM
			;CONTROL RETURNED
			;TO THE MAIN PROGRAM.

SOURCE: Courtesy Teledyne Semiconductor.

reads the 8700 when the new information becomes available; the rest of the time is spent in processing activities.

Interfacing the ADC3511/3711 with the 8080 Microprocessor*

This section describes techniques for interfacing National Semiconductor's ADC3511 and ADC3771 to 8080A microprocessor systems.

A complete A/D port is seen in Figs. 3-72 and 3-73. Figure 3-72 shows a dual–polarity converter and Fig. 3-73 a positive-only polarity converter. Each port contains an A/D converter, Tri-State bus driver, and two gates to control input/output. This A/D port is easily used in single- or mul-

* Portions of this section reprinted from Jake Burma, "CMOS A/D Converter Chips Easily Interface to 8080A Microprocessor Systems," National Semiconductor Application Note AN200, National Semiconductor Corp., Santa Clara, CA, 1978. Used with permission.

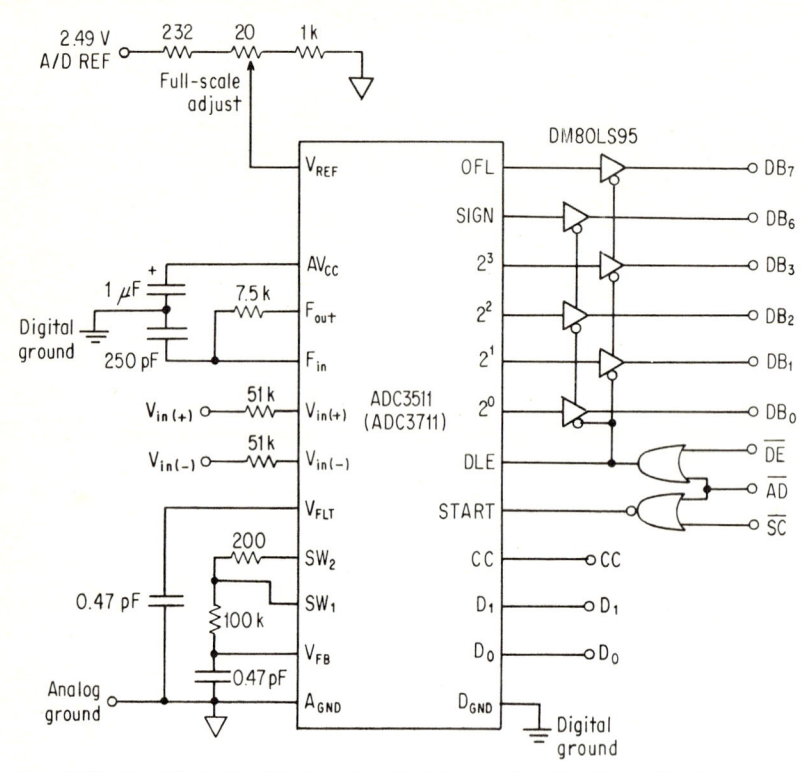

148 Data Conversion and Telecommunications Circuits

Fig. 3-72 Dual-Polarity A/D Requires that Inputs Are Floating with Respect to the Supply. Input range is ±1.999 V. *(Courtesy National Semiconductor Corp.)*

tichannel systems. In multichannel systems, a converter is used on every channel allowing digital multiplexing of the outputs.

Data from the A/D converter in a single-channel system is easily processed using an OUT command to start a conversion and IN commands to read the data after the microprocessor has been interrupted by a CONVERSION COMPLETE.

Figure 3-74 shows the block diagram of a single-channel A/D port that uses a 6-b bus comparator to decode its assigned peripheral address from the lower address bits of the 8080A address bus.

When an interrupt is received, the present status of the processor is stored on the stack memory by a series of PUSH commands. The interrupt is serviced by reading digit 4 [most-significant digit (MSD)] into the processor and checking the overflow (OFL) bit. If the overflow bit is high, the converter input has exceeded its range and an error signal is gener-

Data Conversion Circuits   149

ated, indicating that scaling must be done to attenuate the input signal. If the OFL bit is low, the sign bit is then checked to determine the polarity of the conversion. If the sign bit is low, a 1 is added to the MSB of digit 4. Since this bit would normally be low (maximum converter range allows MSB ≤ 3 or 0011), a 1 in this position is used to denote a negative input voltage. The 4 b of digit 4 which now include the sign are shifted into the upper half of the first byte and the 4 b of digit 3 are packed into the lower half. Similarly, digits 2 and 1 are packed into the second byte and both bytes stored in memory.

Figure 3-75 and Table 3-9 are the flow chart and assembly language routine used to implement this action.

The basic A/D port can easily be expanded to multiple-channel systems. An 8-channel system is seen in Fig. 3-76. This system interrupts the processor when one of the CONVERSION COMPLETE outputs goes high. The processor saves the current status and reads the status word of

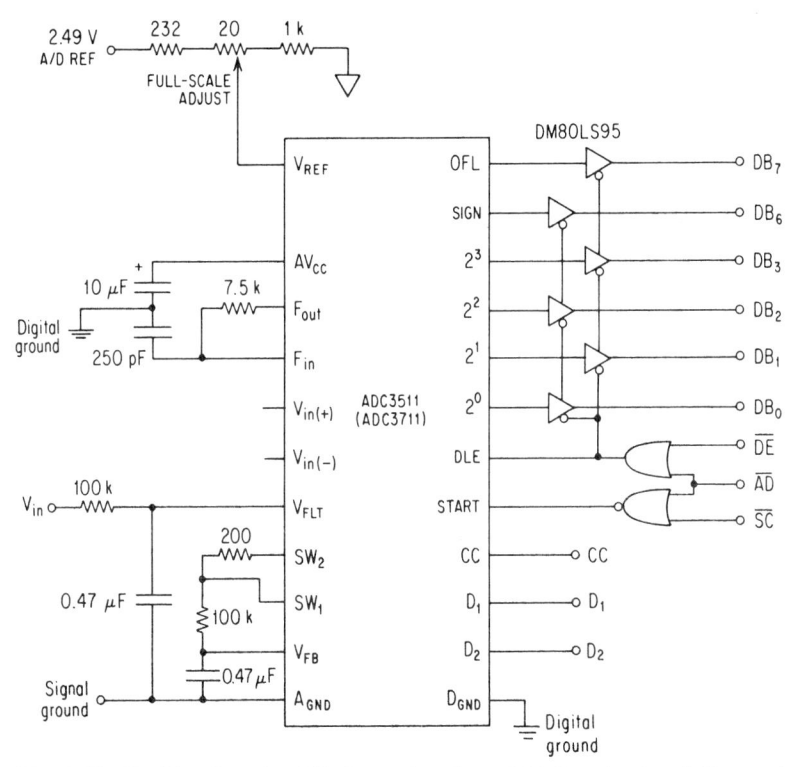

Fig. 3-73 Positive-Polarity A/D Operating from 5-V Supply. Input Range Is +1.999 V. *(Courtesy National Semiconductor Corp.)*

150 Data Conversion and Telecommunications Circuits

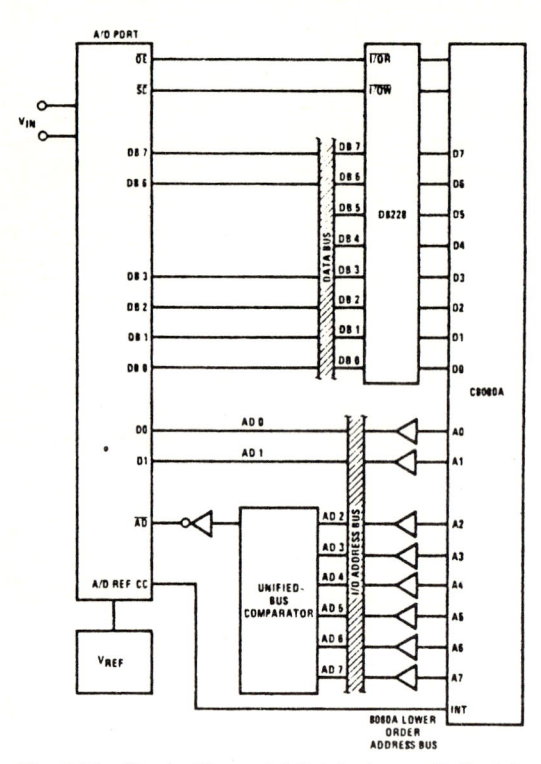

Fig. 3-74 Single-Channel A/D Interface with Peripheral Mapped I/O. *(Courtesy National Semiconductor Corp.)*

the A/D system. The status word is then compared to a priority table. Each level in the table corresponds to a priority level, with high-priority converters being first in the table. If two or more converters have the same priority and are ready at the same time, the converter with the highest number gets serviced first.

The program determines which service routine to use by the bit position of 1s in the status word. The routine loads the address pointer to digit 4 of the selected converter. The program then calls a subroutine which goes through the process of checking overflow and sign and packs BCD digits into 2 bytes. These 2 bytes are then stored in a table in the memory directly above the converter address. After a channel is serviced, the original processor status is restored and the interrupt enabled. If additional channels need service, they immediately interrupt so the new status word is then read and a new priority established.

TELECOMMUNICATIONS CIRCUITS*

As a first step toward understanding telecommunications, it is best to define the term and understand what it is. Basically the function of a communications system is to send messages from point A to point B. Telecommunications includes telegraph, telephone, television, telemetry, facsimile, and telecontrol.

Present telecommunications systems can be represented by the block diagram of Fig. 3-77. The telephone company equivalent of Fig. 3-77 is shown as Fig. 3-78. The messages may be symbols (text or data), voice, pictures, control signals, etc. We can expect in the future an increasing diversity of message types, sources, and uses. For example, electronic technology continues to decrease the cost and increase the number and diversity of printed materials.

* Portions of this section taken from T. J. Mroz, "Designing with Codecs: Know Your A's and μ's," *EDN*, May 5, 1978, pp. 109–114. Used with permission.

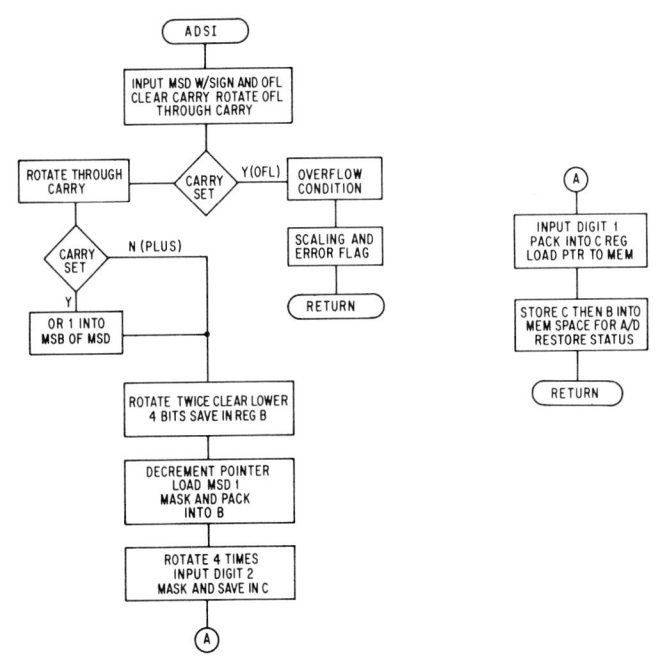

Fig. 3-75 Flow Chart for Single-Channel A/D Converter. *(Courtesy National Semiconductor Corp.)*

TABLE 3-9 Single-Channel Interrupt Service Routine

Label	Op code	Operand	Comment
ADIS:	PUSH	PSW	;A/D interrupt service
	PUSH	H	;save
	PUSH	B	;current status
	IN	ADD 4	;input A/D digit 4
	IN	ADD 4	;delay
	ORA	F0	;reset carry
	RAL	BA	;rotate OFL thru carry
	JC	OFL	;overflow condition
	RAL		;rotate sign thru carry
	JC	PLUS	;positive input
	ORI	20H	;OR 1 into MSB neg input
	RAL		;shift
	RAL		;into position
	ANI	F0	;mask lower bits
	MOV	BA	;save in B
	IN	ADD 3	;input digit 3
	IN	ADD 3	;delay
	ANI	OF	;mask higher bits
	OR	B	;pack into B
	MOV	B,A	;save in B
	IN	ADD 2	;input digit 2

Label	Op code	Operand	Comment
	IN	ADD 2	;delay
	RAL		;rotate
	RAL		;into
	RAL		;upper
	RAL		;4 bits
	ANI	F0	;mask lower bits
	MOV	C,A	;save in C
	IN	ADD 1	;in digit 1
	IN	ADD 1	;delay
	ANI	OF	;mask upper bits
	OR	C	;pack
	MOV	C,A	;save in C
	LXI	H,ADMS	;load ptr to A/D Mem, space
	MOV	M,C	;save C in memory
	INX	H	;point next
	MOV	M,B	;save B in memory
	OUT	ADD 1	;start new conversion
PLUS:	POP	B	;restore
	POP	H	;previous
	POP	PSW	;status
	EI		;enable interrupts
	RET		;return to main program

SOURCE: Courtesy National Semiconductor Corp.

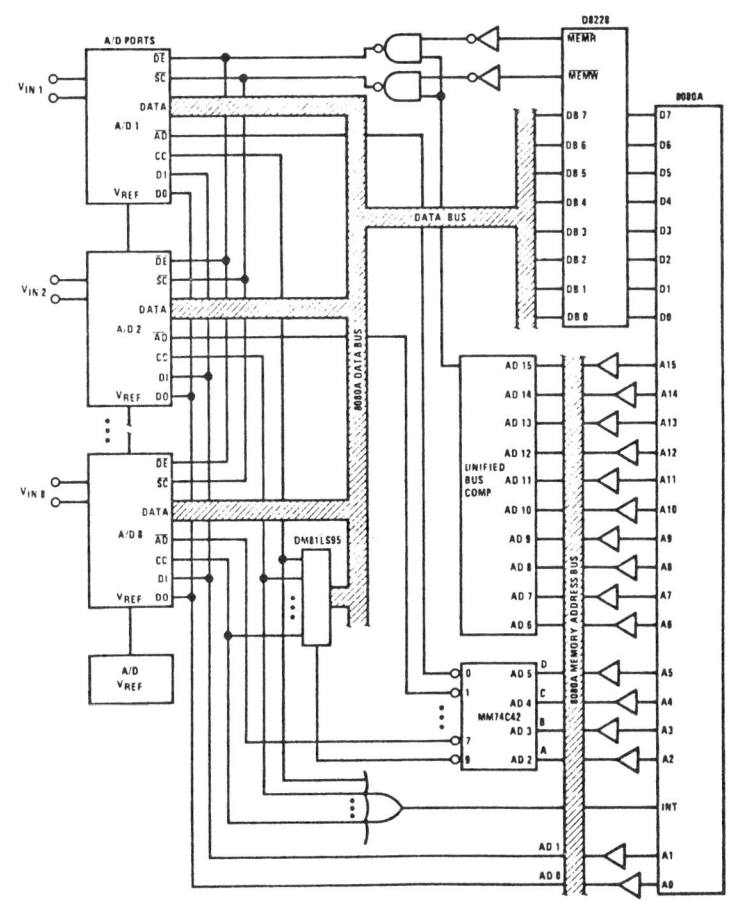

Fig. 3-76 An 8-Channel A/D System with Maskable Priority Interrupt Using Memory-Mapped I/O. *(Courtesy National Semiconductor Corp.)*

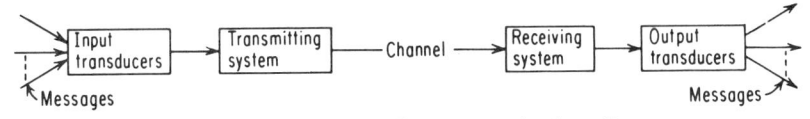

Fig. 3-77 Block Diagram of a Typical Telecommunications System.

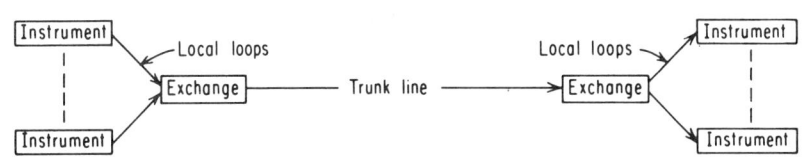

Fig. 3-78 Block Diagram of a Typical Telecommunications System Using Telephone Company Terminology.

153

154 Data Conversion and Telecommunications Circuits

The input transducers shown in Fig. 3-77 convert the messages to a (voltage) form suitable for transmission. Such conversion may include frequency translation, digitization, multiplexing, switching in telephone networks, etc.

The transmitting system matches the voltage equivalent of the messages to the channel. Amplifiers suffice for telegraph, telephone, and television signals sent on open wire or coaxial cable lines. Carrier modulators and RF (radio frequency) power amplifiers and antennas are required in radio links. A light source and optical modulator are part of the transmitting system for optical channels.

The majority of telecommunications applications prior to 1978 were concerned with audio or video signals. The availability of digital links is of considerable importance to the future growth of the telecommunications industry. The key feature of a DCS (digital communications system) is that it only sends a finite set of messages, in contrast to an analog communication system which can send an infinite set of messages.

A message is simply the output of some information source. In general, it might be a symbol or a group of symbols, letters of the alphabet, numbers, or special-purpose characters. A character may be given varying numbers of binary digits (bits) in different systems.

In the context of a DCS, a digital waveform is taken to mean a sinusoidal voltage or current signal representing a message or symbol. The waveform is endowed with specially chosen amplitude, frequency, or phase characteristics which allow the selection of a distinct waveform for each of the messages.

In a DCS, the objective at the receiver is not to reproduce the waveform with precision (since the set of possible waveforms is known exactly and the fidelity of the received waveform is not a criterion for message accuracy). The goal instead is to determine, from the received signal perturbed by noise, which of the finite waveforms has been sent by the transmitter.

The receiver is assumed to be always time-synchronized with the transmitter. Receivers have clocks that track transmitter clocks (or vice versa) with enough accuracy to permit them to "know" when to expect a message from the transmitter. This is called *synchronous detection*. In some cases, the receiver can estimate the transmitter's carrier wave phase reference. When the receiver exploits phase reference knowledge to detect the signals, the process is called *coherent detection;* when it does not have phase reference information, the process is called *noncoherent detection.*

Today's telephone systems employ mazes of switches and filters for both analog and digital signals. One small part of these complex networks, the channel bank, transforms analog voice signals at a local exchange into easily transmitted digital signals. It is then easy to compress, transmit, and repeat these digital signals in cable and microwave transmissions.

Circuits termed "encoders" sample the analog voice signals generated by user phones, convert them to digital bytes, and shift them serially out of the channel bank. On the receiving end, "decoders" accept these bytes and recreate the analog voice signals initially generated by the phone user. The term "codec" applies to a complete encoder-decoder set.

Previously, these conversions were achieved by high-speed A/D and D/A converters multiplexed over many analog channels. But now the advances in large-scale integration have made it economically feasible to encode each channel separately and multiplex the resulting digital signals.

Channel banks, located in telephone company central offices, use codecs to handle many phones, encoding analog voice signals into serial data streams for transmission to various receiving banks. One bank can also decode incoming calls and redistribute analog voice signals to user phones (Fig. 3-79). Both codecs and channel banks follow specifications in the μ Law 255 or the A Law that not only govern transfer characteristics but also define formats for framing, synchronization, and signaling.

The μ Law 255 and A Law differ fundamentally in the transfer characteristics associated with their A/D and D/A conversions (Figs. 3-80 and 3-81).

PCM (pulse code modulation), the technique channel banks employ to transmit and receive information, allows them to digitally sample analog inputs at a fixed rate by the codec at discrete uniform intervals, and then to perform an A/D conversion that quantizes the sample into an 8-b sign-plus-magnitude word. These intervals are determined by the Nyquist sampling theorem, which states that any signal can be ultimately reconstructed if it is sampled at a rate equal to twice its highest frequency of interest. The universally accepted rate for a telephone conversation is 8000 samples/s because voice data transmission up to 4000 Hz has been found to have enough fidelity for practical purposes.

The voice signal must be band-limited at 4 kHz by a low-pass filter to prevent it from being distorted by the codec's A/D and D/A conversions. Ultimately, semiconductor manufacturers will incorporate the filter into the codec itself, and there is some feeling that even the full-duplex hybrid, which allows two-way conversions, will end up in the codec.

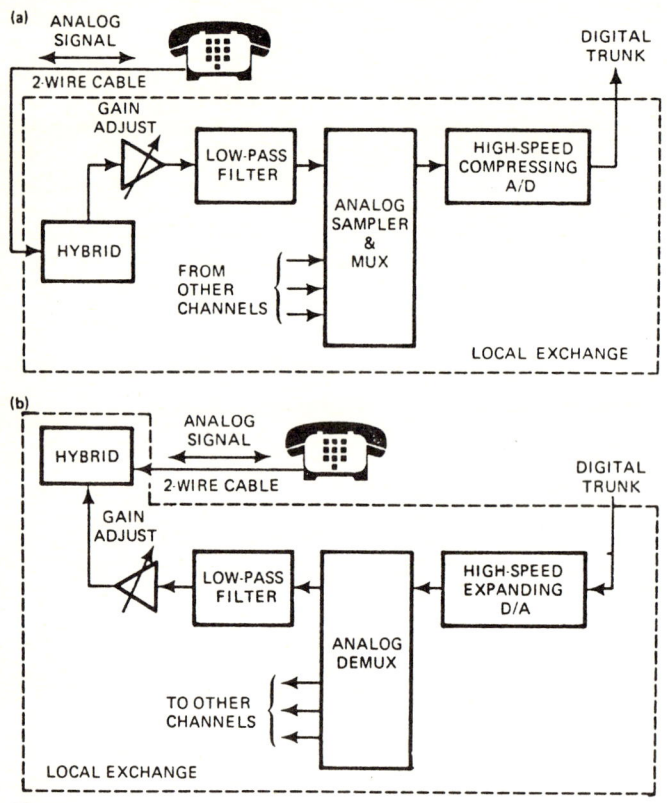

156 Data Conversion and Telecommunications Circuits

Fig. 3-79 Most Existing Phone Systems Employing Pulse-Code Modulation Use Just One High-Performance Data Converter in the Transmitter (a) or receiver (b). Multiplexers Make These Converters Available to All the Channels the Exchange Handles. In These Diagrams, a Hybrid is a Device that Converts Two-Wire Phone Signals to Four-Wire Signals, thus Eliminating Cross Talk between Incoming and Outgoing Data.

The digitally encoded information is transmitted in serial form. This technique allows data transmission from 24 voice channels into 24 time slots on a single, serial transmission highway or bus. One 8-b code word from each voice channel is transmitted in its associated time slot and the bus format groups the twenty-four 8-b time slots into a 192-b block.

Overall synchronization is provided by adding 1 b, and the resulting 193 b are designated a "frame." This last frame synchronization bit defines the boundary between time slots 1 and 24 and serves to keep the time-slot counter of the far-end receiver synchronized with that of the near-end transmitter. The 8-kHz frame repetition rate then produces the characteristic 1.544 Mb/s data used in all AT&T T1 systems. The 8-b

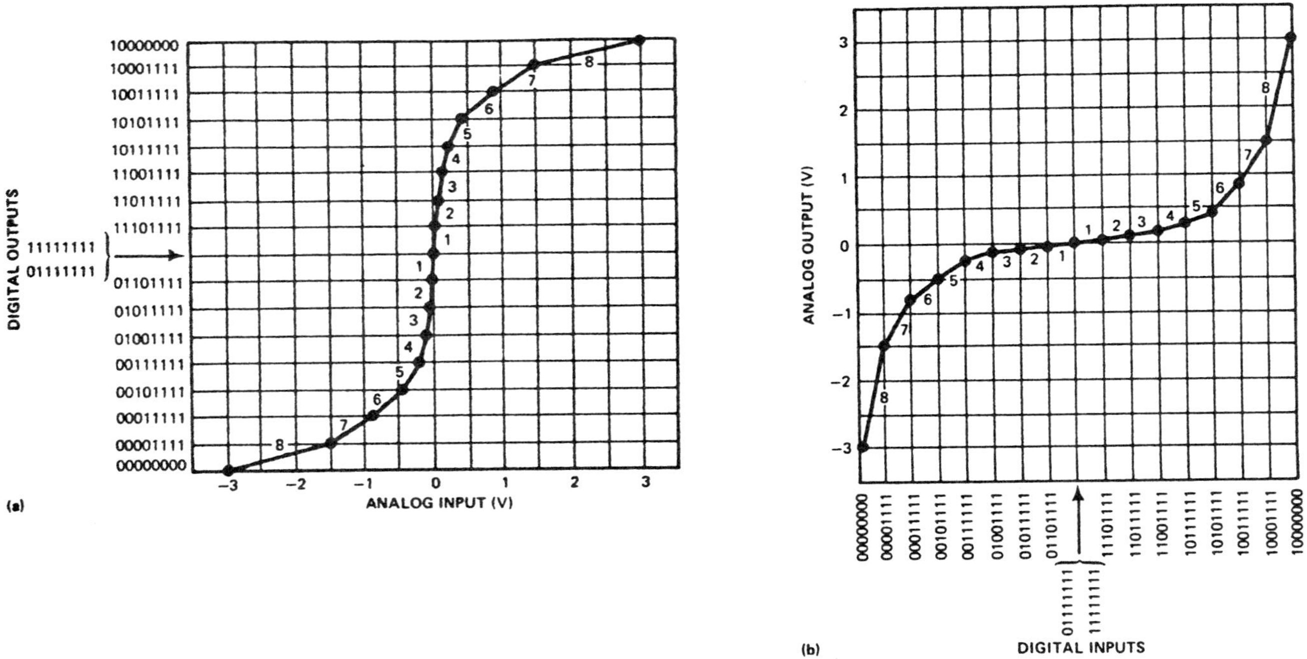

Fig. 3-80 The μ Law 255 Transfer Characteristics for Coders (a) and Decoders (b) Consist of Piecewise Linear Approximations of the Desired Curve. Note that Two Digital Values Correspond to the Origin because the Sign Bit Can Have Either Value at Zero. *(From EDN, May 5, 1978. Reprinted with permission.)*

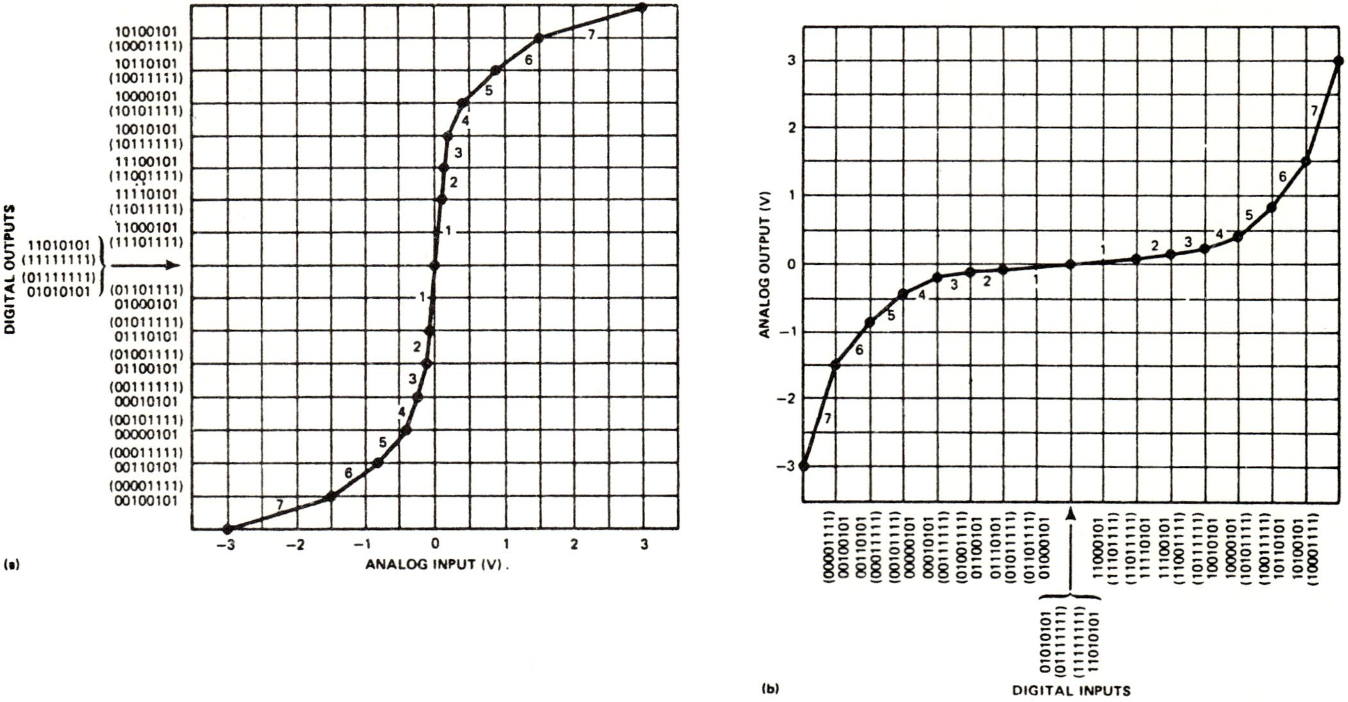

Fig. 3-81 The A Law Transfer Characteristics for Coders (a) and Decoders (b) Reduce the Number of Chords Required in μ Law 255 Curves by Making Chords near the Origin Colinear. Note that Values Listed in Parentheses are the Corresponding μ Law 255 values. (*From EDN, May 5, 1978. Reprinted with permission.*)

data block corresponding to one sample of a given voice channel is demultiplexed by time-slot identification at the receiver.

The critical step in the process of forming the digital PCM signal is the assignment of a binary code to each analog sample as it is presented to the codec. This is usually done by a system of nonlinear quantization referred to as *companding* and performed by a *compander* (compressor-expander).

In a typical companded transmission system, the analog signal is passed through the compressor part of the transmitting compander before being transmitted. At the receiver, the analog signal is passed through the expander part of the receiver compander. The process boosts low-level signals, making them better able to compete with the system noise, and attenuates high-level signals, preventing them from saturating the system.

A digital realization of the desired companding law is obtained by segment or chord approximation to the curve shape. North American PCM systems use digitally realizable approximations to the so-called "μ Law 255" (Fig. 3-80). For this encoding law, there is a total of 8 chords for each input polarity, with each chord having 16 equal steps. The step size within each chord is constant, but doubles in size from one chord to the next, starting with the chord nearest the origin.

European telephone systems also use a chord or segment approximation, but in accordance with the so-called "A Law" (Fig. 3-81). A total of 8 segments is again used for each polarity. For the 4 chords nearest the origin, however, the step size remains the same and they are merged into a single chord. After the first 2 chords, the steps double in size from one chord to the next. Thus the A Law employs a 13-segment line.

In practice, there is not much difference between the two laws. Controversy over which is "better" exists, but involves various nontechnical arguments concerning the place of origin. In any case, most codec manufacturers are making both laws available for the two markets.

Encoder and decoder nonlinear characteristics maintain relatively constant S/D (signal-to-distortion) levels over a wide range of analog input levels. Because the channel bank compresses high input levels and expands low levels, a person speaking softly into a telephone is nearly as intelligible as someone speaking loudly. Thus codecs are termed *companding A/D and D/A converter sets*.

IC Codecs

While codecs have been around in discrete form for years, there has been much difficulty in converting them to IC form because of the past incompatibility of linear and digital functions in a single-chip technology.

Now, however, the semiconductor industry has, almost as a unit, discovered how to adapt the MOS process (already highly developed for digital techniques) for linear operations. Thus the door is opened to all kinds of futuristic applications involving interfacing digital equipment with analog signals. Many of these are likely to concern voice equipment such as speech scramblers, voice-operated equipment, or any type of speech recognition or conversion equipment, as well as any type of voltage-operated equipment—or the reverse operations, D/A conversion. In addition, the new process should significantly enhance the range of applications of microprocessors, as in automatically controlled engine functions in automobiles.

The heart of most IC codecs is the companding D/A converter whose nonlinear transfer characteristics are used to *compress* and *expand* analog signals. As mentioned before, different companding laws, developed to satisfy telephone digital voice-transmission requirements govern the United States (the Bell μ Law 255) and Europe [the CCITT (International Telephone and Telegraph Consultative Committee) A Law]. Both are piecewise linear approximations of exponential curves. More segments are used in the low-level signal region to yield a better S/N ratio. Both laws dictate a total of 8 segments or chords of each polarity (using 3 address bits), and each chord breaks down into 16 uniformly spaced steps (addressed with 4 b); the sign bit determines the positive or negative quadrant.

Because of these nonlinear transfer functions, circuitry for companding DACs is more complex than for linear versions (see Fig. 3-82). The chord generator produces the total current for each segment of the curve; the pedestal generator supplies the starting point for each chord; the step generator outputs the proper current for each chord; the chord decoding logic controls inputs to the pedestal and step generators; and the output switching matrix sums step and pedestal currents and routes them to the proper output node.

Companding DACs act as PCM decoders (receivers) because they perform both decoding and the D/A conversion. When one needs encoding or compression for digital transmission of an analog signal, one can tie an SAR, comparator, and logic to a companding DAC to perform the additional A/D functions.

A true codec consists of a complex mixture of analog and digital functions, including (as a minimum) a band-limiting filter, S/H (SAMPLE-

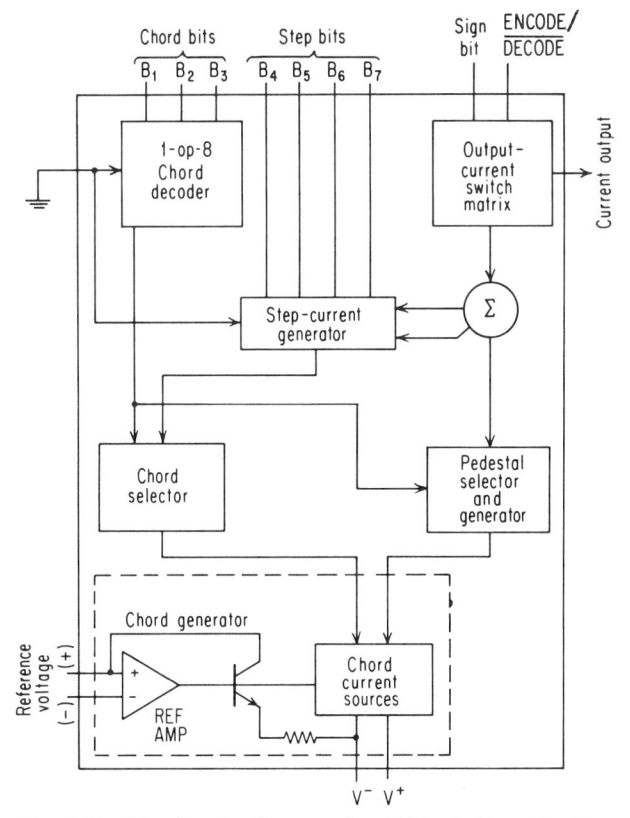

Fig. 3-82 Circuitry for Companding DACs Is Complex Because of Those Devices' Nonlinear Transfer Functions.

AND-HOLD) circuit, and companding DAC encoder-decoder. No device currently available incorporates all these features in one chip.

However, popular usage has pretty well redefined the term "single-chip codec" to mean a unit with just S/H-circuit and encoder-decoder functions.

Table 3-10 summarizes the salient features of presently commercially available CMOS codecs.

National Semiconductor's CMOS/BIFET codecs are separated into digital functions using CMOS circuitry and linear functions using BIFET (bipolar field-effect transistor) technology. The parts consume 250 mW in the operating mode and less than 10 mW in standby.

Separate analog and digital grounding pins on the 28-pin DIPs, plus autozeroing, serve to reduce channel noise and cross talk. Each device also includes a stable reference. In addition National offers a CMOS

Data Conversion and Telecommunications Circuits

TABLE 3-10 Summary of Commercially Available Codec Characteristics

Manufacturer	Mostek	Motorola	American Microsystems	National Semiconductor	Siliconix
μ Law Part(s)	MK5151 MK5116[b]	MC14407[a] MC14406[b]	3501 Coder 3502 Decoder	MM58100 LF3700	DF331 Coder DF332 Decoder
A Law Part(s)	MK5156	MC14407[a] MC14404		MM58150 LF3700	DF341 DF342
Number of Chips	1	1	2[c]	2	2[c]
Filter on Chip?	No (5201)	No (14413)	Yes	No (AF133) (AF134)	No
Number of Power Supplies (Not Including Reference)	2	1	2	2	2
Voltage Supply Values	±5	10–16	±5	±12	±7.5
Mode of Operation	Asyn/Syn	Asyn/Syn	Asyn/Syn	Asyn/Syn	Syn
Number of Pins	24/16 16	28/24 24	18 Coder 16 Decoder	28/20 22/20	14/14 14/14
Voltage Reference on Chip?	No	No	No	Yes	No
Power Dissipation (Typ): Active (mW) Standby (mW)	30 Not available as Option	78/108[d] 33/88[d]	125[d] 30[d]	250/520[d]	80 11
Signaling	A/B (5151)	8th Bit (14406)	A/B +CCIS Opt.[e]	8th Bit	A/B
Data Rate (Mbytes/s)	0.064–2.1	0.128–3.088	0.056–3.152	0.064–2.1	1.25–3.0
Gain Tracking	±0.4/−40[d] ±0.8/−50 ±2.5/−55	±0.3/−40[f] ±0.6/−50 ±2/−55	±0.25/−40 ±0.7/−50 ±2/−55	±0.25/−37 ±0.5/−50	±0.25/−40 ±0.5/−50 ±1.5/−55
S/D Ratio (dB) at (dBmO)	39/−30 34/−40 29/−45	37/−30 31/−40 26/−45	37/−30 33/−40 30/−45	77/−30 30/−40 25/−45	39/−30 34/−40 30/−45
Cross Talk		−80	None	−83 (dBmO)[g]	None
V_{REF} Required (V)	−2.5, +2.5	65–125-μA Current Ref.	−3	On-Chip	−3, +3

NOTE: Mostek's MK5151 replaces the earlier MK5150 codec.
[a] Pin-selectable conversion law.
[b] No-signaling version.
[c] Separate coder and decoder.
[d] Includes filter.
[e] CCIS = common-channel interoffice signaling.
[f] End-to-end codec and filter.
[g] dBm0 = the measure of power with reference to zero dBm at the reference transmission level.

PCM channel filter with transmit, receive, and 50/60 Hz rejection filters and 600-Ω output drivers.*

The chip's operation is truly asynchronous; upon receipt of an incoming 8-b PCM word, it automatically interrupts the ENCODE cycle, completes the decoding process, and returns to encoding. As is usually the case, the required filter is extra.

AMI is offering a two-chip codec system with on chip filtering. The S3501 is an encoder-plus-transmitting filter, while the S3502 is a decoder-plus-receiving filter.

Mostek's MK5116/5156 codecs do not have a power-down mode, but with 30-mW typical power dissipation they do not need it. These 16-pin CMOS parts also do not incorporate an internal reference or filter, but they do have an autozeroing circuit and separate analog/digital grounding pins.

To complement these parts, Mostek offers the μ Law 255 MK5151. Identical in function to the MK5116, this codec facilitates insertion and recovery of signaling information in the data stream. Furthermore, it offers this feature on two channels. Unfortunately, the MK5151's 24-pin package precludes direct replacement and pin compatibility with the MK5116. Fairchild is second-sourcing the MK5151 codec.

Motorola's single-chip codecs, the 24-pin MC14406 and 28-pin MC14407, offer pin selection of μ or A Law companding. The latter unit includes signaling capabilities similar to those of the MK5151.

Motorola utilizes metal-gate linear or bipolar processing and digital CMOS processing to produce the only wide-range, single-power supply parts (10 to 16 V). Operating power dissipation equals less than 75 mW, and a power-down input reduces that figure to 1 mW. Also included on-chip are autozeroing, zero-code suppression, and full-duplex operation to 3.088 MHz.

Motorola's 14413 16-pin PCM active filter, when used in conjunction with the 14406 (μ Law), the MC14407 (A or μ Law), or the MC14404 (CCITT format) codecs, meets or exceeds all Bell System D3 voice frequency requirements and objectives and is compatible with G.712CCITT recommendations. The MC14413 filter has both transmit and receive low-pass filters on-chip, and on-hook power use is only 1 mW. With the MC3419 linefeed and 2-to-4 wire conversion circuit, one is able to design a minimum 3-chip SCU (subscriber channel unit) (Fig. 3-83) including subscriber line interface.

* W. C. Black et al., "A CMOS PCM Channel Filter," *1980 IEEE ISSCC Proceedings,* San Francisco, CA, February 1980.

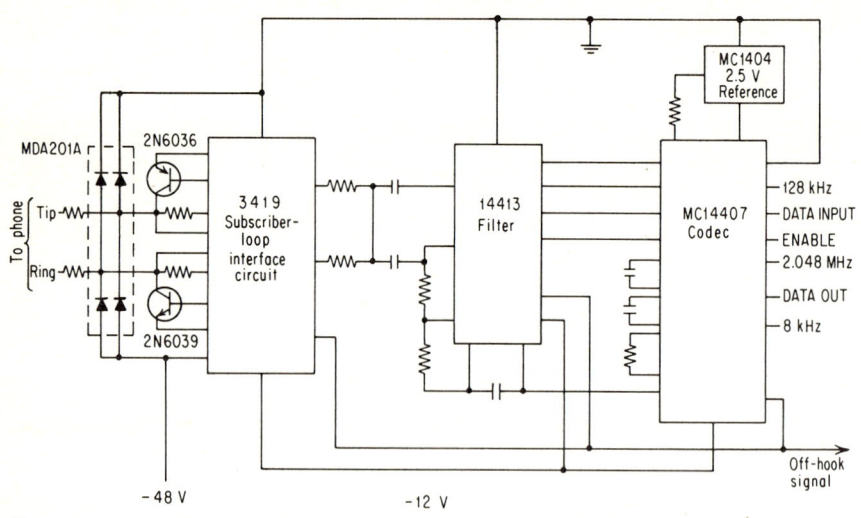

164 Data Conversion and Telecommunications Circuits

Fig. 3-83 Motorola's 3-Chip Subscriber Channel Unit Uses CMOS for Its Filter and Codec and Has a Bipolar Subscriber-Loop Interface Circuit (SLIC). The SLIC's Outboard Components Provide Current Drive for the Telephone Line and Protect against Transients.

The subscriber-loop interface circuit, or SLIC (or "hybrid" in Bell System vernacular), which is fabricated using bipolar technology because of the extreme voltage requirements, takes decoded analog signals from the filter chip and conditions them for driving the noisy telephone lines; it can withstand extremely tough environmental conditions.

Although Motorola uses bipolar technology for the SLIC, it builds its codec and filter chips with CMOS technology. Many manufacturers lean toward a similar three-chip approach, though the choices encompass N-channel and complementary MOS and bipolar.

CMOS was used for the codec and filter by Motorola primarily because of its low power consumption. The MC14407 codec and the MC14413 five-pole elliptic filter each draw less than 100 mW and power-down to less than 1 mW. Motorola has used that inherent quality of CMOS to give its three-chip system a telecommunications selling point—minimal power consumption when the phone is on the hook. Motorola's entire system, as it is now designed, will power-down to less than 10 mW.

Sharing the same level of complexity as the "single-chip" codecs is American Microsystems' 2-chip codec set. Like the single-chip parts, it allows one to implement a minimum system with only two ICs; the functions are partitioned differently, however.

AMI's S3501 encoder with filter and S3502 decoder with filter are partitioned into logically and physically independent chips for transmission and reception channels. The advantages of this type of partitioning are the following:

- They eliminate any potential cochannel cross talk that could be caused by sharing conversion circuitry, or by leakage or coupling between same-chip components.
- One chip can be operated synchronously or asynchronously without regard to the other.
- When, under μ Law 255, either A/D or D/A conversion is required but not both (say, for digital signal processing or tone reception), only one chip is required to do it.
- Interconnections made externally in other codec sets are eliminated; as a result, the encoder comes in an 18-pin DIP and the decoder in a 16-pin DIP—packages that are small enough to be machine-insertable.

Another reason for the set's low number of pins is an on-chip phase-locked loop. This derives all internal timing—except for the buffer-register shift clocks of the encoder output and the decoder input—from a single input, the 8-kHz sampling-strobe signal. Not only does on-chip timing generation hold down the pinout, it eliminates the need for a smoothing filter between the transmission filter and the A/D converter. That is because both the filter and conversion clocks are harmonics of the 8-kHz strobe.

The PLL timing generation brings other benefits such as the lack of offsets among the strobe and filter clocks, which means no output distortion due to intermodulation products. Furthermore, eliminating the usual encoder S/H circuit results in an important system improvement—less group delay.

Deriving the timing signals from an on-chip PLL also simplifies powering down. Gating the 8-kHz strobe signal off when the channel is idle causes the PLL to detect an unlocked condition and automatically power-down all the active circuitry on its chip. In this way a 2-chip set's power drain drops from 200 mW while operating to 25 mW in standby.

Independence is definitely the cornerstone of this system. With each buffer register shift clock working independently not only of each other but also of the PLL-derived timing, the conversion rate and the PCM data rate of either a single chip or the two chips working in tandem are also independent. Both chips can accommodate any data rate from 56 kHz to 3.2 MHz, while sampling at the internationally regulated 8-kHz rate.

Siliconix similarly separates coder (DF331/341) and decoder (DF332/342) functions, but does not include a filter on-chip—a strategy that allows this firm to offer the smallest part on the market: a 14-pin DIP. This codec family is second-sourced by Nitron, which produces the μ Law NC331/332 and A Law NC341/342.

Two single-chip CMOS codec circuits have appeared. Intermetall, GMBH, is developing a one-chip CMOS filter codec. Intended for one channel in pulse-code modulation telephone systems, the very large-scale integrated circuit will come in two designs, one for Europe and the other for the U.S., to accommodate the different approximation methods in those two regions. The chip's two converters work on the charge-redistribution principle, and the two low-pass filters are switched-capacitor types, providing a minimal signal-delay time. Frequency response, delay distortion, idle-channel noise, signal distortion, cross talk, and gain variations exceed CCITT recommendations. The devices, housed in a 24-pin dual in-line plastic package, operate off $+5$ and -5 V and consume less than 150 mW.

The AMI S3505 contains a fully asynchronous PCM filter and codec on one chip. Switched capacitor technology was used not only for the filter but also for the chip's op amps and DACs, using charge redistribution in a binary-weighted capacitor array.

The S3505 operates from a ± 5-V power supply, dissipates less than 200 mW, and is encased in a 28-pin DIP that has inputs for clock, reference trim circuitry, etc. The codec has an on-chip bandgap reference voltage which has two off-chip timers and exhibits an accuracy of 60 ppm/°C. Four package pins are dedicated to the references.

Normally one would expect a single-chip codec to be susceptible to cross talk between encoder and decoder. However, the S3505 has a cross-talk separation of 80 dB between encoder and decoder, which has been reduced by separately busing each section. The analog bus ground line is split and the power lines are split close to the pad to reduce common impedance.

The input of the S3505 consists of a cosine filter, followed by a low-pass filter and then a high-pass filter—all in front of the ADC. The output consists of a low-pass filter and an interpolative filter. The filter capacitors are formed from double polysilicon with thin silicon dioxide dielectric and have values of less than 1 pF lower than one would expect for an off-chip clock signal of 128 kHz.

The switched capacitor cosine filter clocks on each transition of the 128-kHz signal, thus giving an effective switching rate of 256 kHz. Rejecting all of the 128-kHz signal prevents aliasing of the clock signal. In addition, the low-pass filter passes a band of dc to 3.4 kHz. Since the signal is to be encoded at 8 kHz and the frequency to be encoded is just 3.4 kHz, a high-pass filter that can be clocked at 8 kHz can be fabricated with very small capacitors. The high-pass filter removes low-frequency and power-line noise. Its output is connected to a C (capacitor) ladder for A/D conversion.

In the receiving section the output of a sample and hold circuit is applied to a low-pass filter, which removes the high-frequency noise introduced by the 8-kHz sample-and-hold operation and equalizes the output. The interpolative filter, using the 128-kHz signal to clock at an effective rate of 256 kHz, interpolates between two consecutive samples and inserts a stop between them, effectively doubling the output sampling rate.

Close behind are Motorola's CMOS single-chip codec series (MC14400/01/02) and National's and RCA's single-chip CMOS codecs.

Many telecommunications systems are destined for completely internal use and therefore need not conform to either μ or A Law criteria. Such systems can take advantage of devices like Harris Semiconductor Corp's HD15530/15531 and HD6409/6469 CMOS Manchester encoder-decoders.

The HD15530/15531

Both meet MIL-STD-1553 TDM (time-division multiplexing) specifications and offer 1.25-Mb/s data rates, sync identification and lock-in, clock recovery, and 50-mW power dissipation using one ±5-V supply.

The 24-pin HD15530 employs a 16-b data-frame format with added sync and parity bits. A programmable frame length of up to 32 b forces the HD15531 into a 40-pin DIP.

The HD6490 has a digital phase-locked loop for clock recovery, an oscillator-clock generator, and complete encoder and decoder sections for 1 Mb/s data rate operation. Encased in a 20-pin DIP, the HD6409 dissipates 25 mW at a 5-V supply level over a temperature range of −40 to +85°C. The HD6469 is housed in a 14-pin DIP and consists of a repeater function only.

Short-term applications of codecs are likely to be concentrated in central stations and in larger private exchanges (Fig. 3-84). But for the longer term, because codecs will take time to impact the fiscally and technologically conservative telephone industry, applications are ex-

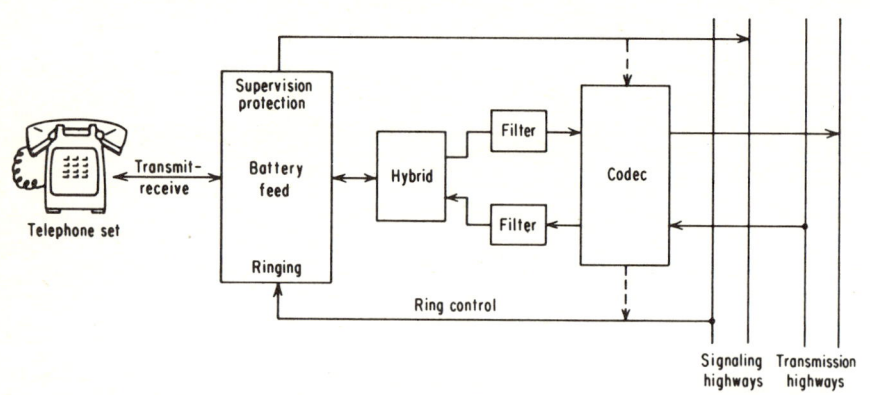

168   Data Conversion and Telecommunications Circuits

Fig. 3-84 Use of a 1-Chip Codec without On-Board Filters in a Private Branch Exchange or Central Office Switching System.

pected in transmission systems, local and tandem (interoffice trunk switching) offices, private automatic branch exchanges, and, ultimately, local subscriber loops and telephones.

MK5150 CODEC*,†

In the MK5150 coder-decoder, the use of a single D/A converter in the coding section and another in the decoding section has several advantages. For one thing, it improves performance because the isolation between the transmitting and receiving circuitry greatly exceeds that of the more common shared-converter approach. For another, it makes the device easier to use, because the transmitting and receiving sections are separate and can even be clocked at different rates.

Metal-gate CMOS is used to fabricate the MK5150 for several reasons. The process makes for high-quality matched capacitors of minimum size. It builds analog circuits well—particularly high-gain amplifiers and comparators. And it can be applied with confidence to high-volume production.

The 5150 does require a ±5-V supply, however, but it minimizes power dissipation by running its logic section from the +5-V line to ground. The entire 10-V swing drives only the analog section, which is by

---

* K. Bohni and M.J. Callahan, "CMOS Codec Splits Transmitting, Receiving Sections," *Electronics*, Sept. 28, 1978, pp. 141–144. Copyright © McGraw-Hill, Inc. Reprinted with permission.

† Recently the MK5150 was replaced by the MK5151 pin-compatible codec upgrade.

far the smaller portion of the chip. The codec typically dissipates 30 mW at room temperature.

The block diagram of the MK5150, as shown in Fig. 3-85, reveals that the device uses not one but two D/A converters for encoding and decoding data, making it easy to use and giving it a system isolation unattainable with a shared approach. What is more, the D/A converters are of the capacitive type and so require no external S/H capacitors. Nor is an external filter needed for autozeroing, as it would be with a shared converter.

The block diagram also shows how few elements are required to handle analog signals. In fact, there are only two on the entire chip: a comparator and a single op amp. The reason is that the data-compression/expansion scheme carried out according to telephone company specifications is implemented with a digital 8-to-13-b converter which allows the use of a linear 13-b D/A converter. The analog compo-

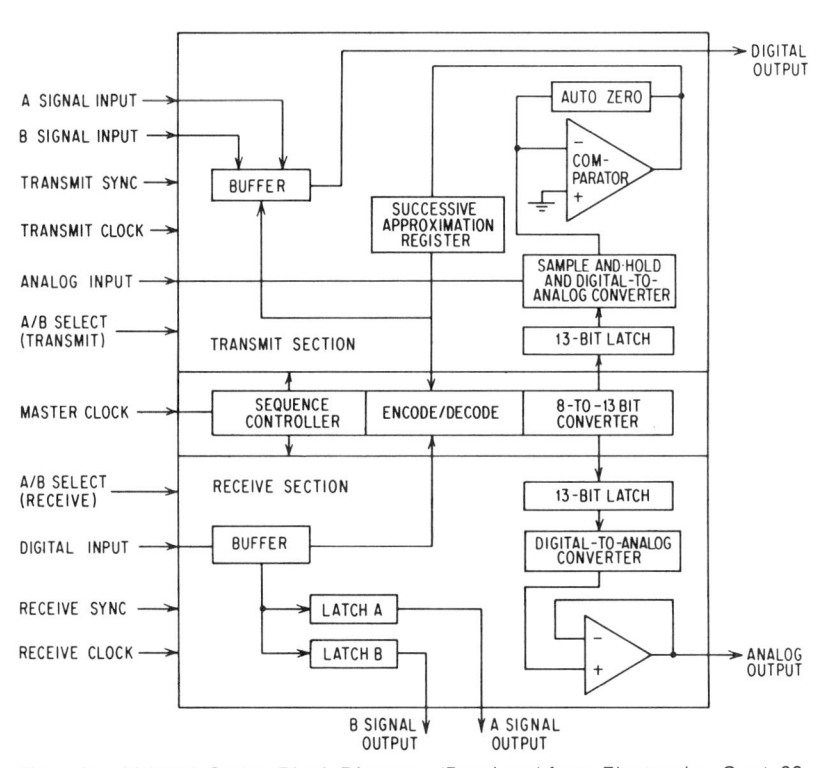

Fig. 3-85 MK5150 Codec Block Diagram. *(Reprinted from Electronics, Sept. 28, 1978. Copyright © 1978 by McGraw-Hill, Inc. Used with permission.)*

nents use the full 10V swing of the power supply, so having as few of them as possible on chip means lower power consumption.

Because the receiving and transmitting sections operate independently, the chip can use synchronous or asynchronous modes at various input and output clock rates.

The timing diagram for the MK5150 is shown in Fig. 3-86. In the receiving mode, the input data is shifted serially into the input buffer at the receive-clock rate and then in-parallel into the decoding section for as long as the RECEIVE-SYNCHRONIZING line is high. After the falling edge of the RECEIVE-SYNC pulse, the encoding process is halted for about 2 to 5 μs while the data is processed through the 8-to-13-b converter. The result then is latched into the 13-b RECEIVE latch, which updates the output of the receiving D/A converter with a 100 percent duty cycle. The receiving converter acts as an S/H circuit and its output is buffered by the unity-gain operational amplifier.

During each signaling frame (which is every sixth 125 μs period in the 12-frame standard data format), only a 7-b decoding operation is performed, and the eighth data bit is latched into either the A or the B signal output latch—whichever is selected by the A/B select input in the receiving section. This eighth bit is a signaling bit and is assigned the analog value of $\frac{1}{2}$ step; the result is a lower S/D ratio than if it were set arbitrarily to either 1 or 0.

In the transmitting mode of operation, the analog signal is sampled by the input S/H circuit, which simultaneously performs the autozero function, and is encoded by a successive-approximation technique. The D/A

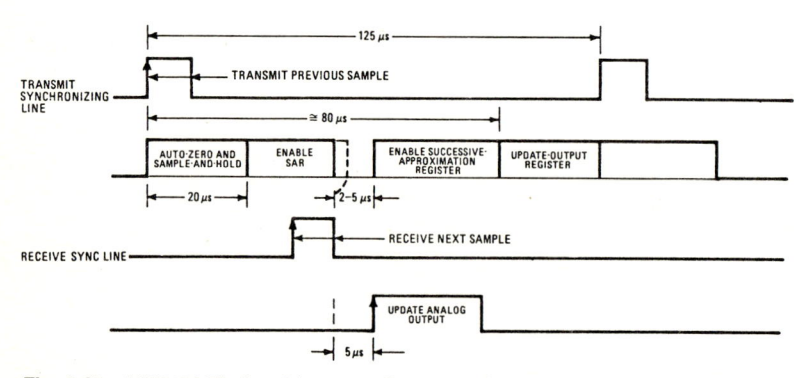

Fig. 3-86 MK5150 Timing Diagram. *(Reprinted from Electronics, Sept. 28, 1978. Copyright © 1978 by McGraw-Hill, Inc. Used with permission.)*

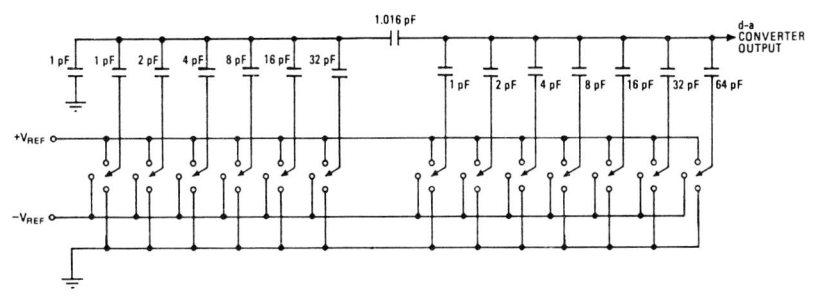

Fig. 3-87 The MK5150 uses a pair of switched-capacitor-type D/A Converters. Since companding is done digitally with an 8-to-13-b Converter, a Linear D/A Converter with Full 13-b Resolution Is Used. *(Reprinted from Electronics, Sept. 28, 1978. Copyright © 1978 by McGraw-Hill, Inc. Used with permission.)*

converter in the transmitting section operates much like the one in the receiving section.

Once the encoding process is complete, the output of the SAR is loaded into the output buffer. The data is transmitted serially at the output clock rate for as long as the TRANSMIT-SYNC line is high. A or B signaling information is inserted during the signaling frame into the output bit stream in place of the eighth data bit as determined by the A/B SELECT input line in the transmitting section of the codec.

The sequence controller, which is driven by a master clock rate ranging from 1.544 to 2.048 MHz, keeps the system in step with itself. All timing signals for S/H successive approximation, and so on, are generated in that section. To ensure proper encoding, a decoding interrupt is allowed only when the SAR clock line is low. The decoding interrupt interval therefore varies from 2 to 5 μs.

The 8-to-13-b converter provides a one-to-one translation between 8-b companded code at its input and 13-b linear code at its output. The D/A converters are therefore 13-b linear types. Shown in Fig. 3-87, each converter operates on the charge-distribution principle of a binary-weighted capacitor ladder.

The capacitor ladder has two sections—7 MSBs and 6 LSBs—connected by a 64:1 capacitive divider. The D/A converter's output is equivalent to the output of a 13-b D/A converter with an output capacitance of 128 pF.

In the encoding section, this equivalent capcitor of 128 pF is also employed to perform autozero and S/H, thereby eliminating the need for external capacitors.

American Microsystems, Inc., has fabricated a CMOS switched

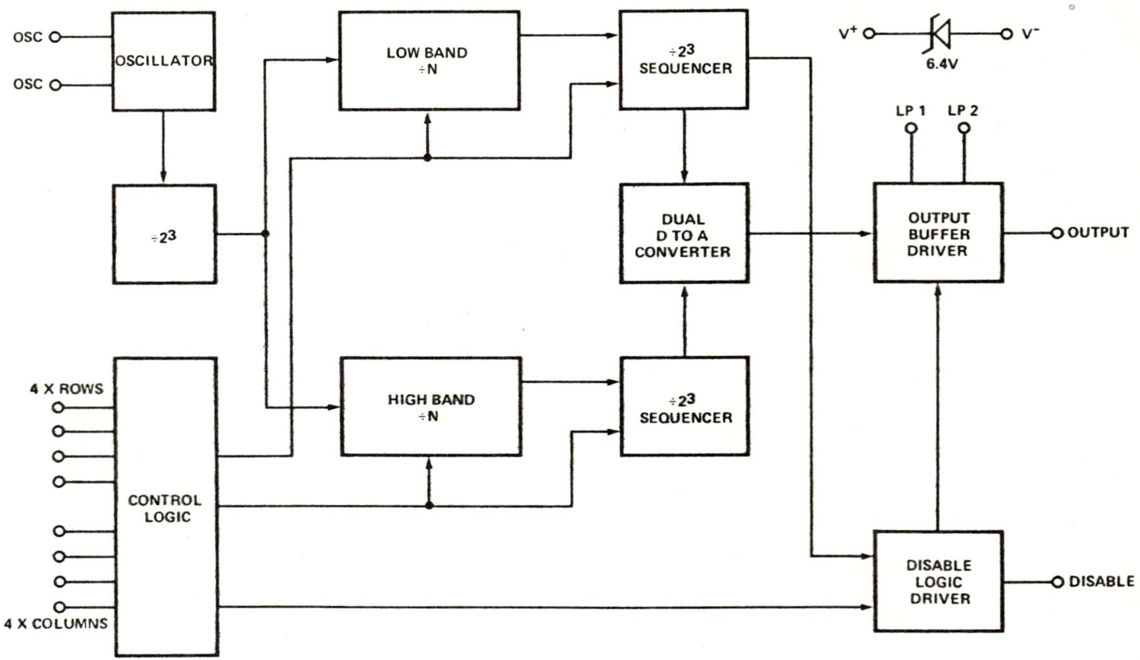

Fig. 3-88 Touch-Tone Encoder Schematic Diagram. *(Courtesy Intersil.)*

Telecommunications Circuits 173

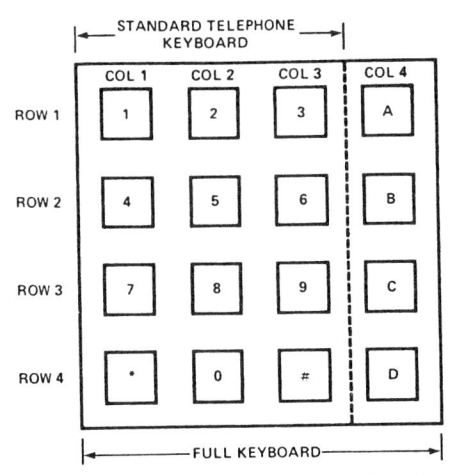

KEY	LOW BAND FREQ. Hz	HI BAND FREQ. Hz
1	697	1209
2	697	1336
3	697	1477
4	770	1209
5	770	1336
6	770	1477
7	852	1209
8	852	1336
9	852	1477
*	941	1209
0	941	1336
#	941	1477
A	697	1633
B	770	1633
C	852	1633
D	941	1633

Fig. 3-89 Keyboard Frequencies. *(Courtesy Intersil.)*

capacitor filter utilizing CMOS operational amplifiers for application in PCM voice codecs* as have National, Mostek, Motorola, and others.

Touch-Tone Encoder

The ICM 7206 is a 2-of-8 sinewave tone encoder for use in tone dialing in telephone systems. The circuit (Fig. 3-88) consists of a high-frequency oscillator, two separate programmable dividers, a D/A converter, and a high-level output driver. The seven or eight inputs are originated from either a 4 × 3 or 4 × 4 single-contact keyboard (Fig. 3-89).

The reference frequency is generated from a fully integrated oscillator requiring only a color TV (3.57-MHz) crystal. This frequency is divided by 8 and is then gated into two divide-by-N counters (possible division ratios 1 through 128) which provide the correct division ratios for the upper and lower bank of frequencies. The outputs from these two divide-by-N counters are further divided by 8 to provide the time sequencing for a four-voltage level synthesis of each sinewave. Both sinewaves are added and buffered to a high-current output driver. Provisions are made for up to two external capacitors for LP filtering, if so desired. Typically the total output harmonic distortion is 20 percent with no LP filtering. The value may be reduced to 2 to 3 percent with filtering.

* See R. Gregorian et al., "CMOS Switched Capacitor Filter for a Two Chip PCM Voice Codec," *1979 IEEE ISSCC Digest of Technical Papers*, Philadelphia, PA, February 1979, for a complete discussion.

The output drive level of the tone pairs will be approximately −3 dBV into a 900-Ω termination. The skew between the high and low groups is typically 2.5 dB without LP filtering.

The ICM7206 will operate with supply voltages as low as 3 V and dissipates less than 5.5 mW at 5.5 V. The device requires only nine additional components to become a minimum-part, low-cost tone generator. The chip contains a zener diode, and the addition of a single external resistor makes the CMOS circuit immune to latch-up.

The chip can generate either single or dual tones. Oscillator start-up time is 5 ns and multiple-key lockout is provided.

The ICM7206 is encased in a 16-pin DIP for use with keyboards having two contacts per key connecting both rows and columns to a common line (positive supply line).

The ICM7206 is fabricated using a standard low-voltage metal-gate CMOS process. Custom options are possible using different quartz crystal frequencies, two-contacts-per-key-type keyboards, and any combination of the output frequencies defined in Fig. 3-89.

The circuit techniques used in the design have been developed over many years for watch and clock applications and are characterized by wide operating supply voltage ranges and low power dissipation.

To minimize chip size, all diffusions used to define source-drain and field regions are butted up together. This results in approximately 6.3-V zener breakdown between the supply terminals and between all components on chip. As a consequence, the usual CMOS static charge and handling problems are not experienced with the ICM7206. In fact, it can be handled with procedures similar to those used for standard TTL gates.

The oscillator consists of a medium-size CMOS inverter having on-chip a feedback resistor and two capacitors of 14 pF each, one at the oscillator input and the other at the oscillator output. The oscillator is followed by a dynamic ÷ 2^3 circuit which divides the oscillator frequency to 447,443 Hz. This is applied to two programmable dividers capable of division ratios of any integer between 1 and 128. Each programmable counter is controlled by a ROM. The outputs from the programmable counters drive sequencers (divide by 8) which generate the eight time slots necessary to synthesize the four-level sinewaves.

The control logic block recognizes signals on the row and column inputs that are only a small fraction of the supply voltage, thereby permitting the use of a simple matrix single-contact-per-key keyboard instead of the more usual two-contacts-per-key type having a common line.

Telecommunications Circuits   175

The row and column pull-up resistors are equal in value and connected to the opposite supply terminals (ICM7206 only; for the ICM7206A all pull-up resistors are connected to the V− terminal). Therefore, connecting a row input to a column input generates a voltage on those inputs equal to one-half the supply voltage.

The ICM7206 employs a unique but extremely simple D/A converter. This D/A converter produces a four-level synthesized sinewave having an intrinsic total harmonic distortion level of approximately 20 percent. Figure 3-90 shows a single-channel D/A converter. The current sources Q_2 and Q_3 are proportioned in the ratio of 1:1.414. During time slots 1 and 8 both S_1 and S_2 are off, during time slots 2 and 7 only S_1 is on,

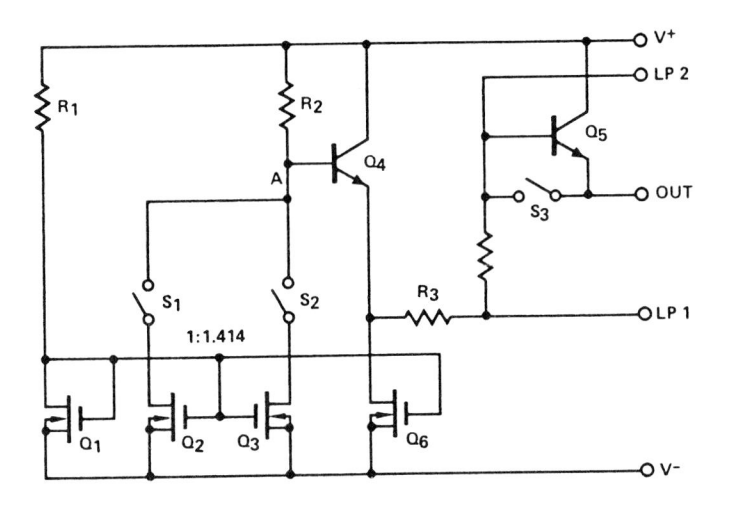

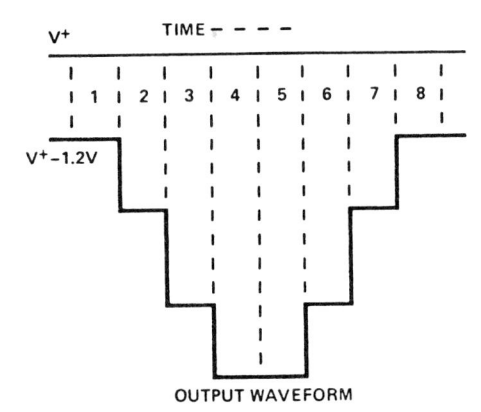

Fig. 3-90 D/A Converter and Output Buffer. *(Courtesy Intersil.)*

during time slots 3 and 6 only S_2 is on, and during time slots 4 and 5 both S_1 and S_2 are on. The resulting currents are summed at node A, buffered by Q_4 and further buffered by R_3, R_4, and Q_5. Switch S_3 allows the output to go into a high-impedance mode under quiescent conditions.

The synthesized sinewave has negligible even-harmonic distortion and very low values of third and fifth harmonic distortion, thereby minimizing the filtering problems necessary to reduce the total harmonic distortion to well below the 10 percent level required for touch-tone telephone encoding. Figure 3-91 shows the LP filter characteristic of the ICM7206 output buffer for $C_1 = 0.0022\,\mu F$ and $C_2 = 0.022\,\mu F$. A small peak of 0.4 dB occurs at 1100 Hz, with sharp attenuation (12 dB per octave) above 2500 Hz. This type of active filter produces a sharper and more desirable knee characteristic than would two simple cascaded RC networks.

Most junction-isolated CMOS ICs, especially those of moderate or high complexity, exhibit a latch-up phenomena whereby they can be triggered into an uncontrollable low-impedance mode between the supply terminals. This can be caused by gross forward biasing of inputs or outputs (with respect to the supply terminals), by high voltage supply transients, or more rarely by an exceptionally fast rate of rise of supply voltages.

The ICM7206 is no exception and precautions must be taken to limit the supply current to those values shown in the "Absolute Maximum Ratings" listing. As an example, do not use a 6 V very-low-impedance supply source with the ICM7206 in an electrically extremely noisy environment unless a 500-Ω current-limiting resistor is included in-series with the V− terminal. For normal telephone encoding applications, no problems are envisioned even with low-impedance transients of 100 V or more if circuitry similar to that shown in the next section is used.

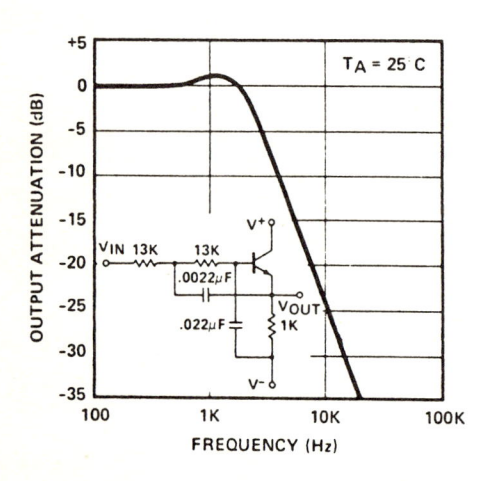

Fig. 3-91 Frequency Attenuation Characteristics of the ICM7206 Output Buffer. *(Courtesy Intersil.)*

TELEPHONE HANDSET

A typical encoder for telephone handsets is shown in Fig. 3-92. This encoder uses a single-contact-per-key keyboard and provides all other switching functions electronically. The diode connected between pins 8 and 15 prevents the output from going more than 1 V negative with respect to the negative supply V−. This circuit operates over the supply voltage range from 3.5 to 15 V on the device side of the bridge rectifier. Transients as high as 100 V will not cause system failure, although the ICM7206 will not operate correctly under these conditions. Correct operation will resume immediately after the transient is removed.

The output voltage of the synthesized sinewave is almost directly proportional to the supply voltage ($V^+ - V^-$) and will increase with increased supply voltage until zener breakdown occurs (at approximately 6.3 V between pins 8 and 16), after which the output voltage remains constant.

PORTABLE TONE GENERATOR

The ICM7206A requires a two-contact-per-key keyboard with the common line connected to the positive supply (pin 16). A simple diode matrix may be used with this keyboard to provide power to the system whenever a key is depressed, thus negating the need for an ON/OFF switch. In Fig. 3-93, the tone generator is shown using a 9-V battery. However, if instead a 6-V battery is used, diode D_4 is not required. It is recommended that a 470-Ω resistor still be included in-series with the negative supply to prevent accidental triggering or latch-up.

Dual-Tone Multiple-Frequency Receiver*

A fully integrated DTMF (dual-tone multiple-frequency) receiver on a single metal-gate CMOS chip featuring the use of switched-capacitor filters has made it possible to reduce the total silicon area while requiring only three external passive elements. Central office quality performance is achieved without the need for trimming. Receiver specifications are listed in Table 3-11.

The block diagram of Fig. 3-94 shows the chip's overall architecture. The incoming telephone signal tone pair is high-pass-filtered to remove 60 Hz and any dc component it might have. It is then passed through a preemphasis network to the two bandsplit filters. The output of each

* B. J. White, et al., "A Monolithic Dual-Tone Multi-frequency Receiver," *1979 IEEE ISSCC Digest of Technical Papers,* Philadelphia, PA, February 1979.

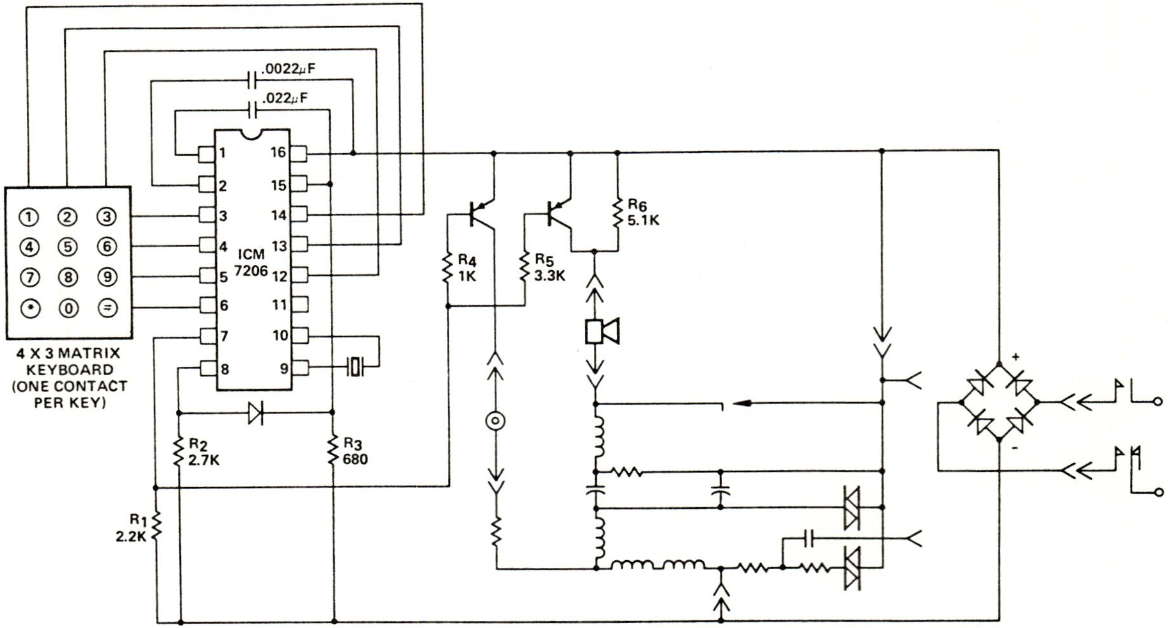

Fig. 3-92 Telephone Handset Touch-Tone Encoder. (*Courtesy Intersil.*)

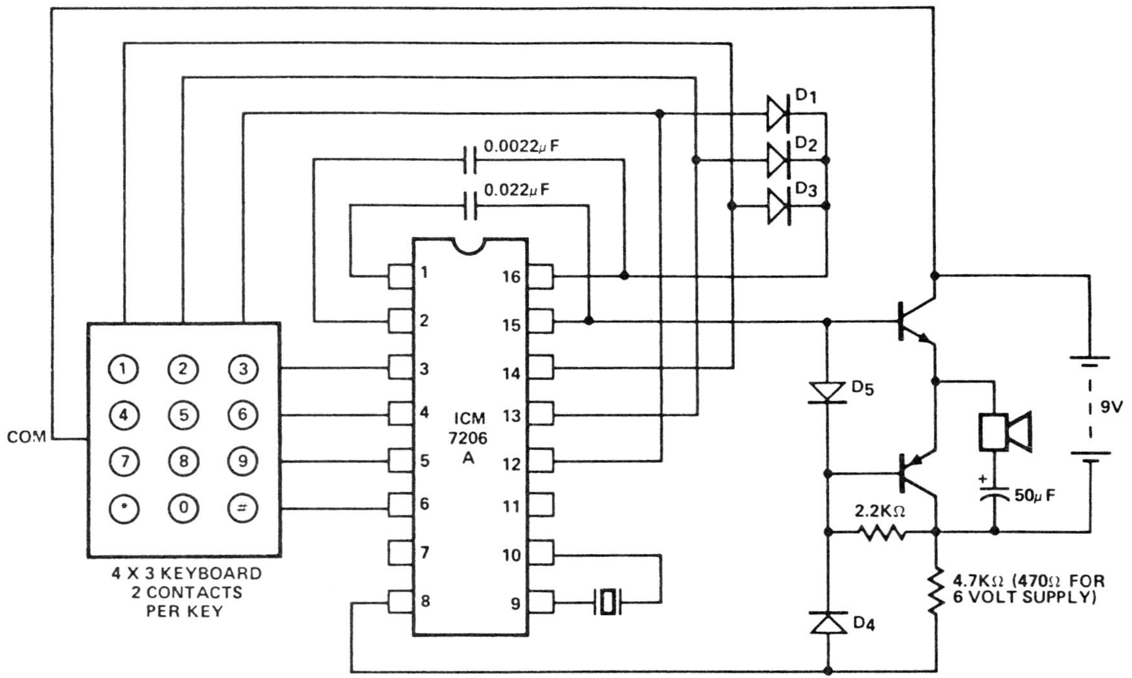

Fig. 3-93 Portable Tone Generator Schematic Diagram. *(Courtesy Intersil.)*

180   Data Conversion and Telecommunications Circuits

TABLE 3-11   System Level Specifications for DTMF Receiver

Specification	Must Accept	Must Reject
Low-Frequency Bandwidth (697,770,852,941)	±(1.5% + 2 Hz)	±3%
High-Frequency Bandwidth (1209,1336,1477,1633)	±(1.5% + 2 Hz)	±3%
Minimum Amplitude Each Frequency	−24 dBm @ 600 Ω*	
Maximum Amplitude Each Frequency	+6 dBm @ 600Ω	
Twist (High ÷ Low)	+4 to −8 dB	
Tone Valid Time	≥40 ms	≤13 ms
Pause Time	≥40 ms	≤35 ms
Talkoff Performance	≤2 Hits on Mitel Tape (Typ)	
60 Hz Tolerance	2 V rms if Full-Scale Input is not Exceeded	
Outputs	Three-State B2/8 or Hexidecimal	

* dBm = the measure of power level in decibels with reference to a power of one milliwatt.

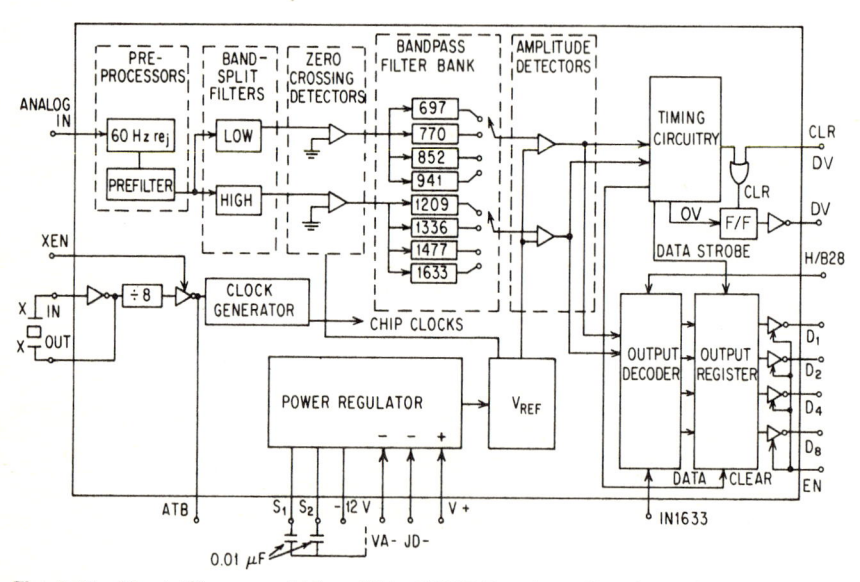

Fig. 3-94 Block Diagram of Monolithic DTMF Receiver. *(Reprinted from 1979 IEEE ISSCC Digest of Technical Papers, Philadelphia, PA, February 1979. Copyright © 1979 by the Institute of Electrical and Electronics Engineers, Inc. Used with permission.)*

bandsplit filter will be one of the two tones and will cause a square wave of fixed amplitude to be sent to the bank of bandpass filters from the zero crossing detector. If the square wave frequency is close enough to one of the bandpass center frequencies, the amplitude detector will indicate to the timing and decode circuitry that a valid tone has been detected. If both tones are simultaneously present for sufficient time, the data outputs are activated and DATA VALID is raised.

One of the most difficult requirements of a DTMF receiver is adequate talkoff performance or speech immunity. Since the hard-limiting zero crossing detector provides a constant rms (root mean square) output for any ac signal input, a 50 percent DUTY cycle square wave output will contain the highest amplitude frequency component. If any interference is present at the input to the zero crossing detector (as in the case of interfering speech), the output DUTY cycle will deviate from 50 percent, thereby reducing the amplitude of the fundamental frequency and prohibiting the detection of a tone pair.

To enhance further the speech immunity of the receiver, the bandsplitting is accomplished with bandstop filters to ensure that interfering signals outside of the DTMF band are permitted to reach the zero crossing detector. Talkoff performance as measured by a speech tape is 3 hits or less.

The bandstop filters are 6-pole Chebyshev designs implemented with second-order biquadratic sections. This circuit was chosen for its small size yet adequate component tolerance.

Exploiting the extreme stability and accuracy of on-chip capacitor ratios, the frequency detection circuit is implemented with eight 2-pole bandpass filters. Bandpass filters have the advantage that inherent in the filtering mechanism is an averaging of the zero crossings, which increases the system tolerance to noise and interference. Whereas digital schemes require more than 16 dB of separation in the bandsplit filter if no twist or noise is present, an analog filter bank permits the rejection specification to be only 8 dB. Additionally, the center frequencies of the bandpass filters can be established wherever necessary, instead of choosing the nearest frequency whose period is a multiple of a master clock period.

The eight precision bandpass filters are biquadratic sections implemented with switched capacitor integrators.

The CMOS operational amplifiers employed throughout the receiver have measured open-loop gains of 3000 typical, unity gain bandwidth of 2 MHz, power consumptions of 6 mW with a 12-V supply, and 0.1

percent settling time in the integrator configuration of 3 μs with a 20-pF load.

The DTMF circuit contains 40 operational amplifiers, 156 precision capacitors, and 250 digital gates, and is encased in a standard 22-pin DIP.

Although one- and several-chip approaches to a DTMF receiver already exist, a two-chip version is available from Mitel Semiconductor. Aided by a few external components, this Mitel design—separate filter and decoder chips designated the MT8865 and MT8860, respectively—affords an optimum trade-off between cost-effectiveness and flexibility.

Like the other chips, the MT8865/8860 offers both the classic telephone functions and many important new ones. It can be used to construct a single-ended or differential DTMF receiver, update older rotary-dial phones by means of tone-to-pulse conversion, and create a call restrictor in an individual telephone or a private branch exchange. But more significantly, it will in the future also serve in many remote control and remote data-entry systems. Examples are the remote control of test measurement and industrial equipment and eventually (in the office or home of the future) the remote control of home appliances, alarm systems, and unattended message recorders.

The MT8865 DTMF analog bandsplit filter consists of two bandpass filters plus zero-crossing detectors for the front-end functions of bandsplitting, dial-tone rejection, and limiting. The MT8860 DTMF digital decoder detects and decodes the tones. External components tailor the chips for different applications.*

IC Telephone Dialers[†]

Improvements in IC technology allow telephone dials to be simpler and less costly than their predecessors. Two circuits from Mostek allow telephone designers to utilize keyboard dialing and signal the central office with either pulses or DTMF signals. Each circuit is designed to interface either into existing hybrid telephone networks or into electronic networks with minimal interfacing and few external components. Designing with either circuit requires an understanding of the telephone system relating to the subscriber set.

---

* For a detailed discussion of the circuit, see R. A. Bromfield et al., "Making a Data Terminal out of the Touch-Tone Telephone," *Electronics*, July 3, 1980, pp. 124–129.

† Charles B. Johnson, "Subscriber Sets Advance with IC Dialers," Session 38, *1978 WESCON Proceedings*, Los Angeles, CA, 1978. Used with permission.

THE MK5098 PULSE DIALER

The MK5098 is a pulse dialer circuit described in the block diagram of Fig. 3-95.

The MK5098 uses a low-cost 455-kHz resonant-mode ceramic resonator to generate an accurate reference frequency from which the pulse signal is derived. The keyboard interface consists of seven inputs from a single-contact keyboard. Keyboard inputs are stored in a FIFO (first-in first-out) buffer constructed with a sequentially addressed memory allowing entries at 40 ms each. This allows the inputs to be entered as fast as the user wishes and the IC ensures that the pulse rate is acceptable to the system requirements. The digits are retrieved one at a time from the FIFO and entered into a down counter. If a digit 3 were needed, the down counter would generate three breaks of the pulse output, 60 ms each spaced 40 ms apart. Each digit is separated from the previous digit by 800 ms, which is the required interdigit time. Before the first digit, an 800-ms pause is generated, which allows proper system operation. The mute output provides muting during pulsing, with no muting during the 800-ms pause. On-chip overvoltage protection is provided with the volt-

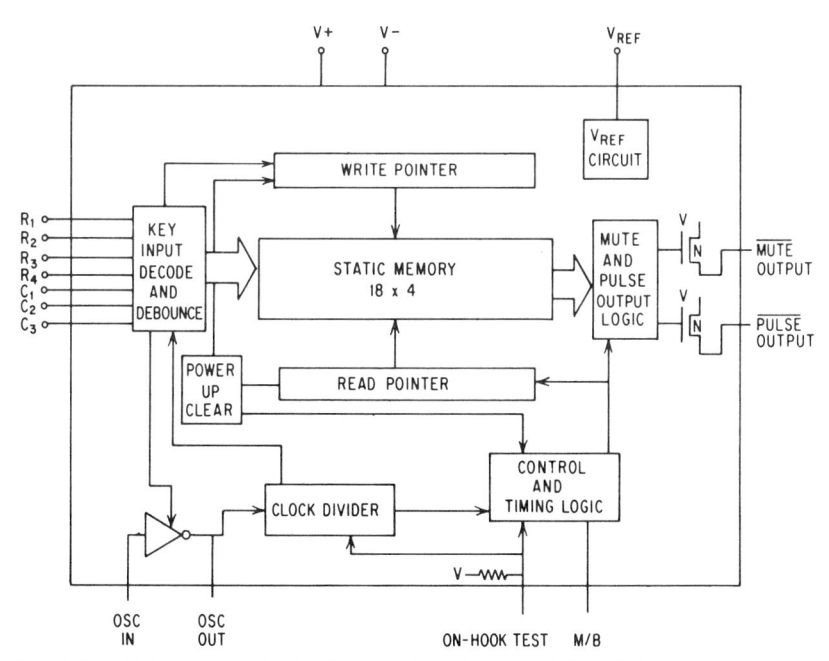

Fig. 3-95 Pulse Dialer Block Diagram. *(Courtesy of WESCON.)*

age reference output. This output is an open circuit if the supply voltage is at the minimum operating voltage of the circuit. When the voltage increases above this minimum, the output sources increase amounts of current as the voltage increases. The typical *IV* curve of this voltage reference is shown in Fig. 3-96.

When used in conjunction with a constant-current source, this output regulates the supply voltage to the circuit under varying telephone conditions including lightning surges. The on-hook input allows the circuit to be quickly tested and empties any unused digits in less than 300 ms if the user terminates a call. Functionally this input speeds up most circuit functions by 100 times the normal rate. The MAKE/BREAK input allows the percentage MAKE/BREAK to be either 40%/60% or 33%/67%. This selection allows the same telephone to be used in systems requiring either ratio.

Installing the MK5098 into the existing hybrid telephone is accomplished by the connection shown in Fig. 3-97.

The speech network remains connected in the same manner as when using a rotary dial. A diode bridge is installed between tip and ring and the network to ensure that the proper polarity voltage is available to power the circuitry. The current source is used to supply the minimum current required to power the IC. The current source is designed to normally source 500 μA. The compound Darlington pair formed by Q_1 and Q_2 is used to make and break the current drawn by the telephone network. The base current required to turn on Q_1 is available through R_3, except when the pulse output is sinking the current through the pulse output transistor. When the pulse output is not sinking current, the network is connected through Q_2 to the line, and, when the IC is sinking

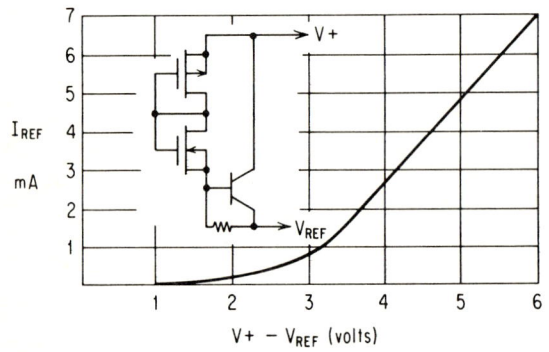

Fig. 3-96 Voltage Reference Output. *(Courtesy of WESCON.)*

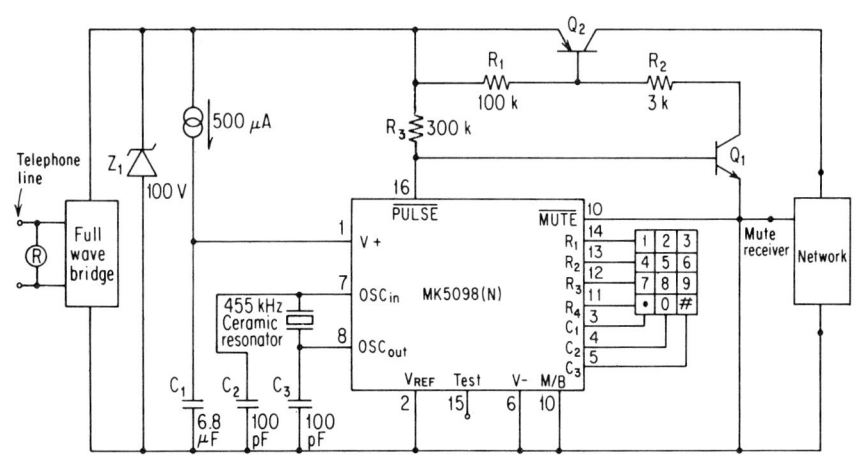

Fig. 3-97 Pulse Telephone Schematic Diagram. *(Courtesy of WESCON.)*

current through the pulse output, the loop current is broken. The mute output is used to turn on a Darlington transistor which shorts the network and attenuates the clicks reaching the receiver.

THE MK5087 TONE DIALER

Designed to generate the DTMF dialing frequencies, the MK5087 utilizes CMOS technology to implement the circuitry shown in the diagram of Fig. 3-98.

A 3.579545-MHz crystal is used to produce a high-accuracy clock frequency. This crystal is widely used as the color-burst oscillator standard in color television and is available at low cost. The 3.58 MHz is divided by 4 and supplied to the ROW and COLUMN divide-by-N counters. The counter divide ratio is determined by the keyboard inputs. These counters supply a pulse train which is equal to the sampling rate of the output fundamental. These two pulse trains each increment D/A converters, which generate a stair-step sinusoidal approximation (Fig. 3-99).

This waveform is a simulation of a sinusoid accurate enough to meet the 10 percent distortion requirement of the U.S. telephone system. Both D/A converter outputs are combined by summing their output currents into an on-chip operational amplifier. This amplifier uses matched N-channel transistors to build a differential pair input stage with a constant-current source, utilizing current mirrors and a bipolar transistor. The second stage utilizes a P-channel gain stage followed by an NPN bipolar transistor used in an emitter-follower configuration which is the

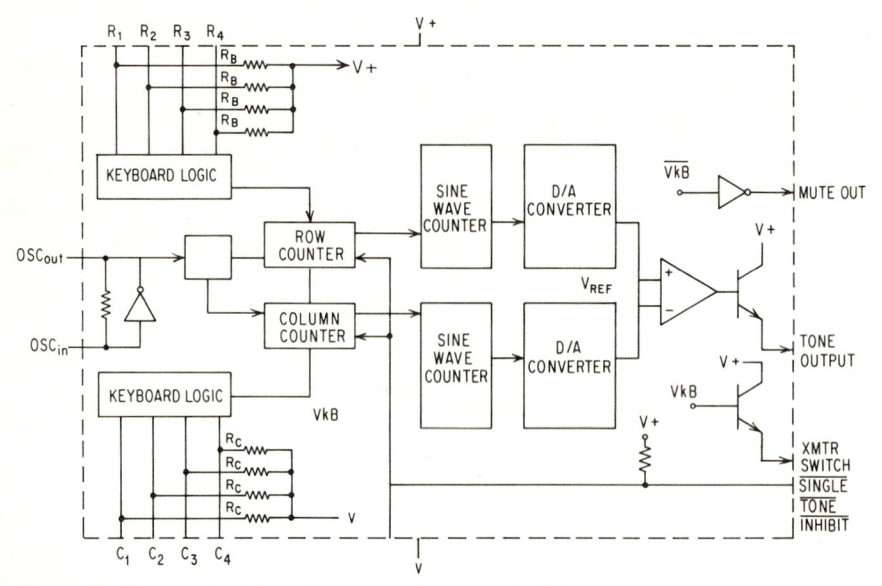

186 Data Conversion and Telecommunications Circuits

Fig. 3-98 Tone Dialer Block Diagram. *(Courtesy of WESCON.)*

output buffer. The bipolar transistor used is inherently available in the CMOS technology. The emitter is formed by the N+ diffusion inside a P well with a P well forming the base and the collector formed by the starting N wafer, as is shown in Fig. 3-100.

The entire op amp exhibits 60 dB of gain at low frequencies, with unity gain occurring at approximately 1 MHz. This op amp is schematically shown in Fig. 3-101.

Both circuits (MK5098 and 5087) are designed to be directly powered from the telephone line. This requires each circuit to derive its power from the 48-V central office battery and signal over tip and ring which

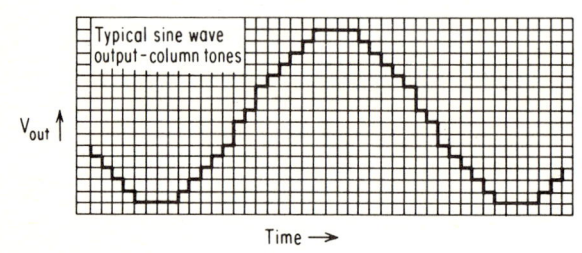

Fig. 3-99 Tone Dialer Sinewave. *(Courtesy of WESCON.)*

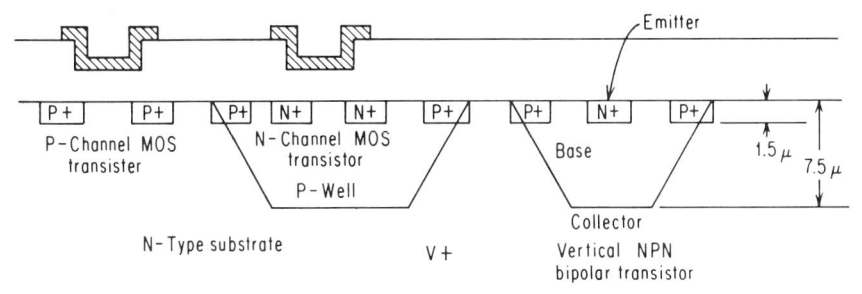

Fig. 3-100 Transistors Available in CMOS. *(Courtesy of WESCON.)*

supply the operating current. System economy is obtained by maximum integration of system functions and through the use of inexpensive external components. Both circuits utilize single-contact keyboards which allow less expensive calculator-type keyboards to be used in the telephone. Inexpensive frequency references are used by each circuit, and both circuits drive the telephone line with a minimum of interfacing.

Figure 3-102 shows how the MK5087 is used in a telephone application.

When using the MK5087 tone dialer either mechanical or electronic switching may be employed to turn on and off the transmitter and receiver. The application of Fig. 3-102 shows only the toning section of the tone telephone. The 2500 network is used in conjunction with the MK5087 to drive tip and ring. The varistor V_2 draws the necessary loop current. The on-chip bipolar transistor drives tip and ring through the 180-Ω resistor while keyboard entries are present. The tone dialer circuit

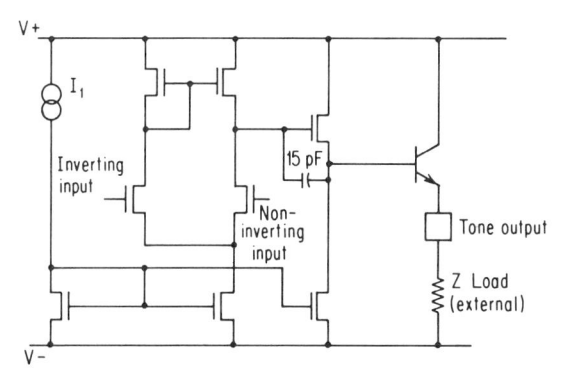

Fig. 3-101 Tone Dialer Op Amp. *(Courtesy of WESCON.)*

188 Data Conversion and Telecommunications Circuits

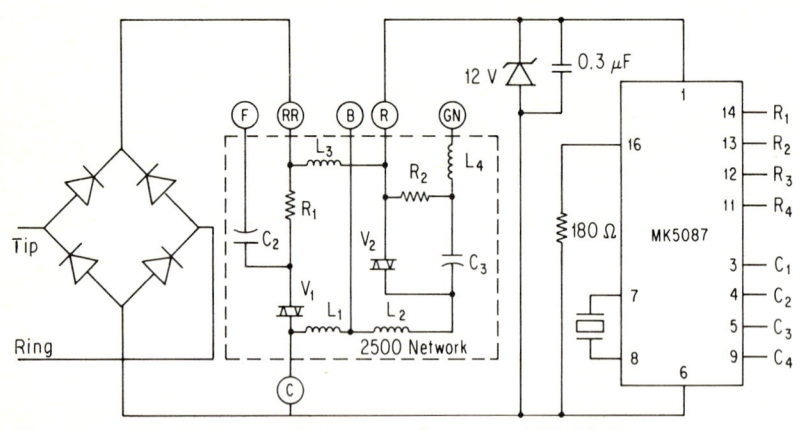

Fig. 3-102 Tone Telephone Schematic. *(Courtesy of WESCON.)*

is powered directly from tip and ring and is designed to operate over the entire voltage excursions present when toning. During toning, the circuit operates from 3.0 to 10.0 V, with the DTMF signal superimposed on the dc voltage supplied from the central office. The diode bridge ensures that the circuit will operate with the voltage reversed on tip and ring. The 12-V zener protects the entire telephone from transients produced by lightning. Mostek offers an entire family of Dialers based on these two circuits.

Other Telecommunications Circuits

The day of the computer-controlled telephone has nearly arrived. Of course, phone company central offices and PBXs have long used computers to exercise varying degrees of intimate control over phone conversations and equipment. But now, with the introduction of microprocessors and controller chips, phones with advanced and/or useful functions and features will soon be sitting on one's desk.

Recent FCC (Federal Communications Commission) rulings have opened the floodgates to specially designed and built telephones for connection to the Bell System. Some of these units have already appeared in this new and promisingly large consumer market.

These custom phones utilize specially programmed microcomputer or microcontroller chips in addition to the more familiar tone and pulse dialer chips. And at least three semiconductor houses have introduced "standard-feature," off-the-shelf phone controllers.

Usually called *repertory dialers,* these ICs can recall and dial, at the push

TABLE 3-12 S2562 Characteristics

Manufacturer	AMI	Manufacturer	AMI
Part No.	S2562	Pulse Dial Output	10 or 20 pps*
Process	CMOS		
Supplies	+4.5–5.5 V	Access Pause Capability	Yes
Current	25–500 μA		
Preferred Number Memory	S5101 256 × 4 CMOS RAM	Oscillator Type	RC
Maximum Number Storage	32 8-Digit or 16 16-Digit	Power-Failure Detection	Yes
		Package	40-Pin DIP
Display Outputs	1 Digit	Remarks	Single-Digit Display Output Needs 4511 Decoder
Real-Time Clock	No		On-Chip Pulse Dialer with 400- or 800-ms IDP,† 33–66 OR 40–60 Make/Break
Timer	No		
Redial Capability	Yes		Single- or Dual-Closure Keyboard

* pps = pulses per second.
† IDP = interdigit pause (the built-in time between numbers when dialing).

of one or two buttons, a previously stored multidigit phone number from a repertory of several dozen. However, their versatility goes far beyond mere memorization and recall of numbers. The repertory dialers' added features portend what one can expect in the way of phone-controller sophistication.

AMI's S2562 remembers 8- or 16-digit numbers, automatically dials and redials, and displays the dialed number. Key features of the S2562 are summarized in Table 3-12.

The circuit contains a PLA (programmable logic array) which receives the keyboard inputs and controls the chip's latches and counters. Although one gives up features like clocks, timers, and calculator interfacing, the S2562 provides options on break/make ratios, interdigit pause times, and dial-pulse rates. In addition, one can also arrange the storage of sixteen 16-digit or thirty-two 8-digit numbers in a 1-kb CMOS RAM. The S2562 also interfaces easily to AMI's S2559 tone generator.

AMI also sells the 2561 tone ringer, which sounds like a regular electromechanical bell but has a novel sequential amplitude feature. Driving a small speaker-transformer combination, the chip gradually increases the tone during the first three rings, then continues to put out maximum-level tones.

Further Reading

1. H. Wurzburg and S. Kelley, "Three LSI Circuits Simplify Digital Switching Systems," *Electronic Design,* October 25, 1980.
2. V. Godbole et al., "One-Chip Codec/Filter Eases Subscriber—Line Interface," *Electronic Design,* December 6, 1980.

CHAPTER FOUR
CMOS Memories

This chapter discusses the use of bulk CMOS silicon and CMOS/SOS technologies to fabricate RAMs, ROMs, and PROMs. The elusive SOS process, still in its infancy, has already proven to be of great value because the substrate is an insulator, thereby eliminating parasitic capacitances that decrease speed and increase power consumption. Because SOS memory devices typically have three to five times the speed of corresponding bulk silicon devices, their use is beneficial in high-speed systems that require very low power consumption.

CMOS memory arrays fabricated on either bulk silicon or insulating substrates have outstanding features not achieved by either single-channel NMOS or bipolar technology. To demonstrate this, Table 4-1 compares the salient features of several commercially available CMOS RAMs (both bulk and SOS substrates) with NMOS and bipolar TTL RAMs.

As shown, CMOS RAMs exhibit speed (access time) performance comparable to the popular industry standard static and dynamic NMOS RAMs, while they are superior in terms of power dissipation.

Memory system requirements have evolved to a point where they will affect all semiconductor memories: generalized memory system design in which there exists pin-, functional-, and organizational-compatible memory families so that program storage and data storage memory can be easily interchanged; and byte-wide IC memory organization struc-

TABLE 4-1 Various Memory Types and Processes

Memory Type	Access Time (ns)	Read Cycle Time (ns)	Operating Dissipation (mW)	Standby Dissipation (mW)	Power Supplies (V)	Peripheral Circuitry Required	Output	Temperature Range (°C)	Pulsed Chip Select	Number of Pins
Bipolar Schottky TTL 64 Bit (SN74S189/289)	35	35	525	525	5	None	Three State/Open Collector	0–75 −55 to +125	No	16
Bipolar TTL 1024 Bit Isoplanar (93F415/425)	20, 35	20, 35	656	656	5	Pull-Up Load	Three State/Open Collector	0–75	No	16
Bipolar ECL[a] 1024 Bit (HM2112)	7	7	820	820	−5.2 2.0	Pull-Down Load	Open Emitter	0–70	No	16
Static NMOS 1024 Bit (2102A)	250–650	250–650	174–289	35	5	None	Three State	0–70	No	16
Static NMOS 4096 bit (Intel 2147H)	35, 45, 55, 70	35, 45, 55, 70	990, 880, 770	165, 110, 165	5	None	Three State	0–70 −55 to +125	No	18

Dynamic NMOS 16, 384 Bit (2118)	80, 100, 120, 150	215, 235, 270, 320	157, 137, 121, 121	22	5	128-Cycle, 2ms Refresh Circuitry	Three State Latched	0–70 −55 to +85	Yes	16
Dynamic NMOS 4096 Bit (MK4027)	120	300	500	0.4	12 5 −5	Circuitry for 64 cycle refresh, sense amplfr.	Three State (Output Data Latched)	0–70 −55 to +85	Yes	16
Static Bulk CMOS 1024 Bit (6508/6518A)	200–600	200	25–0.05		5	None	Three State	0–70	Yes	16/18
Static SOS CMOS 1024 × 4 Bit (MWS5114)	650	650	0.25	0.25	5	None	Three State	0–70	No	18
Static Bulk CMOS 1024 × 4 Bit (HM6514)	270, 320	270, 320	30	0.05	5	None	Three State	0–70 −55 to +125	No	18
Static Bulk CMOS 4096 × 1 (HM6147)	35, 45, 55, 70	35, 45, 55, 70	75	10 μW	5	None	Three State	0–70	No	18

a ECL = emitter-coupled logic.

tures. In addition, the major system performance requirements of higher density, higher speed, and lower power dissipation imposed on semiconductor memories due to their widespread use in microcomputers, minicomputers, remote terminals, automotive applications, electronic toys and games, and the like have virtually dictated that the inherent feature of CMOS memories are being specified for all semiconductor memories:

- Operation from a single +5-V power supply
- Low power dissipation, both active and standby
- Operation over wide temperature and power supply variations while maintaining high noise immunity
- Static operation without need for refresh and support circuitry

These system requirements, plus the circuit design advances made over the last several years and the adaptability of NMOS circuit scaling (device geometry shrinking) to CMOS technology, have led to CMOS memories that exhibit electrical performance characteristics superior to those exhibited by NMOS memories while competing aggressively in terms of density and cost. (Already the mask steps required to enhance the performance of NMOS memories are equivalent to those required for CMOS, which removes the objection that CMOS is a more expensive technology than NMOS.) Thus CMOS memories as well as CMOS LSI are at the crossroads of a new era in data storage, manipulation, and transfer and should enjoy wide usage.

WHAT IS A SEMICONDUCTOR MEMORY?

A semiconductor memory, as shown in Fig. 4-1, is a rectilinear pattern of data storage units called *cell bits,* each of which can be changed to and retain either of the two binary states: a high-level (logic 1) or a low-level (logic 0) voltage. Each cell is accessed for programming or reading by energizing the proper row and column. In practice the cells are operated in groups (words) which store information in binary-coded form. The content of cell groups can represent data such as numbers and letters or instructions for manipulating data.

There are two basic types of semiconductor memories: those that can be both programmed (data entered) and interrogated (data read) and those that are permanently or semipermanently programmed and

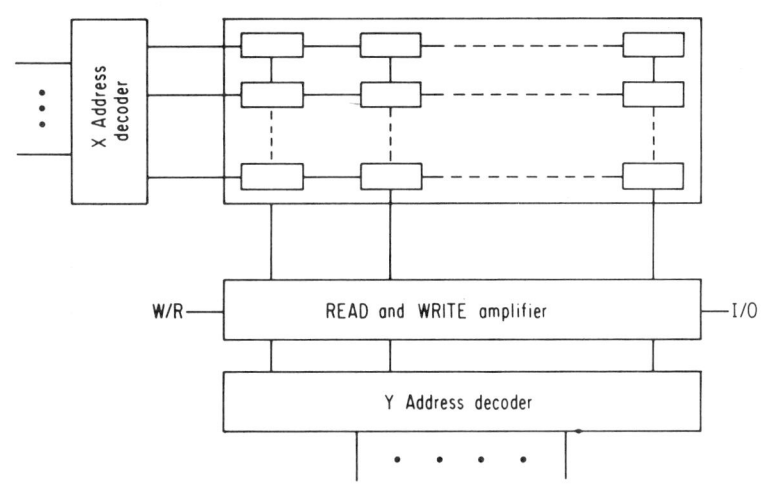

Fig. 4-1 Random-Access Read-Write Memory Block Diagram.

which, during system operation, can be interrogated or read only. The first type is called *read-write* or *RAM* because data can be written into or read from any location in memory without scanning. ROM also can be randomly accessed, but the acronym *RAM* is used to denote only read-write memory.

Since RAM memory cells by virtue of computer architecture must be handled in groups that are selectively accessible for reading and writing purposes, a selection and control system is built into the same chip as the cell matrix. The on-chip control circuitry configuration depends on memory cell structures, organization of the memory matrix, and the intended application. Although the control circuits (as well as the memory array and cell structure) vary from one memory type to the next, they all have several basic features: an address decoder consisting of a row decoder and a column decoder that selectively addresses (energizes) any specified word in memory, at least one CS (CHIP SELECT) or CE (CHIP ENABLE) input that permits the selective addressing of a particular memory chip from among several comprising a memory system, and a READ-WRITE input that permits setting up the proper condition for either reading a particular memory address or writing into the memory. (The READ-WRITE control circuit disables the READ circuit during a WRITE operation and vice versa). Figure 4-1 depicts the block diagram of a typical RAM.

196 CMOS Memories

CMOS MEMORY CELL ARCHITECTURE

RAMs made with MOS technology use two different techniques to store information: MOS RAMs are either static or dynamic. Dynamic MOS circuits make use of the very low leakage associated with the gate circuits of MOS devices. These leakage currents are small enough to permit the circuit's parasitic capacitance to exhibit time constants between milliseconds and seconds. The long time constants can be used to provide temporary storage (as in the three-transistor cell shown in Fig. 4-2), which can be made permanent (or semipermanent) by a charge replacement technique called *refresh*. Thus, the use of dynamic RAMs requires additional circuitry to perform the refresh operation.

Static memory cells retain their information as long as power is applied to the device without need for refresh. All commercially available CMOS memories are static in nature.

Many interrelated factors dictate system use of either static or dynamic memories. The design engineer must, therefore, know the requirements of the particular system and the advantages and disadvantages of each type of memory. The reader should refer to Eugene R. Hnatek, *A User's Handbook of Semiconductor Memories,* John Wiley & Sons, Inc., New York, 1977, for a discussion of static and dynamic memories.

Figure 4-3 displays the memory circuit cell structure for the semiconductor bipolar and MOS memory technologies currently in production for comparative purposes. As can be seen, the CMOS memory cell is very similar to the NMOS static RAM and bipolar TTL memory cells, since both use cross-coupled flip-flops as the data storage medium. The basic CMOS memory cell is static in nature and consists of six MOS transistors connected as a pair of cross-coupled inverters, with pass devices (transmission gates) for selection control, as shown in Fig. 4-4.

The six-transistor CMOS cell has large N-type and small P-type transis-

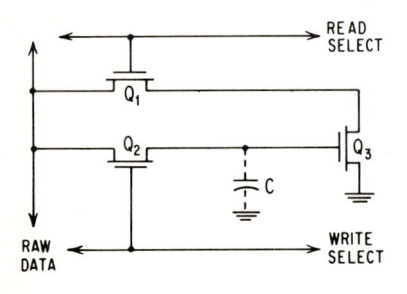

Fig. 4-2 Three-Transistor Dynamic Memory Cell Stores Information (Charge) at the Node Containing Parasitic Capacitor C.

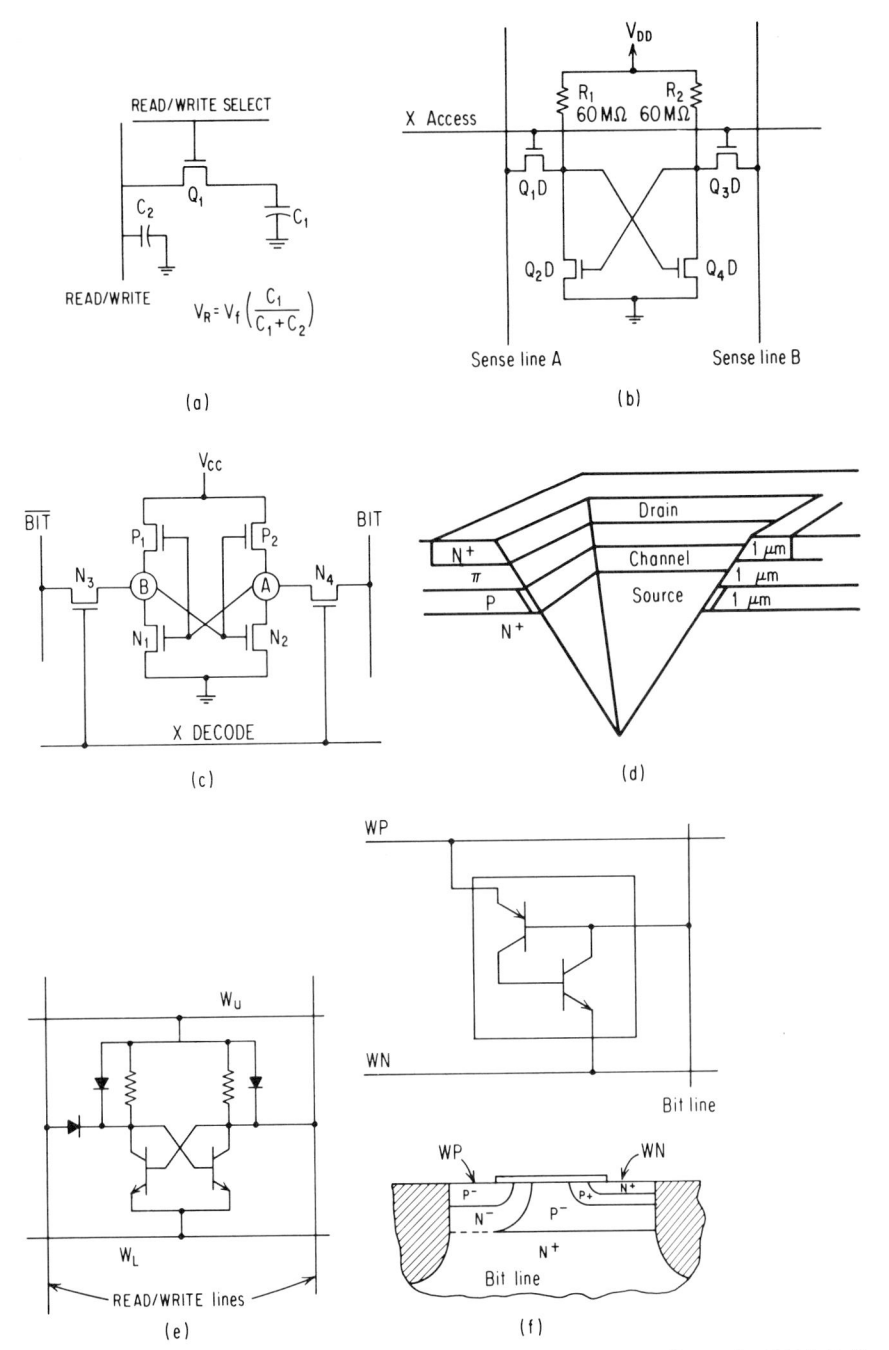

Fig. 4-3 RAM Cell Structures MOS: (a) The One-Transistor Dynamic RAM Cell (Area = 189 μm^2 − 0.697 mil^2); (b) Typical Static NMOS RAM Cell (Area = 0.5 − 2 mil^2); (c) CMOS Six-Transistor (IM6508) with No Buried Contacts (Area = 1.44 − 12 mil^2); (d) VMOS Transistor Cross Section, VMOS Cell. RAM Cell Structures, Bipolar (Area = 1.0 − 1.25 mil^2); (e) Schottky TTL (Cross-Coupled Flip-Flop) (Area = 2.3 − 6 mil^2); (f) Merged Transistor Memory Cell (Area = 0.68 mil^2).

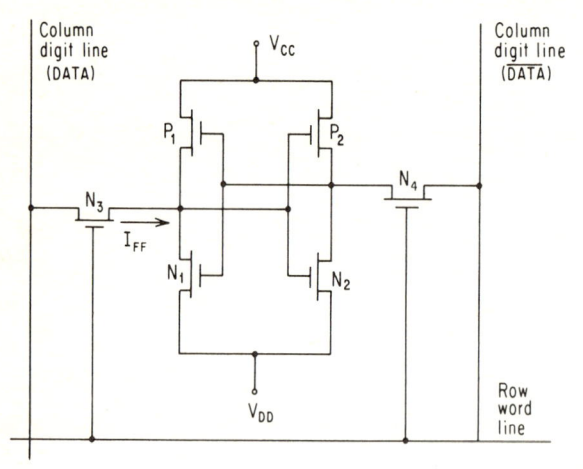

Fig. 4-4 Two CMOS Inverters, Cross-Coupled to Form a Flip-Flop, Comprise the Static Memory Cell.

tors. Such an arrangement prevents a high level on the DATA line or complementary $\overline{\text{DATA}}$ line from disturbing data stored in the cell.

If the pass transistors are N-channel devices (as shown in Figure 4-4), data stored in the flip-flop can more readily be affected by pulling down on a column line; the pass transistor then conducts in common-source mode. Pulling up has significantly less effect because of the weaker source-follower behavior of the pass device. In reading, both DATA and $\overline{\text{DATA}}$ lines are charged high, the cell is selected (decoded), and then the cell's contents are allowed to drive one or the other line low.

In short, the cell is read by a 0 being sensed and is written by a 0 being forced onto one of the lines. Writing into the cell therefore consists primarily of pulling down the required side of the flip-flop. Internal cell regeneration causes most of the rising voltage action on the other side. It thus follows that both DATA and $\overline{\text{DATA}}$ column lines must be held high during row address charge or data can be destroyed.

The typical CMOS inverter is fast and has low power consumption. Since P- and N-channel devices do not conduct simultaneously in the static state, the only power supply current that flows during this state supplies the leakage current in the off devices. This is typically on the order of 1 nA per flip-flop. The data stored is not degraded by this leakage, as is the case with data stored in the dynamic cell. Leakages can vary from cell to cell without resulting in system timing problems or device yield.

When the cell changes state, both inverters conduct for a few nanoseconds. This switching current is, therefore, duty-cycle-sensitive. In the language of the design engineer, a memory said to have a dissipation of so many milliwatts per bit is usually a static memory. Dynamic power consumption in a static memory is sometimes specified for a given cycle time.

There are two kinds of CMOS RAMs available—synchronous or edge-triggered (clocked) and asynchronous (nonclocked).* The synchronous parts exhibit lower operating power and faster access specifications, but require a clock edge to trigger memory operations. Asynchronous RAMs allow and recognize address changes at any time—that is, asynchronously. Thus they need only an address to access data, so that they are easier to use.

Synchronous memories require the address to be valid for only a set-up time period after an edge, while asynchronous RAMs require the address to be valid for the entire signal duration. Both parts can be enabled (during which time they draw what is called *operating current*) or disabled (when they need *standby current*). Because of their clocking schemes, asynchronous RAMs draw operating current as long as they are enabled (Fig. 4-5*a* and *c*), while synchronous statics draw only standby current (even when they are enabled) except at clock edges, when they draw the rated operating current (Fig. 4-5*b* and *c*). So the average operating power of the synchronous part goes up with operating frequency; that is, the user pays for power only when the memory is being used.

An example of an asynchronous RAM is the I5101 256- × 4-b device. The IM6508 is a typical synchronous memory with on-chip address registers, as are the National Semiconductor MM74C920, 921, 929, and 930 RAMs and the Harris/Intersil 6501 RAM. The RCA CD4061A 256 × 1-b RAM and the AMI S2222 512 × 1-b RAM also are synchronous CMOS devices, but without address registers.

Figure 4-6 shows the difference between synchronous and asynchronous designs architecturally—from the method of maintaining proper levels on the DATA and $\overline{\text{DATA}}$ lines. In the asynchronous design (Fig. 4-6a), a P-channel device on each bit line acts as a resistive load during READ operations. Here the P-channel loads are on while the cells are

---

* S. Hume, "Consider 1024-bit CMOS RAMs for Small Static Memory System," *Electronics*, July 24, 1975, pp. 102–105. Copyright © 1975 by McGraw-Hill, Inc. Reprinted with permission.

200 CMOS Memories

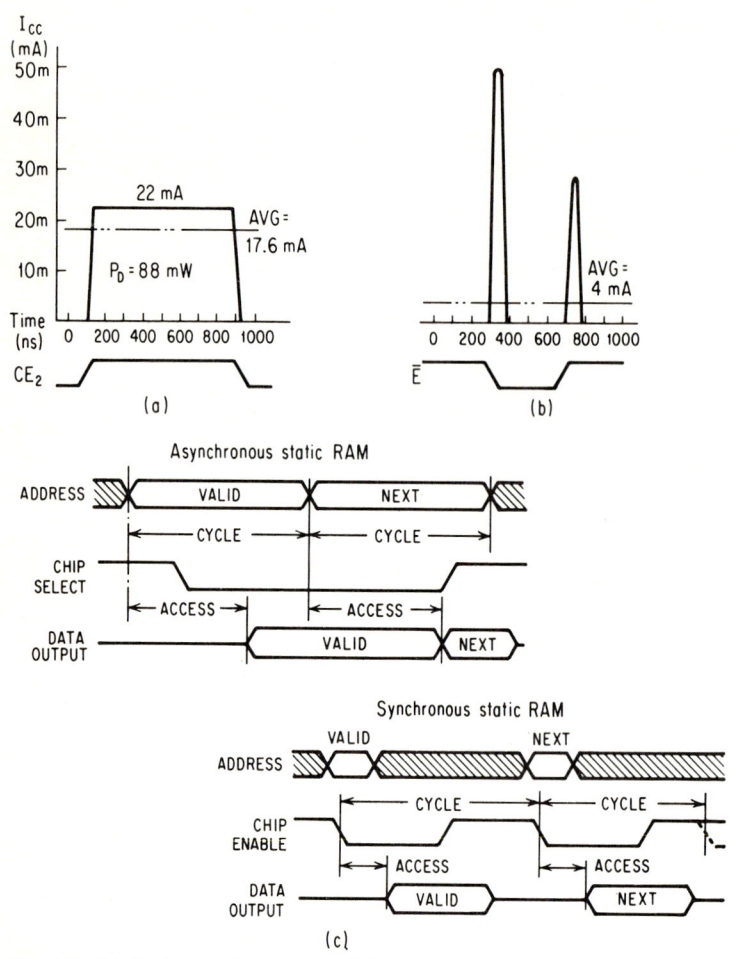

Fig. 4-5 Typical Asynchronous and Synchronous RAM Power Dissipation Curves and Timing Characteristics: (a) Power Dissipation Curve for Asynchronous 5101 Type, (b) Power Dissipation Curve for Synchronous 6501 Type, (c) Timing Characteristics.

being selected, creating a resistive path from the power supply to ground and dissipating about 10 mW of power. When the clock pulses, CS_1 is high, however, all row decoders are low, no cells are selected, and the memory dissipates only nanowatts.

The synchronous design (Fig. 4-6b) also exhibits the nanowatt quiescent power dissipation typical of CMOS memories in all states. A clock is used to disable the row decoders while the columns are being charged

CMOS Memory Cell Architecture

and internal addresses are changing. This arrangement stops cells from being selected while the columns are being charged. In addition, it makes the synchronous cell faster than the asynchronous one, partly because storage cells are not required to drive a resistive load in addition to the capacitive load on the column, and partly too because more sophisticated column sense amplifiers are possible.

Speed of operation is not the only consideration in choosing between synchronous and asynchronous designs. The fact is, while synchronous RAMs do not require on-chip address registers, these can be added with almost no penalty in the die area at significant benefit to the user. For example, most 12- and 16-b microprocessors must have multiplexed

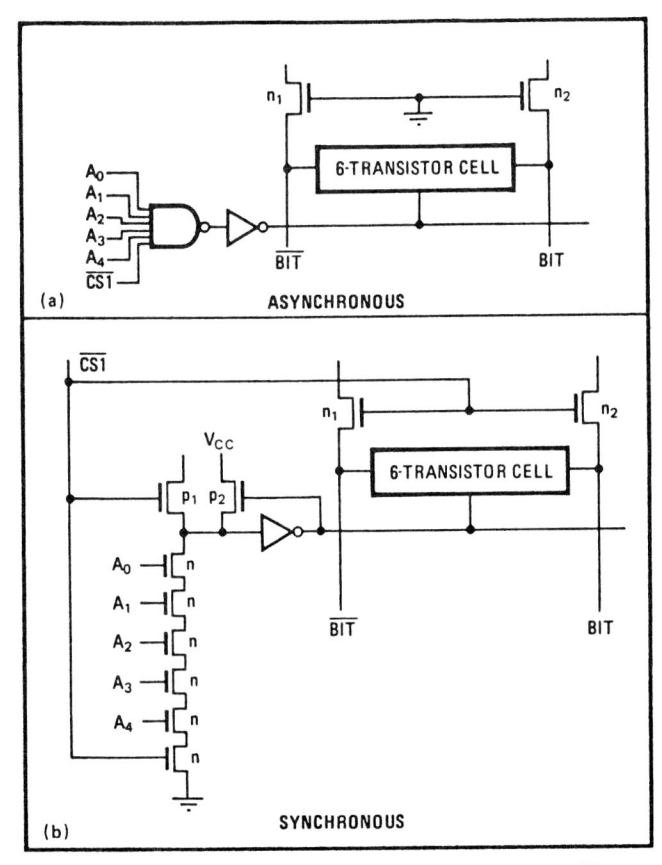

Fig. 4-6 Comparing the Synchronous and Asynchronous Memory Circuit Designs. *(Reprinted from Electronics, July 24, 1975. Copyright © 1975 by McGraw-Hill, Inc. Used with permission.)*

addresses and data on the same line. This requires either on-chip or external address registers.

On-chip address registers give the user more flexible timing, but they add significant propagation delay time to systems—30 ns or so. Since the clock for the synchronous design can be used for control and since double inversion of addresses is required for the decoders anyway, the register function can be implemented without a marked increase in die area. Of course, eliminating external address registers reduces parts count. Also, interconnections in microprocessor-based systems can be reduced by multiplexing addresses and data on the same bus.

MEMORY ORGANIZATION*

Random-Access Memories

A CMOS memory contains all decoding, READ-WRITE logic, and sense amplifier circuitry, in addition to the memory storage array. The static six-transistor memory cell illustrated in Fig. 4-4 repeats and interconnects to form a memory matrix of MxN words M bits by N words, 1 b long. This matrix, with a single representative cell and associated decoding, is shown in Fig. 4-7.

As an illustrative example showing the internal design of CMOS RAMs, Figs. 4-8 and 4-9 show the organization of a 256 × 4- and a 1024 × 1-b RAM, respectively.

The basic structures of the two designs are identical. Each of the designs has address registers controlled by the strobe terminal ($\overline{ST}$ for the 256 × 4-b memories, $\overline{CS_1}$ for the 1024 × 1-b memories). A high level at the strobe terminal allows the external address code to fall through the address register latches and reach the decoder inputs; the decoders, however, are inhibited from selecting any row or column while the $\overline{STROBE}$ input is high. When $\overline{STROBE}$ falls, the address code then contained by the address registers is latched and external code changes have no further effect. $\overline{STROBE}$ LOW simultaneously enables the decoders, causing the selected location(s) to be read, and sends the data to the output latch(es).

---

* K. Rapp, "1 K CMOS RAMs," Application Note in *Memory Applications Handbook*, National Semiconductor Corp., Santa Clara, CA, 1978. Used with permission.

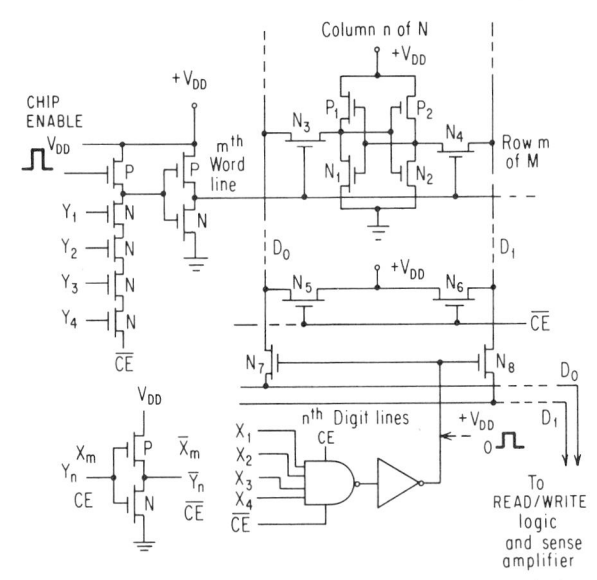

Fig. 4-7 Sixteen Combinations for X or Y Are Possible for This Single Cell in an $M \times N$ Matrix.

While $\overline{\text{STROBE}}$ is low, the data falls through the DATA OUT latch to the Tri-State® buffer input. When $\overline{\text{STROBE}}$ next goes high, the DATA OUT latch locks and stores the data. Note from Figs. 4-8 and 4-9 that $\overline{\text{STROBE}}$ has no control over the Tri-State® buffer. Only the other 2-chip selects and the WRITE ENABLE terminal control it. In WRITE mode (WRITE ENABLE LOW), the output is disabled even though the 2-chip selects are active (low).

Writing into either of the memory types requires that $\overline{\text{STROBE}}$, the other two CHIP SELECTS, and the WRITE ENABLE all be low, i.e., the chip must be selected and WRITE ENABLE be low. In the READ mode (WRITE ENABLE HIGH), however, a chip need not be selected completely for new data to be stored in the DATA OUT latch. Each chip in a system will react as its strobe terminal is brought low, independent of the status of its other CHIP SELECT terminals. New data corresponding to the bus address code will be fetched and stored in the DATA OUT latches. The other CHIP SELECTS choose which chip delivers its data to the data bus. PAGE MODE memory organization thus is available.

The only difference between the two memory types shown in Figs. 4-8

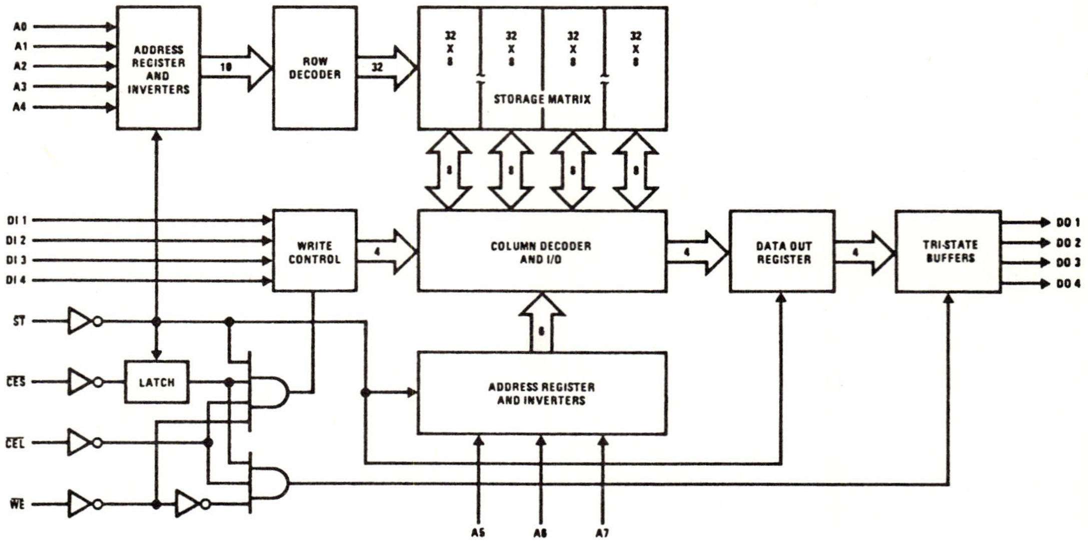

Fig. 4-8 MM54C920/MM74C920 Logic Diagram. *(Courtesy National Semiconductor Corp.)*

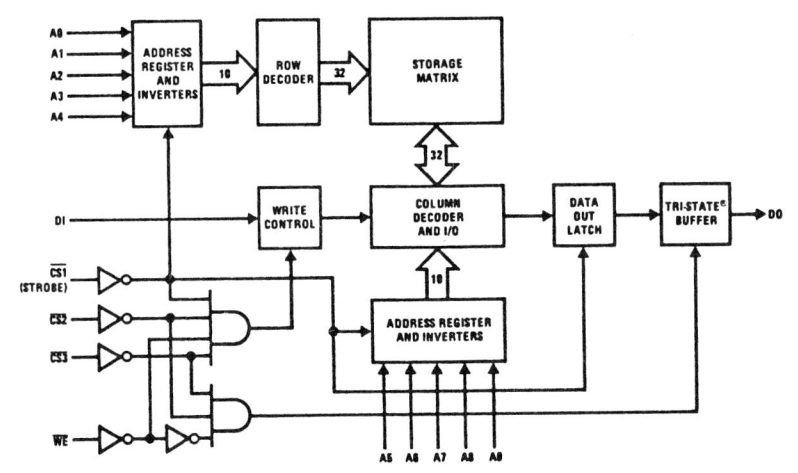

Fig. 4-9 MM54C930 1024 × 1-b RAM Logic Diagram. *(Courtesy National Semiconductor Corp.)*

and 4-9 aside from the word organization lies in the CHIP SELECT input circuits. One CHIP SELECT of the 256 × 4-b RAMs is latched by $\overline{\text{STROBE}}$ exactly as addresses are latched. $\overline{\text{CES}}$ (CHIP ENABLE, STROBED) thus must be established at the proper level prior to the fall of $\overline{\text{STROBE}}$. As with address inputs, set-up and hold time requirements must be met by $\overline{\text{CES}}$ with respect to the negative-going edge of $\overline{\text{STROBE}}$. Once latched by $\overline{\text{STROBE}}$, the captured $\overline{\text{CES}}$ level remains constant until the end of the cycle, i.e., until $\overline{\text{STROBE}}$ again falls at the beginning of the next cycle. The $\overline{\text{CEL}}$ (CHIP ENABLE, LEVEL) control is not latchable. It functions completely independently of $\overline{\text{STROBE}}$. CHIP SELECTS CS_2 and CS_3 on the 1024 × 1-b RAMs both are level-actuated and operate independently of $\overline{\text{STROBE}}$ $\overline{CS_1}$.

Figure 4-10 illustrates how each memory cell fits into the ROW/COLUMN and READ/WRITE selection structure. The memory cells are laid out in a 32 × 32 matrix. Hence, the ROW SELECT signals have a 1-of-32 code; the selected row goes high; the other rows remain low. The pass transistors N_1 and N_2 of all cells on the selected row then conduct. Similarly, selected COLUMN lines go high. The select code is 4-of-32 for the 256 × 4-b memories and 1-of-32 for the 1024 × 1-b memories. A HIGH COLUMN SELECT signal turns on both the WRITE access transistors N_3 and N_4 and the column sense amplifier. If the chip is in WRITE mode, the WRITE circuits pull either the WRITE 1 or WRITE

206 CMOS Memories

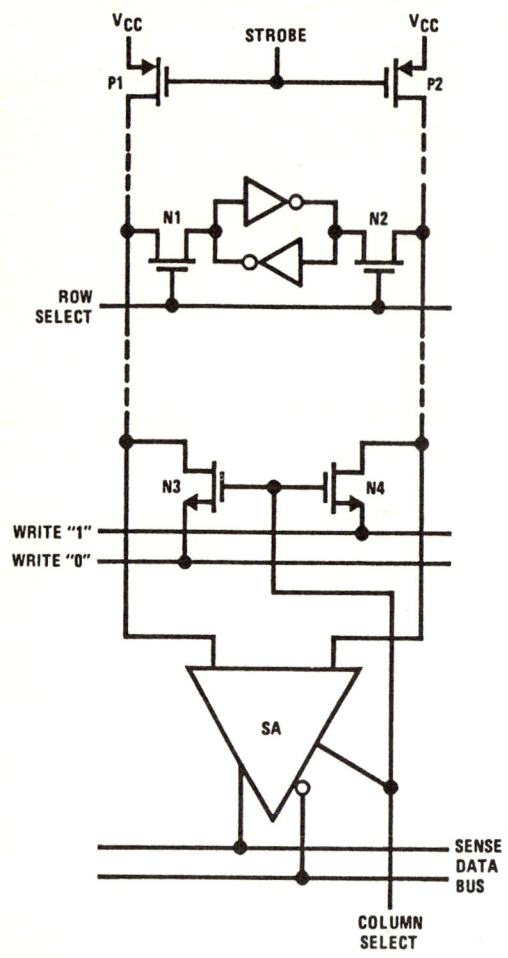

Fig. 4-10 Row/Column Read-Write RAM Structure. *(Courtesy National Semiconductor Corp.)*

0 bus low; the other one floats. In READ mode, both buses float. During selection, column pull-up transistors P_1 and P_2 are turned off (STROBE is high).

ADDRESS DECODERS

Row and column address decoders are logically depicted as shown in Fig. 4-11. Note that $\overline{\text{STROBE}}$ must be low for the output of either decoder type to go high. Address bits A_8 and A_9 are used only for the 1024 × 1-b memories.

Memory Organization 207

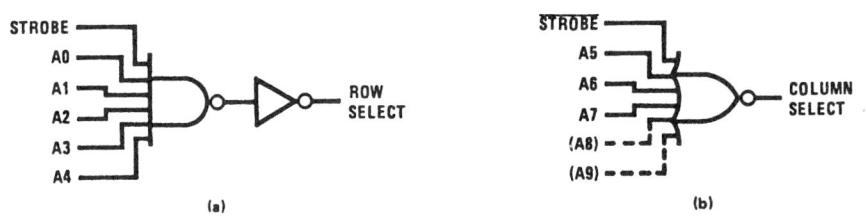

Fig. 4-11 Address Decoders. *(Courtesy National Semiconductor Corp.)*

ADDRESS LATCHES

Address registers comprise fallthrough latches of the type shown in Fig. 4-12. When $\overline{\text{STROBE}}$ is high, the external address code is transmitted directly through to $\overline{A}_K$ and A_K. $\overline{\text{STROBE}}$ latches the code by going low.

$\overline{\text{CES}}$ LATCH

The $\overline{\text{CES}}$ latchable CHIP SELECT on the 256 × 4-b RAMs is implemented using two latches of the type shown in Fig. 4-12 to produce a dual-rank D flip-flop. Positive $\overline{\text{STROBE}}$ transitions thus do not disturb the stored state for $\overline{\text{CES}}$.

WRITE CONTROL LOGIC

Figure 4-13 presents the logic used to control writing. Note that pass transistors N_1 and N_2 conduct only when all input controls are low, thus placing the chip in WRITE mode.

SENSE AMPLIFIER

The high-speed sense amplifier shown in Fig. 4-14 consists of differential amplifier pairs at the end of each column, all connected via the sense data bus to a single pair of load devices, L_1 and L_2. When $\overline{\text{STROBE}}$ falls, current flows in the selected column differential amplifier, causing an amplified voltage differential between the two lines of the sense data bus.

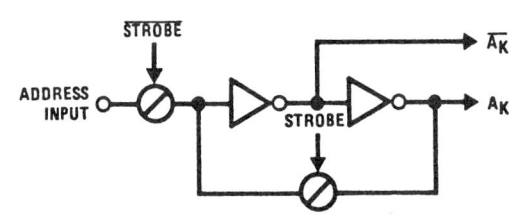

Fig. 4-12 Address Latch. *(Courtesy National Semiconductor Corp.)*

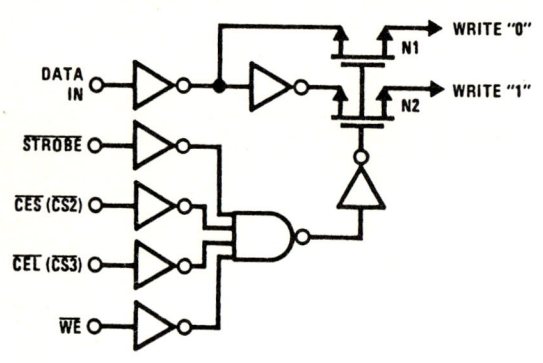

Fig. 4-13 Write Control Logic. *(Courtesy National Semiconductor Corp.)*

DATA LATCH AND OUTPUT BUFFER

Figure 4-15 depicts the fallthrough OUTPUT DATA latch and the Tri-State buffer. When the Tri-State control is low, both P_1 and N_1 turn off and the output floats.

MEMORY TIMING*

In applying any memory the timing, relationships among the various clocks must be understood and adhered to. Consequently this topic is extremely important. The timing diagrams for the MM74C920 and MM74C921 256 × 4-b CMOS RAMs shown in Fig. 4-16 are used as illustrative examples.

Set-up and hold time requirements with respect to $\overline{\text{STROBE}}$ must be met by all address inputs and by $\overline{\text{CES}}$. A minimum downtime is specified

---
* See the article by K. Rapp already cited in reference to the section "Memory Organization" above.

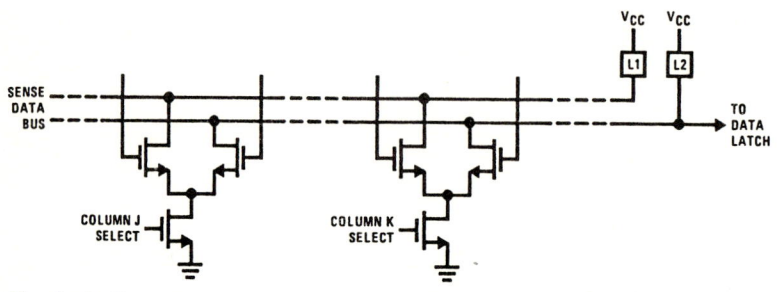

Fig. 4-14 Sense Amplifier. *(Courtesy National Semiconductor Corp.)*

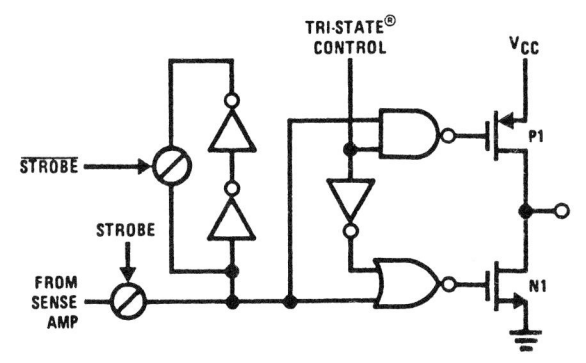

Fig. 4-15 Output Data Latch-Buffer. *(Courtesy National Semiconductor Corp.)*

for $\overline{\text{STROBE}}$ to ensure that new data reach the output latch during READ and to provide sufficient time to reliably enter new data during WRITE. A minimum $\overline{\text{STROBE}}$ uptime is specified to allow for pulling up columns and for precharging sense amplifiers. The sum of these two $\overline{\text{STROBE}}$ time specifications establishes minimum cycle time.

READ Cycle

Access time is specified both from address change and from the falling edge of $\overline{\text{STROBE}}$. Access from address change is the more meaningful measure of the speed with which new data can be fetched. It is important, however, to recognize that access from $\overline{\text{STROBE}}$ will limit if more than minimum address setup time is provided.

Note that the asynchronous character of all auxilliary chip select controls except $\overline{\text{CES}}$ permits them to fall after $\overline{\text{STROBE}}$. Provided that the selection occurs at least t_{OE} (output buffer enable time) before new data are made internally available by $\overline{\text{STROBE}}$, no effect on access time occurs.

WRITE Cycle

WRITE circuits on a chip are activated only when $\overline{\text{STROBE}}$, the other CHIP SELECTs, and WRITE ENABLE are all low. To satisfy the specified minimum write-pulse width, then, these signals must be jointly low for at least the specified t_{WP} (WRITE pulse width, negative).

The first of the above control signals to move high ends the WRITE operation. Input data must be valid a minimum set-up time t_{DS} before the first positive moving control-signal edge and must be maintained a minimum hold time t_{DH} beyond that edge. For successive WRITE cycles,

210 CMOS Memories

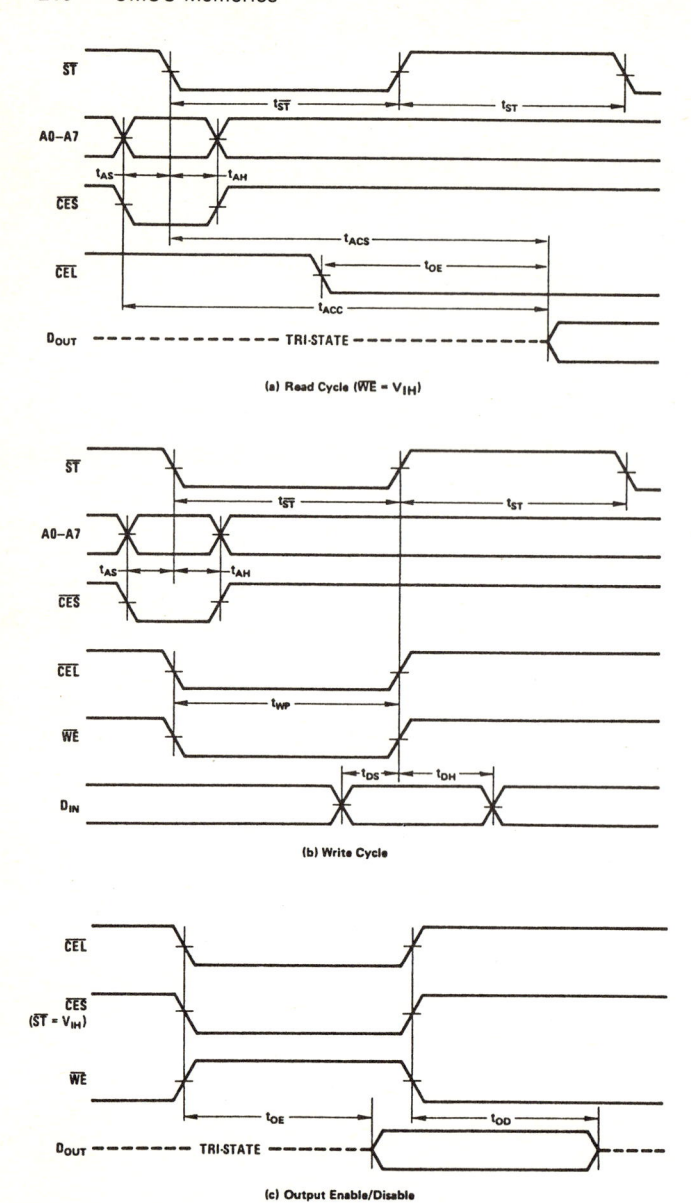

Fig. 4-16 MM74C920/MM74C921 Switching Time Waveforms. *(Courtesy National Semiconductor Corp.)*

note that $\overline{WE}$ and/or the auxiliary CHIP SELECTS may be kept low. STROBE alone then would establish the write period.

Disabling of the output buffer when $\overline{WE}$ (WRITE-ENABLE) is low is useful for common data I/O applications. These CMOS memories have been designed to keep their outputs disabled when $\overline{WE}$ and the auxiliary CHIP SELECTS simultaneously go low. Note, however, that if the CHIP SELECTs fall prior to $\overline{WE}$, the outputs can become briefly enabled until the falling $\overline{WE}$ signal takes control. Conflict between external data drivers and the on-chip data buffers could occur during this interval, causing unexpected data bus transients. No damage to the CMOS RAMs will occur, however.

Read-Only Memory

Figure 4-17 shows the block diagram for a 1024-b mask-programmable CMOS ROM. This device is organized as 256×4 b. Address contents (A_0 through A_7) appear at the four data outputs (B_0, B_1, B_2, and B_3) following the negative-going edge of the clock. When the clock goes high, data present at the outputs are latched. The MEMORY ENABLE may be taken low asynchronously, forcing the data outputs low and resetting the output latches.

The 32 data outputs of the memory matrix must be decoded to provide the required 4-b output word. Three input address lines (A_0, A_1,

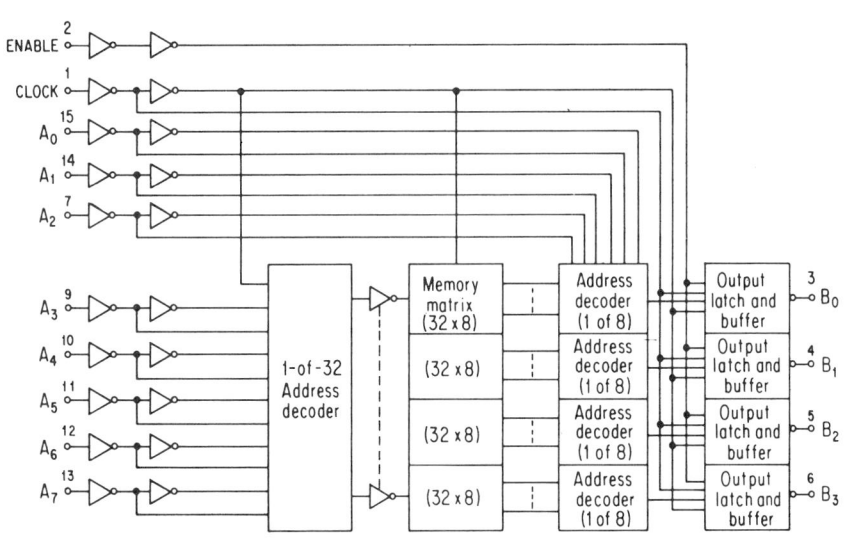

Fig. 4-17 Organization of 1024-b CMOS ROM. *(Courtesy Motorola, Inc., Semiconductor Products Division.)*

and A_2) select which of each group of eight matrix data lines is to be enabled, the remaining lines going to the decoder.

The contents of the memory are determined and fixed during the manufacture of the device.

POPULAR CMOS RAMs

CMOS RAMs substantially reduce the problem of memory volatility because of their extremely low standby power. Data can be kept alive for 3 to 4 months with a standard AA, 3.5-V nickel-cadmium battery for a 2-kb × 8-b memory array. Battery back-up can also be used with static and dynamic NMOS RAMs, but the CMOS standby power drain is far less.

The comparison of CMOS RAMs with bipolar and NMOS RAMs in Table 4-1 showed the ability of CMOS RAMs to operate over wide extremes of temperature and power supply variations while maintaining a high noise immunity and comparable operating speeds. For example, a typical operating range for memories is 3–15 V. Thus, CMOS memory array devices are compatible with the popular bipolar transistor logic circuits without additional interfacing circuits.

Economies in CMOS system design are achieved because the system can operate on a single, noncritical, power supply voltage. Furthermore, static memories, i.e., those that require no refresh memory, also require no high-power clock or preconditioning pulses for proper operation. System noise problems diminish because large transient currents are not developed. System complexity can be less than that required for other semiconductor technologies.

CMOS RAMs are commercially available in densities from 64 to 16,384 b and in a variety of organizations, as summarized in Table 4-2; the most popular organizations are the 256 × 4-, 1024 × 4-, 4096 × 1-, and 2048 × 8-b types. It is worth noting that the 1024 × 4-CMOS types are pin- and functionally compatible with Intel's 2114 NMOS RAM, while the 4096 × 1-b CMOS RAMs are compatible with the Intel 2147 NMOS RAM. This provides system interchangeability to the low-power CMOS version without redesign, or degradation of operating speed—a significant point. The trend of semiconductor RAMs toward byte organization will see many ×8 configurations in the future (256 × 8, 512 × 8, and 1024 × 8 b, for example) in addition to the available 2048 × 8-b types.

The Intersil 6508/6518 1-kb RAM

The IM6508/6518 is a fully decoded and buffered CMOS silicon gate 1024-b static RAM arranged in a 32 × 32 array using the standard six-transistor cell (Fig. 4-18).

It has TTL-compatible inputs and outputs and works directly with bus-oriented microprocessors without additional power supplies or interfaces. It is a pin-for-pin replacement for existing RAMs such as the bipolar TTL 93415.

The IM6508 has on-chip address registers controlled by the CHIP ENABLE line, and is packaged in a 16-pin DIP. A variation, the IM6518, comes in an 18-pin DIP and has three chip enables—two for WRITE ENABLE and output buffering and one for address registers (see Fig. 4-19).

Power is only 5 μW total in standby and 10 mW operating at 1 MHz. Maximum access speed at 5-V V_{CC} is 200 ns. The 6508/18 is available with access times of 200, 300, 350, 460, and 600 ns, while dissipating 0.1 mA, 0.1 mA, 0.5 mA, 0.1 mA, and 0.1 mA respectively.

The operating range is from 4 V ≤ V_{CC} ≤ 11 V for the A version and 5V ± 10% for the basic version. The CHIP ENABLES and WRITE ENABLE are active in the low state and DATA OUT is of the same polarity as DATA IN. Operation of the IM6508 (16-pin version) is equivalent to the IM6518 (18-pin version) with CS_1, CS_2, and CS_3 tied together (Fig. 4-19).

Three CHIP ENABLES (18-pin version) are provided. CS_2 and CS_3 are conventional CHIP ENABLES and disable the internal WRITE circuitry and the output buffer. CS_1 disables internal WRITE circuitry and acts as a clock for the internal address register, but does not disable the output. Addresses are recognized by the IM6508/18 on the negative transition of CS_1.

The Intel 5101 256 × 4-b RAM

The Intel 5101/5101L silicon gate CMOS static RAM combines high density and low power with a fast, fully static 256 × 4-b modular organization that eliminates clocks, interface circuits, and special power supply requirements while minimizing package count. Since the 5101 is fully static, chip enable clocking is not required.

The data is read out nondestructively and has the same polarity as the input data. All inputs and outputs are directly TTL-compatible. The 5101 and 5101L have separate data input and data output terminals. An

TABLE 4-2 CMOS RAM Summary

Memory Size (b)	RCA	AMI	Intel	NSC	Motorola	Fairchild
4 × 8	CD 4036A CD 4039A					
8 × 8	CDP 1826					
16 × 4	CD 40114B			MM54C89		F4710 F4725
64 × 1					MCM14505	
64 × 4					MCM14552	
32 × 8	CD40024 CDP1824D/CD					
256 × 1	CD4061A CD40061A			MM54C200 MM54C910	MCM14537	F4720
512 × 1		S2222/A				
64 × 12						
128 × 8	CDP1823[c]					
256 × 4	CDP1822		5101	MM54C920 MM54C921	MCM145101	
512 × 4						
1024 × 1	CDP1821[a]			MM54C924 MM54C930	MCM146508 MCM146518	
2048 × 1						
1024 × 4	CDP1825 MWS5114[a]			NMC6514		
4096 × 1	MWS5104[a]	S5104		NMC6504	MCM146504	
2048 × 8						
16,384 × 1						

[a] CMOS/SOS
[b] With on-chip registers.
[c] 1800 series devices have input latches for interface with 1802 μP, 10-V power supply range; and wider temperature range.

Popular CMOS RAMs

Harris	NEC	Intersil	Hughes	Hitachi	Toshiba	Fujitsu	OKI
			HCMP1824				
HM6512		IM6512/A					
			HCMP1823				
HM6501 HM6551 HM6561 HM6562	5101	IM6551/A IM6561/A	HCMP1822	HM43101			
HM6513							
HM6508 HM6518	UPD443/ 6508	IM6508/A IM6518/A					
HM6503							
HM6514 HM6533	UPD444/6514 UPD445	IM6514	HCMP1825	HM6148 HM4334	TMM5047 TMM5514	MB8414	MSM5114 MSM5115[b]
HM6504 HM6543		IM6504		HM6147P HM4315	TMM5504	MB8404	MSM5104
HM6516				HM6116	TC5516P		
				HM6167			

[a] CMOS/SOS
[b] With on-chip registers.
[c] 1800 series devices have input latches for interface with 1802 μP, 10-V power supply range; and wider temperature range.

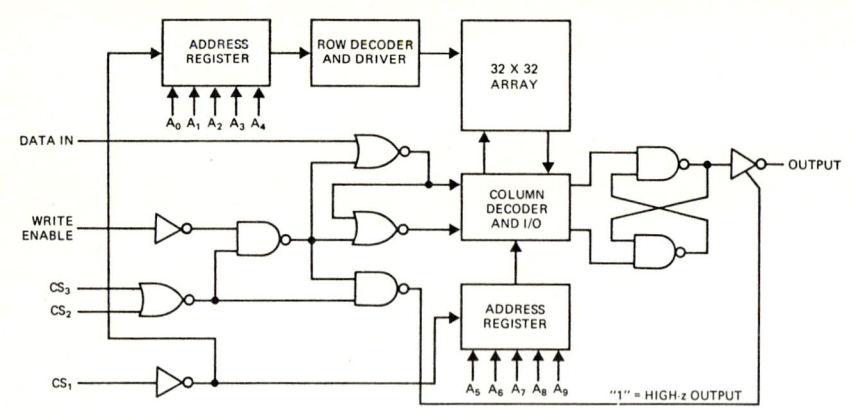

Fig. 4-18 IM6508/6518 Functional Diagram. *(Courtesy Intersil.)*

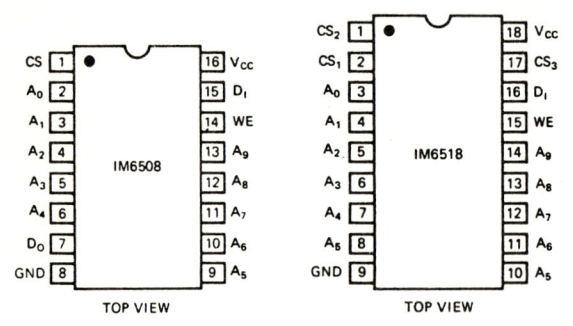

Fig. 4-19 IM6508/6518 Pin Connection Diagrams. *(Courtesy Intersil.)*

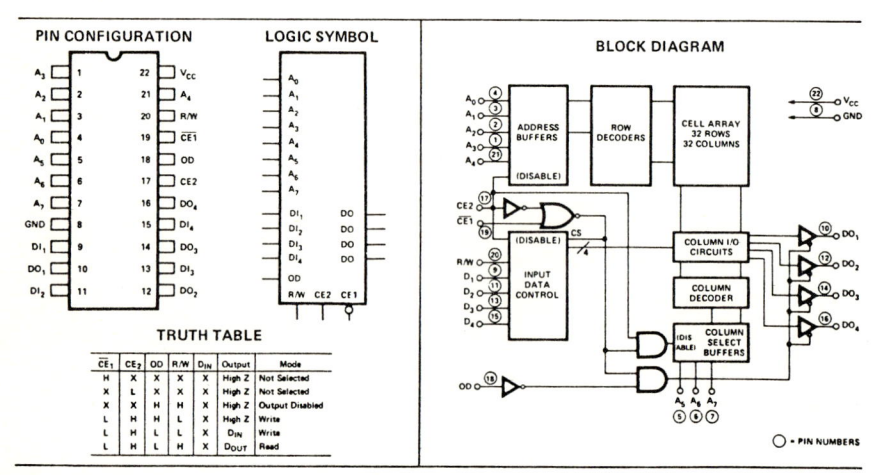

Fig. 4-20 Block and Pin Connection Diagrams for the 5101/5101L 1-kb RAMS. *(Courtesy Intel.)*

output disable function is provided so that the data inputs and outputs may be wire-ORed for use in common data I/O systems.

With the 256 × 4-b configuration, one gets two CHIP ENABLE inputs, four DATA inputs, four three-state outputs with output DISABLE control and READ/WRITE control (see Fig. 4-20) per package. The output DISABLE pin controls the bus state and makes bidirectional logic unnecessary in common I/O buses. The 5101L is identical to the 5101 except that it also has guaranteed data retention at a power supply voltage of as low as 2.0 V.

Minimum standby current is drawn by these devices when CE_2 is at a low level. When deselected, the 5101 and 5101L draw only 15 μA from the single 5-V supply. These devices are ideally suited for low-power applications where battery operation or battery back-up for nonvolatility are required.

The 5101/5101L comes in a 22-pin DIP. At 70°C, the maximum standby current is 15 nA/b, limiting standby power to 75 nW/b. Worst-case access time (and minimum cycle time) is 650 ns over the 0 to 70°C temperature range. However, several different speed and power dissipation versions are available to the user.

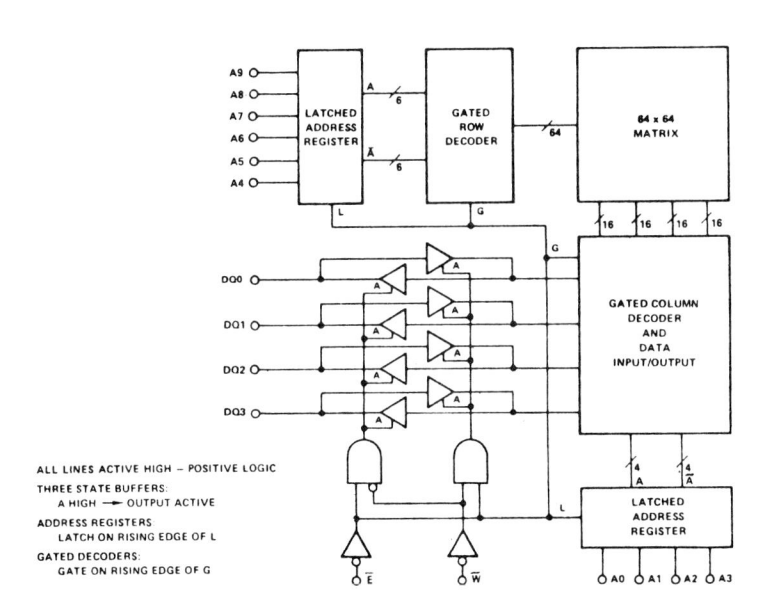

Fig. 4-21 HM6514 Block Diagram. *(Courtesy Harris Semiconductor Products Division.)*

218 CMOS Memories

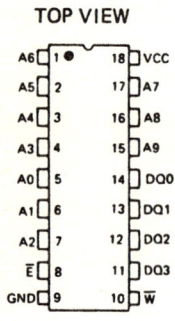

TOP VIEW

```
A6  [ 1●   18 ] VCC
A5  [ 2    17 ] A7
A4  [ 3    16 ] A8
A3  [ 4    15 ] A9
A0  [ 5    14 ] DQ0
A1  [ 6    13 ] DQ1
A2  [ 7    12 ] DQ2
Ē   [ 8    11 ] DQ3
GND [ 9    10 ] W̄
```

Fig. 4-22 HM6514 Pin Connection Diagram. *(Courtesy Harris Semiconductor Products Division.)*

The Harris Semiconductor HM6514 1024 × 4-b RAM

The HM6514 is a 1024 × 4-b static CMOS RAM that utilizes synchronous circuitry to achieve high-performance and low-power operation. A block diagram illustrating it is shown in Fig. 4-21. Data is retained to 2.0 V min.

On-chip latches are provided for the addresses, allowing efficient interfacing with microprocessor systems. The data output can be forced to a high-impedance state for use in expanded memory systems.

The HM6514 is encased in an 18-lead DIP (Fig. 4-22) and is completely pin-compatible with the industry standard Intel 2114 1024 × 4-b NMOS static RAM.

For a READ cycle (Fig. 4-23) the address information is latched in the on-chip registers on the falling edge of $\overline{E}$ ($T = 0$). Minimum address set-up and hold time requirements must be met. After the required hold time, the addresses may change state without affecting device operation. During time $T = 1$ the outputs become enabled, but data is not valid until time $T = 2$.

$\overline{W}$ must remain high throughout the READ cycle. After the data has been read, $\overline{E}$ may return high ($T = 3$). This will force the output buffers into a high-impedance mode at time $T = 4$. The memory is now ready for the next cycle.

The WRITE cycle (Fig. 4-24) is initiated on the falling edge of $\overline{E}$ ($T = 0$), which latches the address information in on-chip registers. If a dedicated WRITE cycle is to be performed and the outputs are not to become active, TWLEL and TEHWH must be met. Under these conditions, TWLDV is unnecessary and input data may be applied at any convenient time as long as TDVWH is still met. If TWLEL is not met, then the outputs may become enabled momentarily near the beginning of the cycle and a disable time (TELQZ) must be met before the input data is applied (TWLQZ = TWLDV). Similarly, if TEHWH is not met, the outputs may enable briefly near the end of the cycle.

The WRITE operation is terminated by the first rising edge of $\overline{W}$ (T = 2) or $\overline{E}$ (T = 3). After the minimum required $\overline{E}$ high time (TEHEL), the next cycle may begin. If a series of consecutive WRITE cycles are to be performed, the $\overline{W}$ line may be held low until all desired locations have been written. In that case, data set-up and hold times must be referenced to the rising edge of $\overline{E}$.

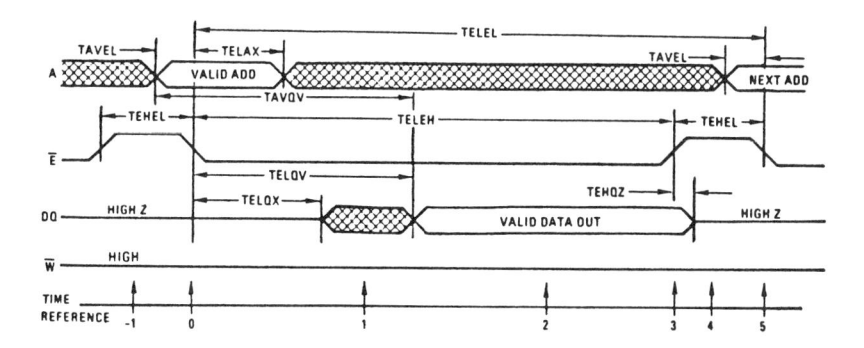

TRUTH TABLE

Time Reference	$\overline{E}$	Inputs $\overline{W}$	A	Data I/O DQ	Function
−1	H	X	X	Z	Memory disabled
0	⇘	H	V	Z	Cycle begins, addresses are latched
1	L	H	X	X	Output enabled
2	L	H	X	V	Output valid
3	⇗	H	X	V	READ accomplished
4	H	X	X	Z	Prepare for next cycle (same as −1)
5	⇘	H	V	Z	Cycle ends, next cycle begins (same as 0)

Fig. 4-23 HM6514 READ Cycle Timing Diagram. *(Courtesy Harris Semiconductor Products Division.)*

220 CMOS Memories

If the pulse width of $\overline{W}$ is relatively short in relation to that of $\overline{E}$, a combination READ-WRITE cycle may be performed (Fig. 4-25). If $\overline{W}$ remains high for the first part of the cycle, the outputs will become active during time $T = 1$. Data out will be valid during time $T = 2$. After the data is read, $\overline{W}$ can go low. After minimum TWLWH, $\overline{W}$ may return high. The information just written may now be read or $\overline{E}$ may return high, disabling the output buffers and preparing the device for the next cycle. Any number or sequence of READ-WRITE operations may be performed while $\overline{E}$ is low providing all timing requirements are met.

The guaranteed electrical characteristics for the primary device are shown in Table 4-3.

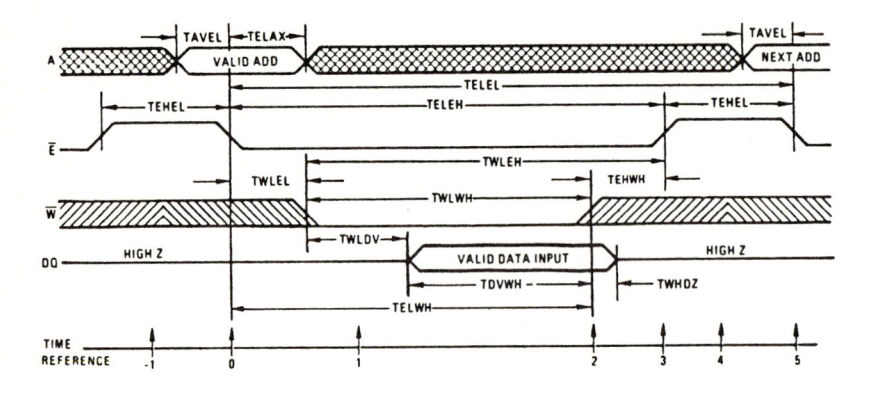

TRUTH TABLE

Time Reference	$\overline{E}$	Inputs $\overline{W}$	A	DQ	Function
−1	H	X	X	Z	Memory disabled
0	↘	X	V	Z	Cycle begins, addresses are latched
1	L	L	X	Z	Write period begins
2	L	↗	X	V	DATA IN is written
3	↗	H	X	Z	WRITE completed
4	H	X	X	Z	Prepare for next cycle (same as −1)
5	↘	X	V	Z	Cycle ends, next cycle begins (same as 0)

Fig. 4-24 HM6514 WRITE Cycle Timing Diagram. *(Courtesy Harris Semiconductor Products Division.)*

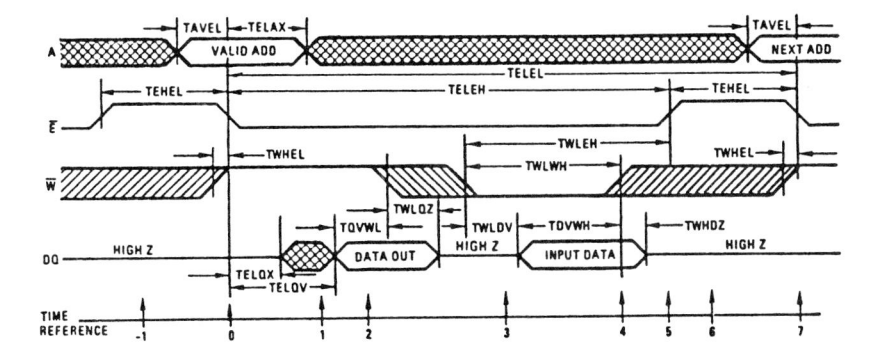

TRUTH TABLE

Time Reference	Inputs $\bar{E}$	$\bar{W}$	A	Data I/O DQ	Function
−1	H	X	X	Z	Memory disabled
0	↘	H	V	Z	Cycle begins, addresses are latched
1	L	H	X	X	READ mode, output enabled
2	L	H	X	V	READ mode, output valid
3	L	L	X	Z	WRITE mode, output high Z
4	L	↗	X	V	WRITE mode, data is written
5	↗	H	X	Z	WRITE completed
6	H	X	X	Z	Prepare for next cycle (same as −1)
7	↘	H	V	Z	Cycle ends, next cycle begins (same as 0)

Fig. 4-25 HM6514 READ-MODIFY-WRITE Cycle Timing Diagram. *(Courtesy Harris Semiconductor Products Division.)*

The Intersil IM6504 4096 × 1-b RAM

The IM6504 is a fully static, synchronous 4096-b CMOS RAM with on-chip address registers. Being synchronous enables the RAM to use faster and more sophisticated sense amplifiers, which means less power dissipation and higher speed than with asynchronous devices. All internal nodes are precharged to a known value before each memory access, which makes access time impervious to address and data patterns.

Address registers are necessary for interfacing the memory chip to microprocessors such as the Intersil IM6100, the Intel 8748, 8048, and

8035, and the Texas Instruments TMS9900, all of which multiplex addresses and data. Having them on-chip simplifies system timing requirements because addresses need only be present for a short hold time (typically 75 ns) following the falling edge of the address strobe. Not only that, but putting registers on-chip saves the two to four IC packages that off-chip address registers would have needed.

The IM6504 dissipates 75 μW at most in a 3-V standby mode, which translates into a substantial cost savings, especially if battery back-up

TABLE 4-3 HM6514 Electrical Characteristics

Symbol	Parameter	Temp. and V_{cc} Operating Range Min	Max	Temp. = 25°C,[a] V_{cc} = 5.0 V Min	Typ	Max	Units	Test Conditions
Direct Current								
ICCSB	Standby Supply Current		50		0.1	10	μA	IO = 0 VI = V_{cc} or GND
ICCOP	Operating Supply Current[b]		7		5	6	mA	f = 1 MHz, IO = 0 VI = V_{cc} or GND
ICCDR	Data Retention Supply Current		25		0.01	5	μA	IO = 0 VCC = 3.0 VI = V_{cc} or GND
VCCDR	Data Retention Supply Voltage	2.0		2.0	1.4		V	
II	Input Leakage Current	−1.0	+1.0	−0.5	0.0	+0.5	μA	GND ≤ VI ≤ V_{cc}
IIOZ	Input/Output Leakage Current	−1.0	+1.0	−0.5	0.0	+0.5	μA	GND ≤ VO ≤ V_{cc}
VIL	Input Low Voltage	−0.3	0.8	−0.3	2.0	1.5	V	
VIH	Input High Voltage	V_{cc} = −2.0	V_{cc} = +0.3	2.5	2.0	5.3	V	
VOL	Output Low Voltage		0.45		0.35	0.4	V	IO = 2.0 mA
VOH	Output High Voltage	2.4		3.5	4.0		V	IO = −1.0 mA
CI	Input Capacitance[c]		8.0		5.0	8.0	pF	VI = V_{cc} or GND f = 1 MHz
CIO	Input/Output Capacitance[c]		10.0		6.0	10.0	pF	VO = V_{cc} or GND f = 1 MHz
Alternating Current								
TELQV	CHIP ENABLE Access Time		300		170	250	ns	[d]
TAVQV	Address Access Time		320		170	270	ns	[d]
TELQX	CHIP ENABLE Output Enable Time		100		50	80	ns	[d]
TWLQZ	WRITE ENABLE Output Disable Time		100		50	80	ns	[d]

Popular CMOS RAMs 223

TABLE 4-3 HM6514 Electrical Characteristics (*Continued*)

Symbol	Parameter	Temp. and V_{CC} Operating Range Min	Max	Temp. = 25°C,[a] V_{CC} = 5.0 V Min	Typ	Max	Units	Test Conditions
			Alternating Current					
TEHQZ	CHIP ENABLE Output Disable Time		100		50	80	ns	d
TELEH	CHIP ENABLE Pulse Negative Width	300		250	170		ns	d
TEHEL	CHIP ENABLE Pulse Positive Width	120		100	70		ns	d
TAVEL	Address Set-Up Time	20		20	0		ns	d
TELAX	Address Hold Time	50		50	20		ns	d
TWLWH	WRITE ENABLE Pulse Width	300		240	150		ns	d
TWLEH	WRITE ENABLE Pulse Set-Up Time	300		240	150		ns	d
TELWH	WRITE ENABLE Pulse Hold Time	300		240	150		ns	d
TDVWH	Data Set-Up Time	200		160	100		ns	d
TWHDZ	Data Hold Time	0		0	0		ns	d
TWHEL	WRITE ENABLE Read Set-Up Time	0		0	0		ns	d
TQVWL	DATA VALID to Write Time	0		0	0		ns	d
TWLDV	WRITE DATA Delay Time	100		80	50		ns	d
TWLEL	Early Output High-Z Time	0		0	−10		ns	d
TEHWH	Late Output High-Z Time	0		0	−10		ns	d
TELEL	READ or WRITE Cycle Time	420		350	240		ns	d

SOURCE: Courtesy Harris Semiconductor Products Division.
[a] All devices tested at worst-case limits. Room temp., 5-V data provided for information—not guaranteed.
[b] Operating supply current (ICCOP) is proportional to operating frequency. Example: Typical ICCOP = 5 mA/MHz.
[c] Capacitance sampled and guaranteed—not 100 percent tested.
[d] AC test conditions: inputs—TRISE = TFALL = 20 nsec; outputs—1 TTL load and 50 pF; all timing measured at $\frac{1}{2} V_{cc}$.

must be provided. The address access time is specified as 150 ns max. The IM6504 comes in an 18-lead DIP and is pin-compatible with the industry standard Intel 2147 static 4096 × 1-b NMOS RAM. It has three-state outputs and TTL-compatible inputs and outputs.

The fabrication of this memory is made possible by the use of a selective oxidation CMOS process (Intersil's SELOXC) which permits smaller geometries and results in smaller die sizes for increased yields (and thus a lower cost per bit); the process typically involves low parasitic capacitances and tends to produce memory access times below 100 ns.

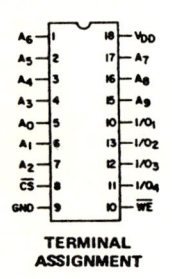

Fig. 4-26 MWS5114 Pin Connection Diagram.

The RCA MWS5114 1024 × 4-b CMOS/SOS RAM

The MWS5114 is a fully static 1024-word by 4-b static RAM fabricated with CMOS/SOS technology that has common data inputs and data outputs and operates from a single 5-V power supply. All inputs and outputs are TTL-compatible and the outputs are three-state.

Encased in an 18-lead DIP (Fig. 4-26), the MWS5114 is pin-compatible with Intel's industry standard 2114 1024 × 4-b NMOS static RAM with a power dissipation of 250 μW maximum. The maximum address access time is 650 ns.

Tables 4-4, 4-5, and 4-6 present the salient electrical characteristics and appropriate timing diagrams for the MWS5114's static electrical characteristics, READ cycle dynamic characteristics, and WRITE cycle dynamic characteristics, respectively. Table 4-7 lists the device data retention characteristics.

TABLE 4-4 Static Electrical Characteristics at T_A = 0 to +70°C, V_{DD} ±5 Percent, Except as Noted

Characteristic	V_O (V)	V_{in} (V)	V_{DD} (V)	MWS5114 MWS6514A Min	MWS5114 MWS6514A Typ*	MWS5114 MWS6514A Max	MWS5114-5 MWS6514A-5 Min	MWS5114-5 MWS6514A-5 Typ*	MWS5114-5 MWS6514A-5 Max	Units
Quiescent Device Current, I_{DD} Max	...	0.5	5	...	20	50	...	...	100	μA
($V_{in} = V_{DD}$ or V_{SS}, $\overline{CS}$ = High Level)	...	0.2	2	...	1	15	...	5	25	μA

TABLE 4-4 Static Electrical Characteristics at $T_A = 0$ to $+70°C$, $V_{DD} \pm 5$ Percent, Except as Noted (*Continued*)

Characteristic	Conditions V_O (V)	V_{in} (V)	V_{DD} (V)	Limits MWS5114 MWS6514A Min	Typ*	Max	MWS5114-5 MWS6514A-5 Min	Typ*	Max	Units
Output Low-Drive (Sink) Current, I_{OL} Min	0.4	0.5	5	2	4	...	2	4	...	mA
Output High-Drive (Source) Current, I_{OH} Min	4.6	0.5	5	−0.4	−1	...	−0.4	−1	...	mA
Output Voltage Low Level, V_{OL} Max	...	0.5	5	...	0	0.1	...	0	0.1	V
Output Voltage High Level, V_{OH} Min	...	0.5	5	4.9	5	...	4.9	5	...	V
Input Low Voltage, V_{IL} Max	0.5,4.5	...	5	...	1.2	0.8	...	1.2	0.8	V
Input High Voltage, V_{IH} Min	0.5,4.5	...	5	2.4	...	...	2.4	...	...	V
Input Current, I_{in} Max	...	0.5	5	...	±0.1	±1	...	±0.1	±1	μA
Three-State Output Leakage Current, I_{out}	0.5	0.5	5	...	±0.5	±5	...	±0.5	±5	μA
Operating Current, $I_{DD,1}$†	...	0.5	5	...	4	...	...	4	...	mA
Input Capacitance, C_{in}	...	...	...	...	5	7.5	...	5	7.5	pF
Output Capacitance, C_{out}	...	...	...	...	5	7.5	...	5	7.5	pF

NOTE: Recommended operating conditions at $T_A = 25°C$. For maximum reliability, operating conditions should be selected so that operation is always within the following ranges: for dc operating voltage 4–6 V; for input voltage, V_{SS} to V_{DD} volts.
* Typical values are for $T_A = 25°C$ and nominal V_{DD}.
† Outputs open-circuited; cycle time = 1 μs.
SOURCE: RCA Solid-State Products Division.

TABLE 4-5 READ Cycle Times

Characteristic		Limits (ns)					
		MWS5114 MWS6514A			MWS5114-5 MWS6514A-5		
		Min*	Typ†	Max	Min*	Typ†	Max
Read Cycle	t_{RC}	650	450	...	650	450	...
Access	t_{AA}	...	450	650	...	450	650
Chip Selection to Output Valid	t_{CO}	...	400	600	...	400	600
Chip Selection to Output Active	t_{CX}	20	100	...	20	100	...
Output Three State from Deselection	t_{OTD}	...	125	175	...	125	175
Output Hold from Address Change	t_{OHA}	50	150	...	50	150	...

NOTE: Dynamic Electrical Characteristics at T_A = 0 to +70°C, V_{DD} = 5 V ±5 percent, t_r, t_f = 10 ns, C_L = 50 pF, and 1 TTL Load.
* Time required by a limit device to allow for the indicated function.
† Typical values are for T_A = 25°C and nominal V_{DD}.
SOURCE: RCA Solid-State Products Division.

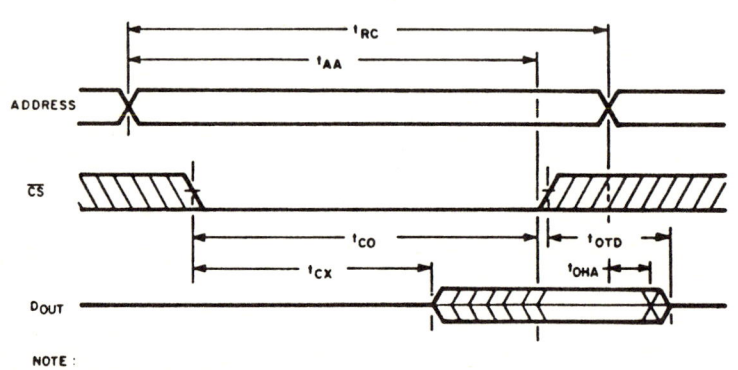

NOTE:
WE IS HIGH DURING THE READ CYCLE.
TIMING MEASUREMENT REF LEVEL IS 1.5 V

The RCA MWS5104 4096 × 1-b CMOS/SOS RAM

The MWS5104 is a 4-kb ×1-b asynchronous static RAM fabricated with CMOS/SOS technology and based on a complementary (buried) contact structure. The increased density stems largely from the use of a five-transistor memory cell (Fig. 4-27). Because of the buried contact technol-

TABLE 4-6 WRITE Cycle Times

| Characteristics | | \multicolumn{3}{c|}{MWS5114 MWS6514A} | \multicolumn{3}{c|}{MWS5114-5 MWS6514A-5} |

Characteristics		MWS5114 MWS6514A Min*	Typ†	Max	MWS5114-5 MWS6514A-5 Min*	Typ†	Max
WRITE Cycle	t_{WC}	500	400	...	500	400	...
Write	t_W	450	350	...	450	350	...
Write Release	t_{WR}	50	25	...	50	25	...
Address to Chip Select Set-Up Time	t_{ACS}	0	0	...	0	0	...
Data to Write Set-Up Time	t_{DSU}	0	0	...	0	0	...
Data Hold from Write	t_{DH}	30	10	...	30	10	...

NOTE: Dynamic Electrical Characteristics at T_A = 0 to +70°C, V_{DD} = 5 V ±5 percent, t_r, t_f = 10 ns, C_L = 50 pF, and 1 TTL Load.
* Time required by a limit device to allow for the indicated function.
† Typical values are for T_A = 25°C and nominal V_{DD}.
SOURCE: RCA Solid-State Products Division.

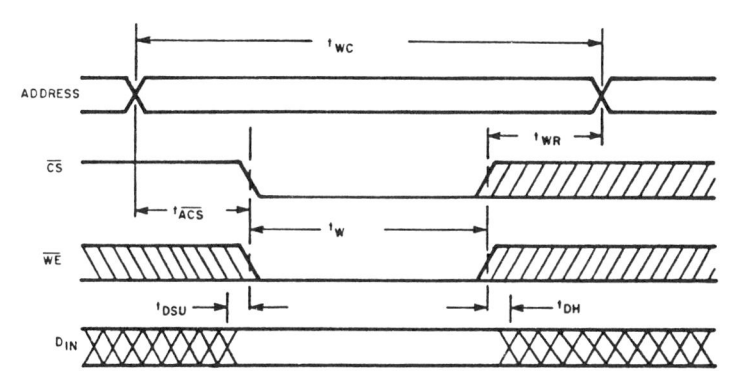

NOTE:
WE IS LOW DURING THE WRITE CYCLE
TIMING MEASUREMENT REF LEVEL IS 1.5 V

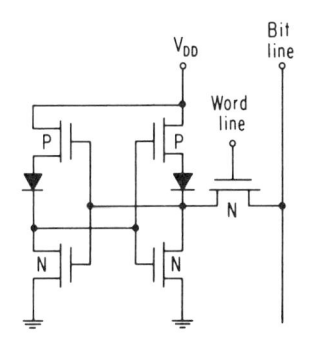

Fig. 4-27 New Five-Transistor RAM Cell with Buried Contact Diodes.

TABLE 4-7 Data Retention Characteristics at T_A = 0 to 70°C; See below

Characteristic		Test Conditions		MWS5114 MWS6514A			Units
			V_{DD} (V)	Min	Typ*	Max	
Data Retention Voltage	V_{DR}			2	...	...	V
Data Retention Quiescent Current, I_{DD}	MWS5114 MWS6514A	V_{DR} = 2 V		...	1	15	μA
	MWS5114-5 MWS6514A-5			...	5	25	μA
Chip Deselect to Data Retention Time	t_{CDR}		5	600	...	...	ns
Recovery to Normal Operation Time	t_{RC}		5	600	...	...	ns

* Typical values are for T_A = 25°C and nominal V_{DD}.
SOURCE: RCA Solid-State Products Division.

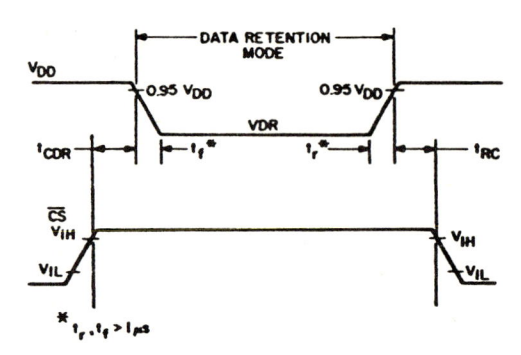

ogy, the cell shown requires just two metal lines, whereas six-transistor cells require three. The diodes, appearing in the contacts to the P⁺ episilicon, are forward-biased at all times during quiescent operation, with less than a 0.2-V drop. The buried contact diodes also help reduce the voltage levels needed for writing.

To operate successfully at high data rates, five-transistor memory cells require carefully designed peripheral control logic. Early CMOS memory designs were invariably based on the more easily writable six-transistor cells. Computer simulations of five-transistor memory cells have, however, shown that writeability of buried contact five-transistor

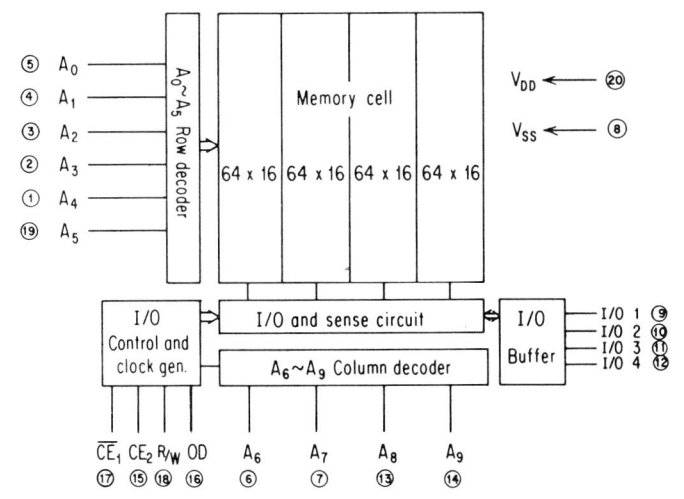

Fig. 4-28 RAM Block Diagram. The Chip is Organized as Four 1-kb RAMs—That Is, 1-kb word × 4-b RAM—and Is Integrated with About 26,600 Elements. *(Reprinted from 1977 IEEE ISSCC Digest of Technical Papers, Philadelphia, PA, February 1977. Copyright © 1977 by the Institute of Electrical and Electronics Engineers Inc. Used with permission.)*

cells is enhanced by the buried contact structure, thereby improving design margins both for WRITE 1 and WRITE 0.

A C²MOS Static Ram*

A C²MOS (clocked CMOS) 4-kb static RAM (Fig. 4-28) has been developed with the following features:

The RAM is static in the sense that the content of the cell is retained statically. However, most circuits are dynamically operated at each transient time; that is, row decoder, sense circuit, and I/O controller are designed with clocked gates and half-bit inverters, both contributing to a reduction in the number of active elements and the pattern area of these functional blocks.

The sense circuit shown in Fig. 4-29 is simple, constructed with only two P-channel FETs, while to each sense line two N-channel FETs are added to precharge it and to retain its signal level statically. It can operate as quickly as more complicated sense amplifiers—for example, a

---

*K. Ochii et al., "C²MOS 4K Static RAM," *1977 IEEE ISSCC Digest of Technical Papers*, Philadelphia, PA, February 1979. Copyright © 1977 by the Institute of Electrical and Electronics Engineers Inc. Reprinted with permission.

230 CMOS Memories

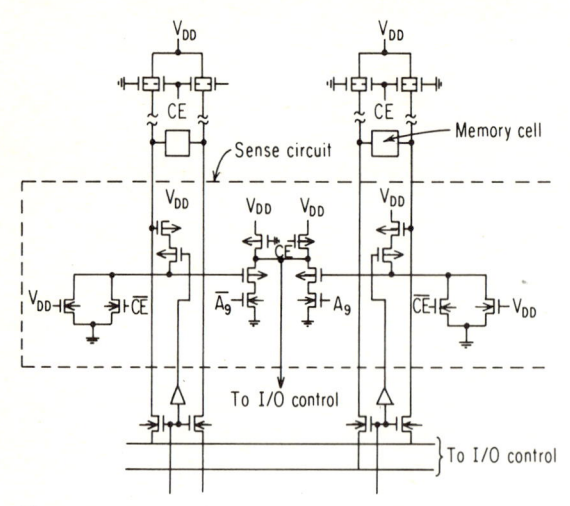

Fig. 4-29 Sense Circuit Schematic Diagram. *(Reprinted from 1977 IEEE ISSCC Digest of Technical Papers, Philadelphia, PA, February 1977. Copyright © 1977 by the Institute of Electrical and Electronics Engineers, Inc. Used with permission.)*

sense latch or a differential amplifier—because a logic circuit threshold can be designed as an element threshold which is very suitable for the sense circuit of a CMOS RAM.

The speed of this sense circuit is more dependent on an element threshold than on the capacitance of a bit line, and thus the larger the scale of this RAM, the more the merit of this sense circuit is apparent.

An OR-type clocked row decoder circuit is shown in Fig. 4-30. Clock

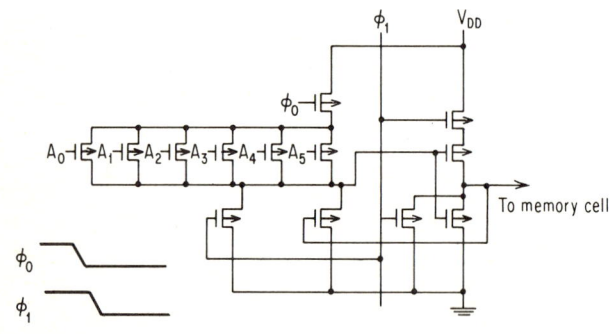

Fig. 4-30 Clocked Row Decoder Schematic Diagram. *(Reprinted from 1977 IEEE ISSCC Digest of Technical papers, Philadelphia, PA. February 1977. Copyright © 1977 by the Institute of Electrical and Electronics Engineers, Inc. Used with permission.)*

pulses ϕ_0 and ϕ_1 are generated internally; ϕ_0 determines the timing of precharge and ϕ_1 of the on-time of the decoder. An adequate timing design of these clocks makes high-speed operation of the decoder possible.

One figure of merit, access time (t_{ACC}) multiplied by power dissipation (P_D) of the RAM, is about 10 nJ, which is a fraction of that of NMOS RAMs. Since the standby current is as small as 0.1 μA, which is comparable to the natural discharge current of the battery, battery back-up operation of the CMOS RAM is easy and inexpensive.

The RAM features an access time of 300 ns at 5-V supply voltage and 200 ns at 8 V. The RAM is assembled in a standard 20-pin DIP, and with a metal mask option can be also assembled in an 18-pin package. The wide range of operational power supply voltage variation, from 4 to 8 V, permits the use of an inexpensive power supply unit, which further contributes to the cost reduction of a total memory system. Table 4-8 summarizes the RAM's operational characteristics and Fig. 4-31 displays its power dissipation as a function of cycle time and power supply voltage.

Using a patterning technology of 4 μm and these design features and circuits, it is estimated that an 8-kb CMOS RAM can be integrated on a 6-mm² chip.

The Hitachi HM6147 4-kb × 1-b

The HM6147 high-speed static CMOS RAM offers total function and pin compatibility with the 2147 industry standard NMOS 4-kb × 1-b 18-pin RAM, but it operates at one-fourth the power: 220 mW worst-case active and 4.2 mW maximum standby, with a low-power version that has a maximum standby power dissipation of 520 μW. For the first time a

TABLE 4-8 Basic Features of the RAM

Supply Voltage	4 to 8 V
Access Time	300 ns at V_{DD} = 5 V
	200 ns at V_{DD} = 8 V
Operational Power Dissipation	50 mW at t_{cycle} = 1 μs
Standby Power Dissipation	0.5 μW
Package	20-Pin DIP
I/O Interface	I/O Common, TTL-Compatible

SOURCE: *1977 IEEE ISSCC Digest of Technical Papers,* Philadelphia, PA, February 1977. Copyright © 1977 by the Institute of Electrical and Electronics Engineers, Inc. Reprinted with permission.

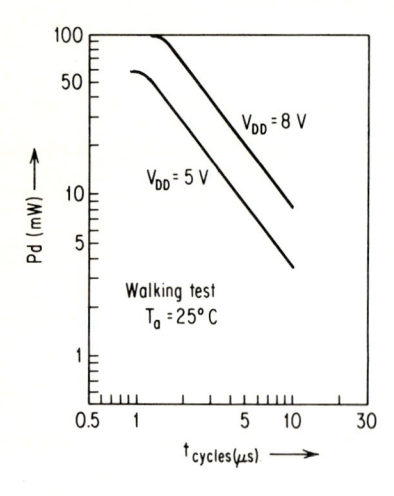

Fig. 4-31 Power Dissipation of the RAM. *(Reprinted from 1977 IEEE ISSCC Digest of Technical Papers, Philadelphia, PA, February 1977. Copyright © 1977 by the Institute of Electrical and Electronics Engineers, Inc. Used with permission.)*

CMOS RAM operates at the same speed as its NMOS counterpart—55 ns max (with a new 35ns/45ns version available)—while providing the advantage of low power. There is no degradation of performance or penalty for using CMOS technology, as there was in the past.

The HM6147 offers direct TTL compatibility in all areas including input, output, and operation from a single +5-V power supply; separate data input and output; three-state outputs; automatic power-down; and completely static operation.

The HM6147 is not a true CMOS part since its memory array is all N-channel flip-flops. But by using polysilicon load resistors with the extremely high value of 50 GΩ in the cells and surrounding the array with CMOS peripheral circuits, Hitachi has achieved combined high-speed and low-power performance characteristics that should be widely copied in second-sourcing the HM6147 and in future designs. In addition, the design uses a bipolar transistor (inherent in the CMOS process) as a pull-up to drive the output. This provides better current-carrying capability than any MOS transistor for a given size.

The electrical characteristics of the HM6147 are summarized in Table 4-9.

Hitachi has followed the HM6147 with the HM6148 1024 × 4-b RAM (a CMOS version of the NMOS 2148) that exhibits the same high-performance standards as the 6147 but again with superior power dissipation to its NMOS counterpart.

Hitachi is also using its CMOS process for a 16-kb static RAM that achieves greater density by eliminating the positive supply contacts to the cells. The HM6116 (2 kb × 8 b) is faster and dissipates less power than

TABLE 4-9 HM6147 Electrical Characteristics

DC and Operating Characteristics (0°C ≤ T_A ≤ 70°C, V_{CC} = 5 V ± 5 percent, GND = 0 V)

Parameter	Symbol	Min	Typ	Max	Unit	Test Conditions
Input Leakage Current	$\lvert IL1 \rvert$	...	...	2.0	µA	V_{CC} = max V_{in} = GND to V_{CC}
Output Leakage Current	$\lvert IL0 \rvert$	...	...	2.0	µA	$\overline{CS} = V_{IH}$ $V_{out} = 0 \sim V_{CC}$
Operating Power Supply Current (1) dc	I_{CC}	...	15	35	mA	$\overline{CS} = V_{IL}$ Output open
Operating Power Supply Current (2) dc	$I_{CC.1}$	...	12	...	mA	$\overline{CS} = V_{IL}$ V_{in} ≤ 0.2 V or V_{in} ≥ V_{CC}−0.2 V
Average Operating Current (3)	$I_{CC.2}$	...	14	...	mA	Cycle 150 ns duty 50%
Standby Power Supply Current (1) dc	I_{SB}	...	5	12	mA	$\overline{CS} = V_{IH}$
Standby Power Supply Current (2) dc	$I_{SB.1}$	...	...	800	µA	$\overline{CS}$ ≥ V_{CC}−0.2 V V_{in} ≤ 0.2 V or V_{in} ≥ V_{CC}−0.2 V
Output Low Voltage	V_{OL}	...	...	0.40	V	I_{OL} = 12 mA
Output High Voltage	V_{OH}	2.4	...	...	V	I_{OH} = −8.0 mA

AC Characteristics (T_A = 0°C to 70°C, V_{CC} = +5 V ±5 percent, unless otherwise noted)
READ CYCLE

Symbol	Parameter	HM6147P-3 Min	HM6147P-3 Max	HM6147P Min	HM6147P Max	Unit
t_{RC}	READ Cycle Time	55		70		ns
t_{AA}	Address Access Time[a]		55		70	ns
t_{ACS}	CHIP SELECT Access Time[a]		55		70	ns
t_{OH}	Output Hold from Address Change	5		5		ns
t_{LZ}	Chip Selection to Output in Low Z	10		10		ns
t_{HZ}	Chip Deselection to Output in High Z	0	40	0	40	ns
t_{PU}	Chip Selection to Power-Up Time	0		0		ns
t_{PD}	Chip Deselection to Power-Down Time		30		30	ns

234 CMOS Memories

TABLE 4-9 HM6147 Electrical Characteristics (*Continued*)

WRITE CYCLE

Symbol	Parameter	HM6147P-3 Min	HM6147P-3 Max	HM6147P Min	HM6147P Max	Unit
t_{WC}	WRITE Cycle Time[a]	55		70		ns
t_{CW}	Chip Selection to End of WRITE	45		55		ns
t_{AW}	Address Valid to End of WRITE	45		55		ns
t_{AS}	Address Set-Up Time	0		0		ns
t_{WP}	WRITE Pulse Width	35		40		ns
t_{WR}	WRITE Recovery Time	10		15		ns
t_{DW}	Data Valid to End of Write	25		30		ns
t_{DH}	Data Hold Time	10		10		ns
t_{WZ}	Write Enabled to Output in High Z	0	30	0	35	ns
t_{OW}	Output Active from End of WRITE	0		0		ns

[a] 35- and 45-ns versions also available.

any other 16-kb RAM. The access time of 100 ns of the HM6116 is better than the NMOS TMS4016's 150 ns, and it can be put into a standby mode where it dissipates only microwatts. The operating power is about one-half that of the TMS4016.

Ingeniously, the HM6116 leaves an N-type island within each cell in the P well and feeds in from the N-type substrate by transistor action—the P-type silicon surrounding each island behaves as the gate of a buried JFET (junction field-effect transistor) (see Fig. 4-32). The current is determined by the size of the island, but it is not critical since all that is necessary to power the cell is for the resistance of the JFET to be far less than that of the loads. The power dissipation of the HM6116 is similar to that of the HM6147. The HM6167 (16 kb × 1-b) offers increased performance over that of NMOS 16-kb dynamic RAMs: 55 to 70 ns for the HM6167 versus 80 to 100 ns for the Intel 2118. Both parts operate from a single +5-V supply, while providing low power dissipation.

16-kb CMOS/SOS RAM*

The feasibility of fabricating a 16-kb CMOS/SOS RAM has been demonstrated by the RCA Technology Center. The circuit makes use of a new

---

*Excerpted in part from R. G. Stewart and A. G. F. Dingwall, "16K CMOS/SOS Asynchronous Static RAM," *1979 IEEE ISSCC Digest of Technical Papers,* Philadelphia, PA, February 1979. Copyright © 1979 by the Institute of Electrical and Electronics Engineers Inc. Reprinted with permission.

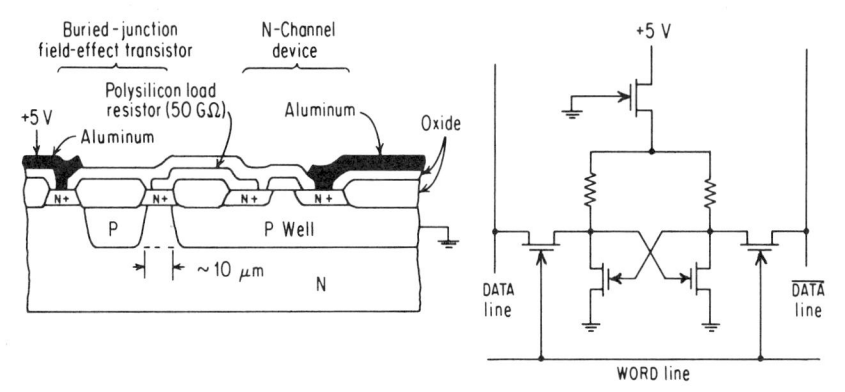

Fig. 4-32 Hitachi's 4-kb RAM Surrounds an N-Channel Array with CMOS Peripheral Circuitry. Density is Increased to 16 kb by Adding a Buried JFET (a) device feeds power to loads (b) while eliminating positive-supply metalization.

five-transistor memory cell in which all internal connections are made via the buried contacts, as was shown in Fig. 4-27. Metal contacts are used only for the external connections between the cell, the bit line, and power supply buses. Contact sharing between adjacent cells is used to reduce further the number of metal contacts to an average of only one per cell.

Also, since the memory cell requires only a single bit line and a single word line, only two metal lines are needed for each column of cells, compared with three for a buried contact NMOS cell or five for a conventional CMOS memory cell.

The organization of the 16-kb RAM is depicted in Fig. 4-33. Asynchronous operation is provided by generating the precharge signal internally whenever transitions are detected on any of the memory address inputs. This makes the precharge totally transparent to the user. During WRITE operations, a 8.7-V level is supplied to the word decoder from a frequency-regulated power supply with a quiescent power dissipation of approximately 10 μA. All inputs and Tri-State outputs are TTL-compatible, and the pinout conforms to the proposed 22-pin JEDEC (Joint Electronic Devices Engineering Council) standard for 4-kb × 4-b RAMs.

This 16-kb RAM contains 87,000 transistors and operates over a voltage range of 3.1 to 12 V, with data retention down to 1.3 V. Quiescent power dissipation at 5 V is typically 0.005 mW during READ and 0.05 mW during WRITE. Dynamic power dissipation is typically 14 mW at 1 MHz. A plot of memory access time versus voltage is shown in Fig. 4-34. Typical access time ranges from 150 ns at 5 V to 80 ns at 10 V.

The developmental pace of 4096 and 16384-b CMOS RAMs is hectic

236 CMOS Memories

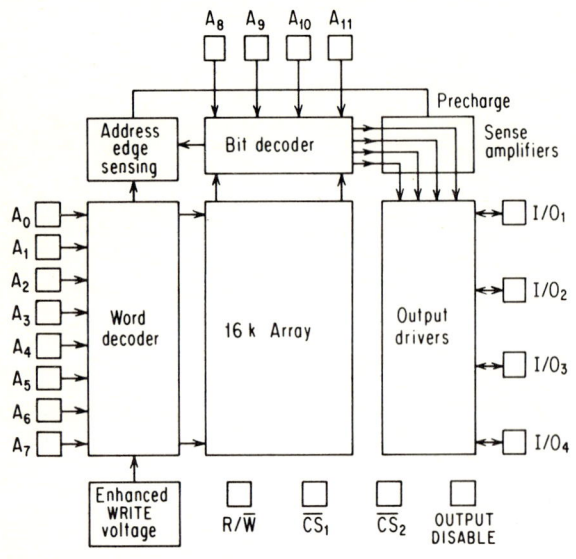

Fig. 4-33 16-kb RAM Memory Organization.

to say the least. Researchers in Japan have developed a high-density CMOS process utilizing E-beam lithography for mask making and dry etch techniques. These efforts have produced two 2048 × 8-b RAMs. One, developed by Toshiba, has an address access time of 95 ns and active power dissipation of 200 mW and 1 μW in the standby mode. The device operates from a single +5-V power supply.* The second 16-kb RAM, developed by Hitachi, has a 74-ns access time and dissipates 200

---

* A. Nozawa et al., "High Density CMOS Processing for a 16K Bit RAM," *IEDM Proceedings*, Washington, DC, December 1979. See also "Static, 16K-Bit Bulk CMOS RAM with 95 ns Access, 200 mW Operation Power, 1 μW Standby," *1980 IEEE ISSCC Proceedings*, San Francisco, CA, February 1980, by the same authors.

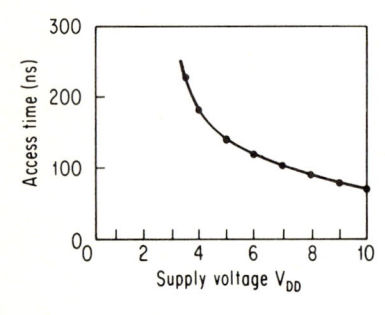

Fig. 4-34 Typical Memory Access Time as a Function of Supply Voltage. Access Time Ranges from 260 ns at 3.2 V to 80 ns at 10 V. *(Reprinted from 1979 IEEE ISSCC Digest of Technical Papers, Philadelphia, PA, February 1979. Copyright © 1979 by the Institute of Electrical and Electronics Engineers, Inc. Used with permission.)*

mW in the operating mode and 25 mW in the standby mode.* Concurrently, a 899-μm^2 memory cell was developed by Toshiba† using variable polysilicon loads. The future portends high-performance 32- and 64-kb CMOS RAMs.

CMOS RAM Electrical Characteristics Summary

For the user's benefit, the primary characteristics of some of the more popular CMOS RAMs listed in Table 4-2 are summarized in Table 4-10. From this table it is seen that there exists a wide variety of available CMOS RAMs with many speed-power-package options available from which one may effect an optimal system design.

POPULAR CMOS ROMs/PROMs

A 1024-b PROM‡

PROMs (programmable read-only memories) have been available for several years. Previous fused-link PROMs have used bipolar technology with its accompanying high power dissipation. MOS PROMs have used charge storage in dual-dielectric or floating-gate structures. The operating range of these memories is limited by charge loss at elevated temperatures. This section describes a fused-link CMOS PROM which overcomes these earlier limitations. The basic features are summarized in Table 4-11.

Figure 4-35 is a block diagram of the PROM, and Fig. 4-36 is a schematic diagram of the memory cell including row and column drivers.

The PROM is fully static with a single CHIP SELECT ($\overline{CS}$). One extra pin ($\overline{PE}$) is required for programming. The output is forced to a high-impedance state when $\overline{CS}$ is high. An 18-pin version has also been made which has latched $\overline{CS}$ and outputs.

The basic problem in the design of a MOS fused-link PROM is to provide the large currents needed to blow the fused link while retaining reasonable speed for the read operation. The memory matrix and pro-

---

* T. Yasui and K. Uchibori, "16-kb (2-kb × 8-b) HCMOS Static RAMs, *1980 IEEE ISSCC Proceedings*, San Francisco, CA, February 1980.

† T. Iizuka et al., "Variable Resistance Polysilicon for High Density CMOS RAM," *1979 IEDM Proceedings*, Washington, DC, December 1979.

‡ J. E. Schroeder and R. L. Goslin, "A 1024-bit, Fused-Link CMOS PROM," *1977 IEEE ISSCC*, Philadelphia, PA, February 1977. Copyright © 1977 by the Institute of Electrical and Electronics Engineers, Inc. Reprinted with permission.

TABLE 4-10 CMOS RAM Electrical Summary

P/N	Supplier	Organization (b)	t_{AA}(ns)	P_{diss}/P_{stdby}(mW)	Power Supplies(V)	TTL-Compatible	Three State	Comm. I/O	Sep. I/O	Output Latch (Disable)	Address Register	Package (DIP) (pins)
MM54C/74C89	National	16 × 4	280(10 V) 650(5 V)	4.5	Single (3–15)	×					×	16
CD4061A	RCA	256 × 1	380(10 V) 750(5 V)	0.1(10 V) 0.025(5 V)	Single (3–15)		×		×			16
MCM14552	Motorola	256 × 1	3150,6300	1.5	Single (−0.5–18)	×	×			×		24
MM54C/74C910	National	64 × 4	700	20	Single (+5 V)	×	×					18
S2222/A	AMI	512 × 1	350,700	25.0/0.002	Single (+10 V)	×			×			16
IM6508/18	Intersil	1024 × 1	200–600	2.5–0.05	Single (+5 V)	×	×		×		×	16/18
MM54C/74C929	National	1024 × 1	240,315	20	Single (+5 V)	×	×		×		×	16
MM54C/74C930	National	1024 × 1	240,315	20	Single (+5 V)	×	×		×	×	×	18
I5101	Intel	256 × 4	450	135,110/ 0.075	Single (+5 V)							22
			650	135,110/ 1.0,0.075	Single (+5 V)							
			800	150,125/ 2.5, 0.25	Single (+5 V)							
IM6551	Intersil	256 × 4	180,360	2.5/0.5	Single (+5 V)	×	×		×	×		22
HM6551	Harris	256 × 4	215,375	15	Single (+5 V)	×	×		×	×		22
IM6561	Intersil	256 × 4	180,360	2.5/0.5	Single (+5 V)	×	×	×				18
HM6561	Harris	256 × 4	215,375	15	Single (+5 V)	×	×	×				18
HM6562	Harris	256 × 4	215,375	15	Single (+5 V)	×	×					16
MM54C/74C920	National	256 × 4	250,325	20,15	Single (+5 V)	×			×		×	22
MM54C/74C921	National	256 × 4	250,325	20,15	Single (+5 V)	×					×	18
HM6513	Harris	512 × 4	270,320	30/MHz/ 50 μW, 5 mW	Single (+5 V)	×	×	×			×	18

Part	Mfr	Organization	Access (ns)	Power	Supply							Pins
HM6503	Harris	2048 × 1	250, 300	20/MHz/ 50 μW, 5 mW	Single (+5 V)	×				×		18
HM6504	Harris	4096 × 1	270, 330	0.03, 0.05	Single (+5 V)	×	×			×		18
IM6504	Intersil	4096 × 1	150	100/≪1	Single (+5 V)	×	×			×	×	18
HM6147P[a]	Hitachi	4096 × 1	34, 45, 55, 70	75/10 μW	Single (+5 V)	×	×			×	×	18
HM6514[f]	Harris	1024 × 4	270, 320	30/0.05	Single (+5 V)	×						18
				2.5				×				
MWS5114[b,f]	RCA	1024 × 4	650	0.25	Single (+5 V)	×	×	×				18
UPD444[f]	NEC	1024 × 4	200, 250	45–95/0.065 μW	Single (+5 V)	×	×	×				18
			300, 450									
UPD445	NEC	1024 × 4	450, 650	45, 75/100 μW	Single (+5 V)	×	×			×		20
HM6533	Harris	1024 × 4	350	35/0.005	Single (+5 V)	×	×	×				22
HM6148[c]	Hitachi	1024 × 4	55, 70	150/0.05	Single (+5 V)	×	×			×	×	18
N/A[b,d]	RCA	4096 × 4	80/150	14/0.005/0.05	Single (+5 V)	×	×	×		×	×	22
TC5516P[e]	Toshiba	2048 × 8	250	70 mA/50 μA	Single (+5 V)	×	×			×		24
HM6516	Harris	2048 × 8	200	5	Single (+5 V)	×	×	×		×		24
HM6116	Hitachi	2048 × 8	100, 120, 150	175/0.02	Single (+5 V)	×	×	×		×		24
HM6167	Hitachi	16,384 × 1	55, 70	200 mW/100 μW	Single (+5 V)	×	×			×		20

NOTE: All electrical parameters guaranteed at 25°C.
[a] 2147 CMOS equivalent.
[b] CMOS/SOS technology.
[c] 2148 CMOS equivalent.
[d] N/A = not available.
[e] Pin-compatible with 2716 EPROM.
[f] Pin-compatible with 2114 NMOS RAM.

239

TABLE 4-11 Summary of Fused-Link CMOS PROM Characteristics

Operating Temperature Range	+55 to 125°C
Supply Voltage, V_{CC}	4.0 to 11.0 V
Input/Output Levels	TTL*
Address Access Time, t_{AA} (25°C)	225 ns*/175 ns†
CHIP SELECT Access Time, t_{AC} (25°C)	250 ns*/200 ns†
Supply Current (−55 to 125°C)	
Standby ($\overline{CS}$ High)	10 µA*/100 µA‡
Enabled ($\overline{CS}$ Low)	10 mA*/30 mA‡

* V_{CC} = 5 V.
† V_{CC} = 10 V.
‡ V_{CC} = 4 to 11 V.

gramming circuitry is P-channel, allowing the use of 30- to 35-V levels in the fuse path during programming. All other circuitry is CMOS for speed and low power.

During programming, the N-channel devices must be isolated from the high voltage levels in the fuse path. For the column lines, this requires one extra transistor per column. In the row drivers, the N-channel device is isolated by two P-channel devices when the row is not being programmed. When the row is being programmed, the N-channel de-

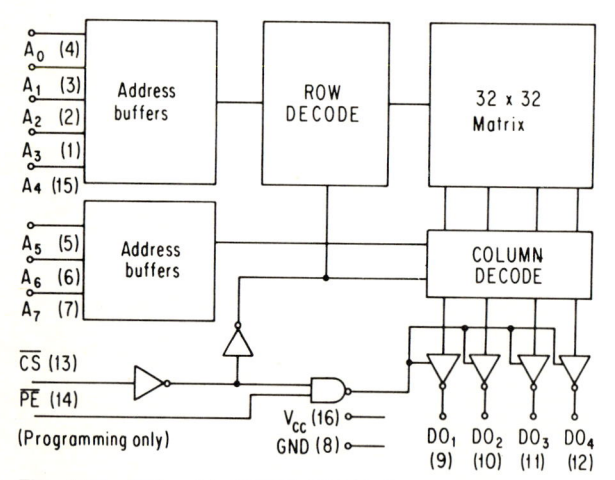

Fig. 4-35 256 × 4-b CMOS PROM Block Diagram. *(Reprinted from 1977 IEEE ISSCC Digest of Technical Papers, Philadelphia, PA, February 1977. Copyright © 1977 by the Institute of Electrical and Electronics Engineers, Inc. Used with permission.)*

Popular CMOS ROMs/PROMs 241

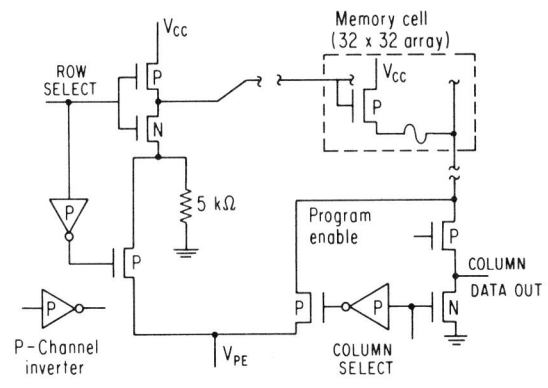

Fig. 4-36 Schematic Diagram of Memory Cell and Row and Column Drivers. *(Reprinted from 1977 IEEE ISSCC Digest of Technical Papers, Philadelphia, PA, February 1977. Copyright © 1977 by the Institute of Electrical and Electronics Engineers, Inc. Used with permission.)*

vice can be pulled negative because they are tied to ground through the 5-kΩ resistor shown.

Figure 4-37 shows the timing for the READ and PROGRAMMING cycles. During READ, $\overline{PE}$ is high. The matrix device, fused link, and column transistors form a voltage divider which is connected through a

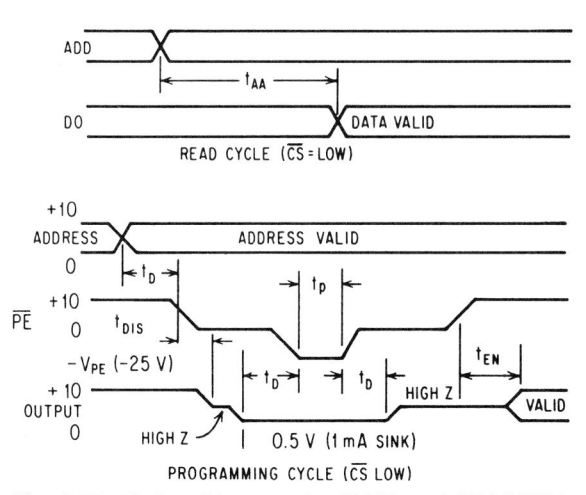

Fig. 4-37 Timing Diagrams for READ and PROGRAMMING Cycles. *(Reprinted from 1977 IEEE ISSCC Digest of Technical Papers, Philadelphia, PA, February 1977. Copyright © 1977 by the Institute of Electrical and Electronics Engineers, Inc. Used with permission.)*

242 CMOS Memories

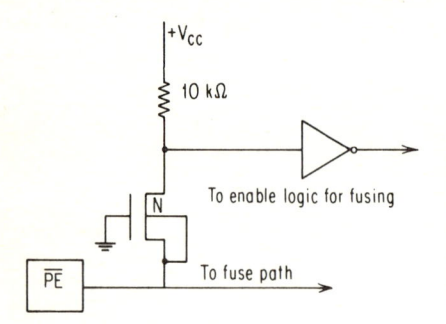

Fig. 4-38 Three-Level Input Buffer. *(Reprinted from 1977 IEEE ISSCC Digest of Technical Papers, Philadelphia, PA, February 1977. Copyright © 1977 by the Institute of Electrical and Electronics Engineers, Inc. Used with permission.)*

transmission gate to the sense amplifier. The only dc current during READ is through this voltage divider path.

Grounding $\overline{PE}$ forces the outputs to a high-impedance state. These pins are then used to select the bits to be programmed in the addressed word. To program the fused links, $\overline{PE}$ is switched to a negative voltage V_{PE}, typically -25 V.

The circuit shown in Fig. 4-38 is used to keep the programming path off until $\overline{PE}$ is at least 2 V negative. Similar circuits are used to address the extra rows and columns included for testing.

The PROM is fabricated using silicon-gate CMOS technology. Polysilicon fuses are fabricated in conjunction with the silicon-gate process, and satisfactory programmability yield and reliability have been demonstrated. All aspects of the circuit design and process are compatible with the polysilicon fuse or with a metal fuse such as nichrome.

The HM6611 1024-b PROM

The HM6611 (Fig. 4-39) is a 256×4-b CMOS PROM, fabricated with self-aligned silicon-gate technology and fusible links. It is specified for interfacing both with standard CMOS and with standard TTL logic families. User benefits are

- Lower-power standby (100 μA)
- Low-power operation (10 mA)
- CMOS RAM pinout, except for $\overline{PE}$
- TTL-compatible input/output
- Three-state output
- High output drive
- Fully static operation

- Fast access time (300 ns)
- High noise immunity
- High reliability
- Easy programming

The HM6611 is packaged in a 16-pin DIP (see Fig. 4-40), and offers all the usual benefits inherent in a field-programmable ROM with the extremely low active and standby power of CMOS.

Convenient memory array expansion can be attained by using the $\overline{CS}$ on the 6611.

Programming can be done on-board, and parallel programming (all desired bits in the selected location) is permissible. Programming conditions are nominally +5 and −25 V, with 30 mA per blown fuse required.

Conditions are simplified so that standard CMOS or open-collector TTL devices can be used to establish all required conditions. The PROM is fabricated with all bits as 1s; blown links create an output low condition when they are addressed.

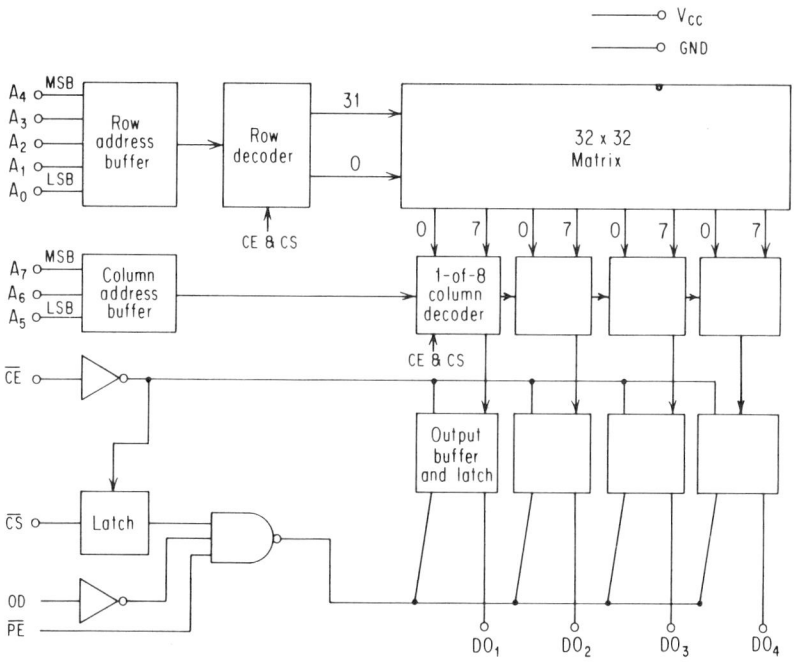

Fig. 4-39 HM6611 Block Diagram. *(Courtesy Harris Semiconductor Products Division.)*

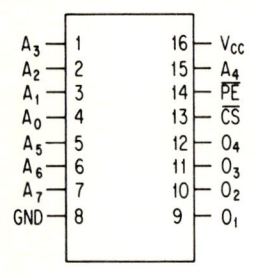

Fig. 4-40 HM6611 Pin Connection Diagram *(Courtesy Harris Semiconductor Products Division.)*

The Intersil IM6603/6604 4-kb PROMs*,†

The IM6603/6604 are 4-kb PROMs fabricated using a low-threshold ion-implanted silicon-gate CMOS process and organized as 1024×4 b for the 6603 and 512×8 b for the 6604 by means of a metal mask option.

The memory elements in the 6603/04 are arranged as two 64×32-b subarrays separated by row address decoding logic, as shown in Fig. 4-41.

* G. Ramachandran, "Single-Supply Erasable PROM Saves Power with CMOS Process," *Electronics,* July 6, 1978, pp. 106–111. Copyright © 1977 by McGraw-Hill, Inc. Used with permission.

† Y. F. Chan, "A 4K CMOS Erasable PROM," *IEEE J. Solid State Circuits*, vol. SL-13, no. 5, pp. 677–680, October 1978.

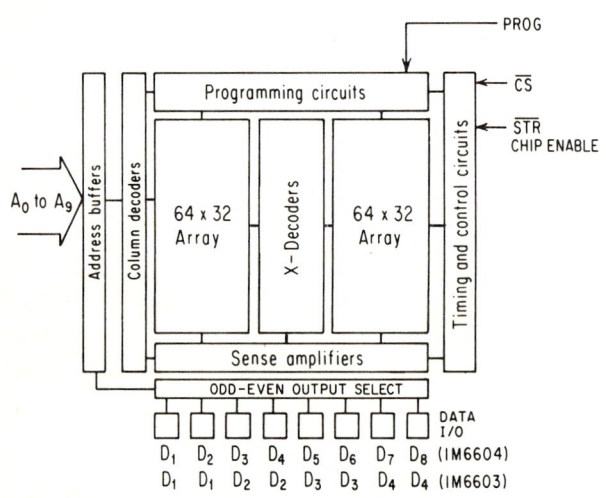

Fig. 4-41 IM6603/6604 Functional Block Diagram. *(Courtesy Intersil.)*

Popular CMOS ROMs/PROMs

Both devices have a 10-b internal address latch. The six address bits, A_3 through A_8, are used by the central row address decoder to select one of eight columns, which are connected to each of the eight sense amplifiers and data I/O buffers.

In the 6603, the 10th bit (A_9) is used as an odd/even select to steer data to and from the array. Since A_4 activates alternate columns and sense amplifiers, only 4 b at a time are read out of the array, making it effectively appear twice as deep: 1024 b instead of two 512-b arrays. The 6604 needs only 9 b for addressing, so it treats the 10th bit as a latched CHIP ENABLE input.

The two devices have an on-chip address register permitting use with bus-organized systems that multiplex data and addresses. As shown in the timing diagram of Fig. 4-42, valid address line levels are required within a hold time following the falling edge of the STROBE line, $\overline{STR}$. Because of the very fast regenerative address buffering, address set-up time is specified to be 0; in fact, it is a negative time—the address can be presented after the STROBE line falls because of the differences in logic delays in STR and address input paths.

The 6604's CHIP ENABLE, treated as an address input, is latched and activates the chip if it is low. In addition, a fast CHIP SELECT circuit

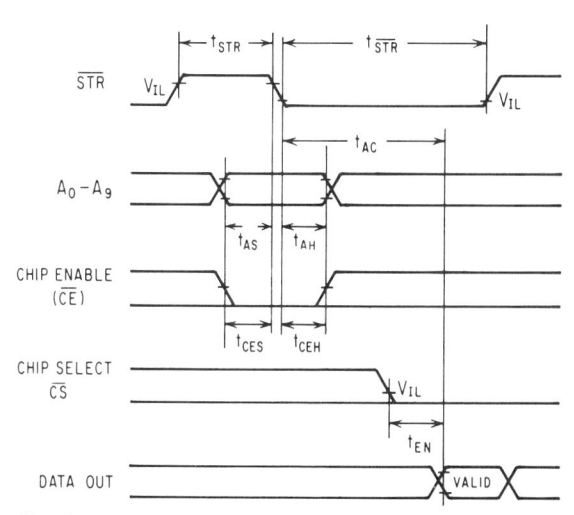

Fig. 4-42 The Fast Regenerative Buffering Used in the 6603/4 EPROM Means that Valid Address Lines Can Be Presented after the Falling Edge of the STROBE Line. Thus, Set-up Time for Addresses Is Actually Negative. *(Reprinted from Electronics, July 6, 1978. Copyright © 1978 by McGraw-Hill, Inc. Used with permission.)*

246 CMOS Memories

ensures that $\overline{CS}$ decoders may be used with an array of erasable PROMs with no sacrifice in access time.

The $\overline{STR}$ line need only be low for the duration of an access. As soon as the microprocessor has acquired the data, the STROBE line can return high and subsequently cause the outputs to go to a high-impedance mode; however, it must remain high long enough for all columns to precharge.

When $\overline{STR}$ is high, the address buffers, row and column decoders, sense amplifiers, and column lines are precharged to appropriate levels. A sequence of internal timing signals is generated on the negative-going edge of $\overline{STR}$. The address buffers are activated and the correct address is latched. Then row and column decoders are activated with the appropriate memory cells selected. The current-sensing amplifiers are then activated and the signal levels are sent to the data I/O buffers to be read out.

The 6603/04 use a P-channel floating-gate fuse technique (Fig. 4-43) similar to that of the 1702 256 × 8-b EPROM (whereas the 2708 and 2716 EPROMs use a dual polysilicon N-channel technique) because of the simpler process.

The READ-PROGRAM-VERIFY sequence illustrated in Fig. 4-44 shows that, in both programming and reading, a cell is addressed by strobing the desired address in. Reading the location verifies that it is unprogrammed. $\overline{STR}$ is left low to ensure that the address does not

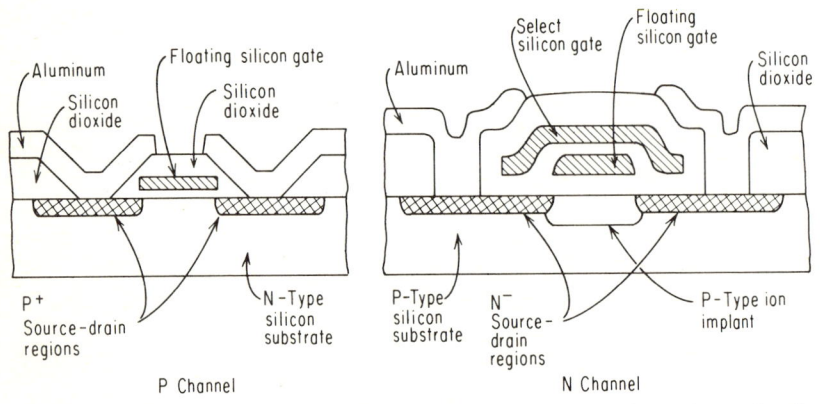

Fig. 4-43 P-Channel Floating-Gate Fuse. Programming is achieved by avalanche injection of electrons into selected P-channel floating gates. High-voltage pulsing is needed only on a single pin (PROG), the power supply need only be at 10 V, and no voltage-supply switching is required for the PROGRAM-READ cycles. (*Reprinted from Electronics, July 6, 1978. Copyright © 1978 McGraw-Hill. Used with permission.*)

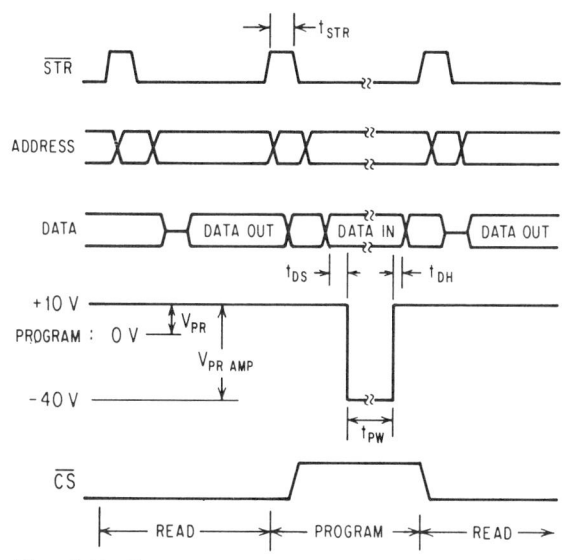

Fig. 4-44 Programming Waveforms. *(Reprinted from Electronics, July 6, 1978. Copyright © 1978 by McGraw-Hill, Inc. Used with permission.)*

change, and then the chip is deselected. When $\overline{CS}$ goes high, the data output buffers turn off and the program data may be presented to the I/O pins.

With data at the I/O pins, the program pin is pulsed negatively from 10 to -40 V, which alters the threshold of the floating-gate transistor in each cell and stores the data. The next positive strobe pulse is especially important, as it precharges the selected line up from operation. Exposure of the array to high-intensity UV light generates photocurrents that discharge the floating gates for reuse.

The CMOS design differs significantly from other erasable PROM configurations, as the simplified schematic of one of the floating-gate memory cells and its associated programming and data-sensing circuitry in Fig. 4-45 shows. When the STROBE line is high, both complementary address register outputs are high for all 10 b (9 address bits and $\overline{CE}$ in the 6604). Decoder Y outputs will be low; $\overline{Y}$ and $\overline{X}$ lines will be high.

These conditions ensure that P-channel devices Q_8 and Q_{11}, which connect the column to the sense amplifier and the programming cell, and the P-channel cell-select transistor Q_9 will all be off, thus isolating all column lines. Because the STROBE line is high, $\overline{STR}$ is low, causing the column lines to be precharged to a known level.

248 CMOS Memories

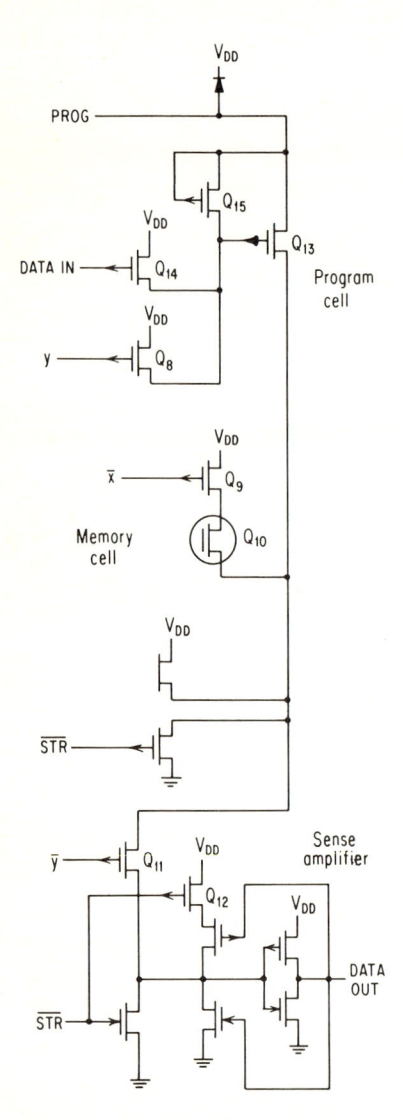

Fig. 4-45 Column Decoders Generate the Complementary Y and $\overline{Y}$ Signals and Row Decoders Produce the $\overline{X}$ Signal. *(Reprinted from Electronics, July 6, 1978. Copyright © 1978 by McGraw-Hill Inc. Used with permission.)*

A minimum positive strobe pulse width is important, since too small a pulse will not allow all lines to be uniformly precharged. If the column has not been allowed sufficient precharge time, a subsequent read from any of its memory cells may not yield valid data if the cell tries to discharge the column toward ground (the cell contained a 0).

On the other hand, the minimum access time is set by the maximum time required to allow a cell to charge a properly recharged column up to the level at which the sense amp can acquire it. This time depends on column line capacitance, sense amp gain, column resistance, transistor thresholds, and other factors.

With the STROBE line high, the column is disconnected from the sense amp. Also, in the sense amp, the feedback loop of the data latch is broken by the turning off of the P-channel transistor Q_{12}, and the internal data output bus is high. Once the STROBE line falls, internally generated clocks time the following series of events:

- The address is latched.
- The address decoders are activated and a single word is selected to drive the columns.
- The columns are connected to the sense amps, and the data is latched internally after an access time.
- If the chip is selected, the output data buffers are turned on, and the internal data latch drives the outside world through transmission gates. If V_{CC} is less than V_{DD}, level translation also takes place.

The buffered address latch can work with many logic levels because it has been designed so that the input switching threshold varies very little with supply voltage. For a supply range of 2 to 12 V, the minimum voltage that is needed to operate the latch varies only from about 1.6 to 2.6 V.

However, the 6603/04 are specified at a minimum of 2.7 V over their standard operating range. This specification allows the user to supply a higher drain voltage for fast access and still drive the address lines with open-collector TTL gates and resistive pull-ups to 5 V. Even TTL gates with totem-pole outputs will drive the address lines satisfactorily. When loaded by the 1-μA CMOS inputs, their output high level is 3.5 V; the 2.4-V level guaranteed by TTL assumes a much heavier load of 10 TTL inputs or 400 μA.

For a set-up in which the address inputs are being driven by TTL, the data output can also be made TTL-compatible by tying pin 24 to a 5-V supply. The output drivers are transmission gates switching either ground or the V_{CC} supply, and they are independent of the drain voltage that powers the memory array. Of course, by connecting V_{CC} and V_{DD} together, an all-CMOS system is also possible. To take full advantage of high-speed operation while maintaining a TTL interface at the inputs and outputs, the STROBE and CHIP SELECT lines should be driven by fast level translators since they require the standard CMOS threshold voltage.

250 CMOS Memories

The main advantage of CMOS is, of course, its low power dissipation. The quiescent power in the 6604/03 is slightly more than 1 nW/b for a 5-V supply. When the devices are cycling at 500 ns, the consumption climbs to 2 and 8 μW/b.

The quiescent and active power dissipations of a variety of erasable PROMs are shown in Table 4-12. The P-channel 1702A, which uses a floating gate similar to the one in the 6604/03, is listed, even though its cycle time is 1500 ns. It is the earliest erasable PROM made and draws the most power. The 2704, 2708, and 2716 use a more complex N-channel stacked-gate cell. What is more, most other erasable PROMs require multiple power supplies, while the 6603/04 require only one.

CMOS/SOS Floating-Gate Avalanche Injection 2-kb EAROM

RCA has designed a nonvolatile 2-kb static, asynchronous EAROM (electronically alterable read-only memory), which dissipates only 50 μW with 5-V operation, retains data for 17.3 yr at 125°C, and can endure more than 300 WRITE and ERASE cycles. The READ access time for this device is 470 ns at 5 V.

The EAROM is based on a storage cell containing two concentric P-type transistors: a floating-gate device for data storage and a conventional device for word line switching. Information is written by hot electron injection from a localized avalanche region that lies along the grain boundaries of the SOS film.

TABLE 4-12 Popular EPROM Power Dissipation Comparison

Device Type P/N	Quiescent Power per Bit (μW)	Active Power per Bit for 450-ns Cycle, 25°C (μW)
1702A (2048 b)	N/A*	160†
6603/04 at 5 V (4096 b)	0.0015	2–8
2704 (4096 b)	N/A	145
TMS2708 (8192 b)	N/A	100
TMS27L08 (8192 b)	N/A	70
2716 (16348 b)	3	17

* N/A = not applicable.
† 1500-ns cycle time.

Popular CMOS ROMs/PROMs 251

With concentric devices, the control and storage gates can function as shields to block the hot electrons from adjacent memory cells.

The 2-kb EAROM chip, measuring about 3.3 × 4.8 mm, has impressive endurance characteristics. During tests, all 2048 b were written into the 1 state and the entire array was simultaneously block-erased with 10-ms pulses, applied to the saturation control line. The storage cells, however, required continually higher erase voltages after each cycle because of electron trapping in the field oxide.

Even so, a typical memory will sustain more than 300 WRITE and ERASE cycles before the erase voltages approach 60 percent of the field-oxide rupture point.

The chip's retention characteristics are similar to those of existing bulk floating-gate avalanche MOS memories. The retention for the weakest 1 state is expected to be 17.3 yr at 125°C or 590,000 yr at 20°C. However, as with bulk FAMOS (floating-gate avalanche-injection metal-oxide semiconductor) memories, simultaneous high voltage and temperature accelerate floating-gate charge losses; as a result, the saturation control gate is grounded except during a WRITE.

Since power dissipation is very low during WRITE and READ operations, the chip can be written with a single pulse, but with few thermal effects.

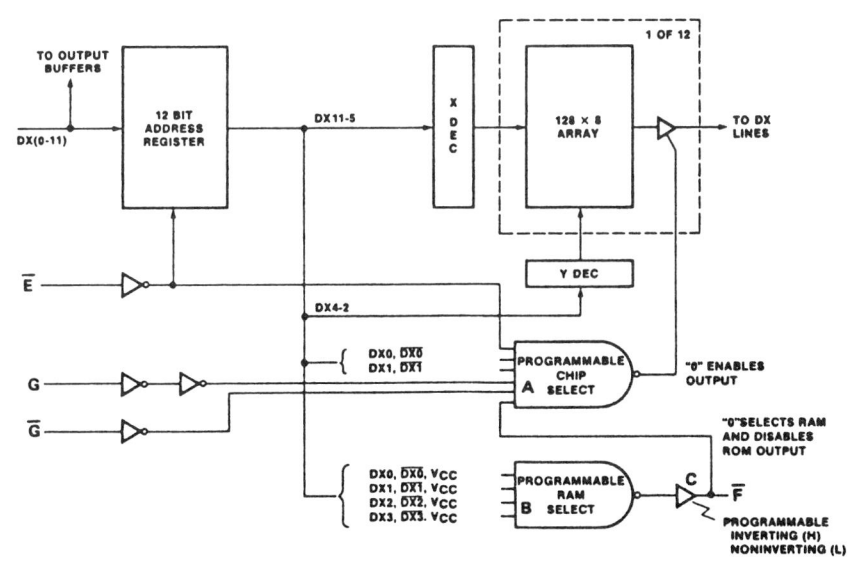

Fig. 4-46 HM6312/A Functional Diagram. *(Courtesy Harris Semiconductor Products Division.)*

The EAROM chip operates from a single power supply of 4 to 6 V, and is compatible with other CMOS, NMOS, or bipolar devices. Further details may be found by referring to the paper by R. G. Stewart, "A CMOS/SOS Electrically Alterable Read Only Memory," *1979 IEDM Proceedings,* Washington, DC, December 1979.

The Harris HM6312/A 1024 × 12-b ROM

The HM6312 and HM6312A (Fig. 4-46) are high-speed, low-power silicon-gate CMOS static ROMs organized 1024 words × 12 b. In all static states, these units exhibit the microwatt power requirements typical of CMOS. The basic part operates at 4 to 7 V, with a typical 5-V, 25°C access time of 350 ns. Signal polarities and functions are specified for interfacing with the HM6100 microprocessor.

The HM6312/A comes in an 18-pin DIP, as shown in Fig. 4-47. Addresses and DATA OUT are multiplexed on 12 lines, DX_0 through DX_{11}. Addresses are loaded into an on-chip register by the falling edge of $\overline{CE}$. DATA OUT, corresponding to the latched address, is enabled when $\overline{CE}$ and OEL are low and OEH is high and the decoded state of DX_0 and DX_1 are true. The RSEL output defines an area in the 4096-word addressing space dedicated to RAM. It can be programmed by DX_0, DX_1, DX_2, and DX_3. This output eliminates a 4-b register and decoder for the high-order address bits to select RAM.

The electrical specifications and switching waveforms are shown in Table 4-13. Combining the HM6312 with the HM6561 256 × 4-b RAMs and the HM6100 microprocessor forms a minimal computing system as shown in Fig. 4-48.

The Supertex 32,768-b CMOS ROM

The CM3200 is the first of a family of high-density CMOS ROMs. It is organized in a 4096 × 8-b configuration and is packaged in a 24-lead DIP. The low-power and low-cost CMOS technology with advanced circuitry makes the CM3200 ideal for microprocessor-based or battery-operated systems. The device operates at high speed and interfaces easily with TTL circuits. It does not require a clock input, yet consumes very little operating or standby power. Each output can drive one standard TTL load.

This device is designed for high-density asynchronous fixed-memory applications such as look-up logic tables and microprogramming. The

Popular CMOS ROMs/PROMs

TABLE 4-13 HM6312/A Electrical Specifications

Absolute Maximum Ratings

Supply Voltage HM6312	+8.0 V
Supply Voltage HM6312A	+12.0 V
Applied Input or Output Voltage	GND—0.3 V to V_{CC} + 0.3 V
Storage Temperature Range	
Industrial HM6312/6312A-9	−40 to +85°C
Military HM6312/6312A-2	−55 to +125°C

DC Characteristics [V_{CC} = 4 to 7 V (HM6312); V_{CC} = 4 to 11 V (HM6312A)]

Parameter	Symbol	Min	Typ	Max	Units	Test Conditions
Logical 1 Input Voltage	V_{IH}	70% V_{CC}			V	
Logical 0 Input Voltage	V_{IL}			20% V_{CC}	V	
Input Leakage	I_{IL}	−1.0		+1.0	μA	0 V ≤ V_{in} ≤ V_{CC}
Logical 1 Output Voltage	V_{OH}	V_{CC} − 0.01			V	I_{out} = 0
Logical 0 Output Voltage	V_{OL}			GND + 0.01	V	I_{out} = 0
Output Leakage	I_O	−1.0		1.0	μA	0 V ≤ V_{in} ≤ V_{CC}
Supply Current	I_{CC}		1.0		μA	V_{in} = 0 or V_{CC}
Input Capacitance*	C_{in}		5.0	7.0	pF	
Output Capacitance*	C_{out}		6.0	10.0	pF	

AC Characteristics (T_A = 25°C, C_L = 50 pF)

Parameter		Symbol	Min	Typ	Max	Units	Test Conditions (V)
Access Time From $\overline{CE}$	(6312)	t_{AC}		500		ns	V_{CC} = 5.0
	(6312A)			250		ns	V_{CC} = 10.0
Output Enable Time	(6312)	t_{EN}		250		ns	V_{CC} = 5.0
	(6312A)			125		ns	V_{CC} = 10.0
STROBE Positive Pulse Width	(6312)	t_{CE}		220		ns	V_{CC} = 5.0
	(6312A)			110		ns	V_{CC} = 10.0
Address Set-Up Time	(6312)	t_{ADDS}		50		ns	V_{CC} = 5.0
	(6312A)			25		ns	V_{CC} = 10.0
Address Hold Time	(6312)	t_{ADDH}		50		ns	V_{CC} = 5.0
	(6312A)			25		ns	V_{CC} = 10.0
Propagation To RAM Select	(6312)	t_{RS}		250		ns	V_{CC} = 5.0
	(6312A)			125		ns	V_{CC} = 10.0

SOURCE: Courtesy Harris Semiconductor Products Division.
* Guaranteed and sampled, but not 100 percent tested.

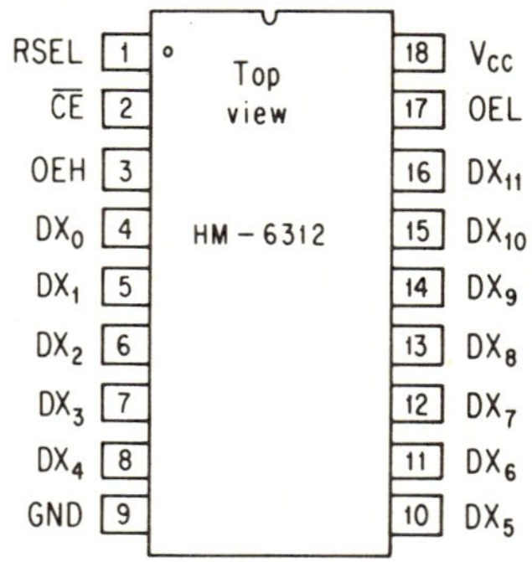

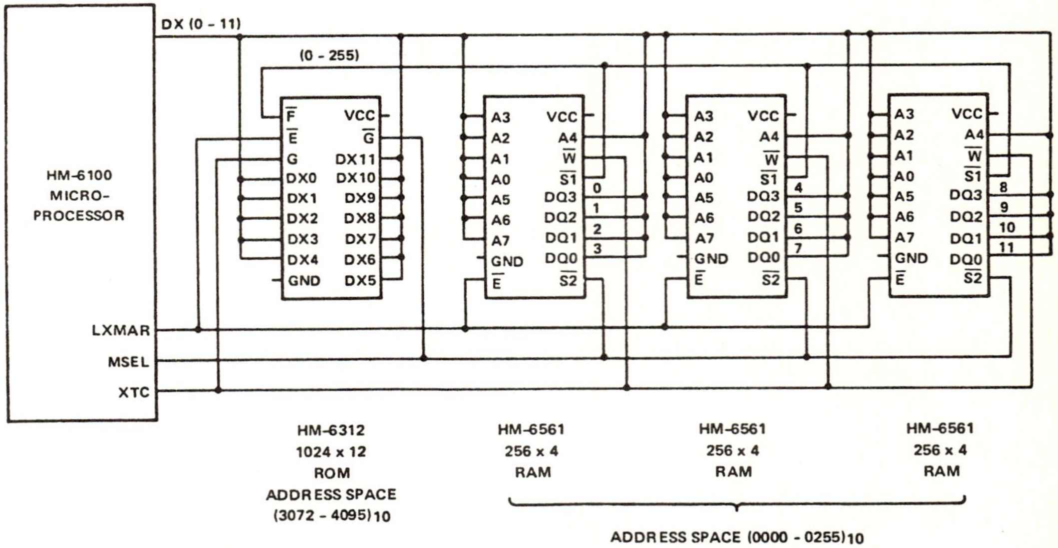

Fig. 4-48 A Minimal CMOS Microprocessor System. *(Courtesy Harris Semiconductor Products Division.)*

Fig. 4-47 HM6312/A Pin Connection Diagram. *(Courtesy Harris Semiconductor Products Division.)*

CM3200 is pin-compatible with the Intel 2708, industry standard 8-kb EPROM and the TI TMS 4732 32-kb EPROM.

The CM 3200 features the following:

- TTL I/O Compatible
- Single 5-V power supply
- Maximum access time—450 ns
- Minimum cycle time—450 ns
- Typical operating supply current—10 mA
- Typical standby supply current—10μA
- Industry standard pin configuration
- Low-power CMOS technology
- Two mask-programmable CHIP SELECTS
- Three state outputs—allow multiple devices on a common bus

Figure 4-49 is a block diagram of the CM3200.

The 12 address inputs are decoded on-chip and select one of 4096 words of 8-b data. Both CHIP SELECT inputs are programmable to be either active high or active low level. When both CHIP SELECT inputs are active, the eight outputs are enabled. But when either of the inputs is not active, the outputs are in a high-impedance state.

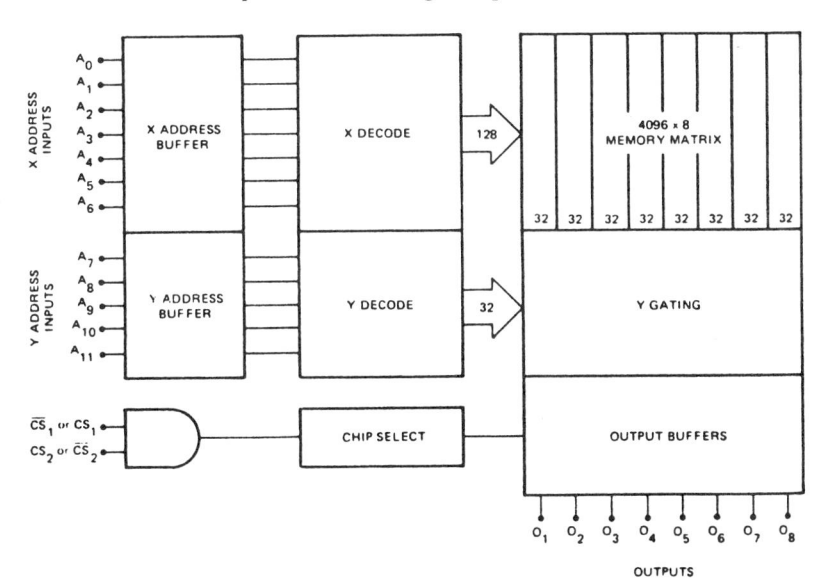

Fig. 4-49 CM3200 Block Diagram.

CMOS Memories

TABLE 4-14 CM3200 Electrical Characteristics

Absolute Maximum Ratings	
Supply Voltage to Ground Potential*	−0.5−+7.0 V
Applied Output Voltage*	−0.5−+7.0 V
Applied Input Voltage*	−0.5−+7.0 V
Power Dissipation	1.0 W
Operating Temperature (Ambient)	0−+70°C
Storage Temperature (Ambient)	−55−+155°C

DC Electrical Characteristics ($T_A = 0$ to $+70°C$; $V_{CC} = +5.0$ V ± 5 percent)

Symbol	Parameter	Min	Typ	Max	Unit	Condition
I_I	Input Current			10	µA	$V_{CC} = 5.25$ V; $V_{in} = 5.25$ V
I_{LKC}	Output Leakage Current	−10		10	µA	$V_O = 0.4$ V to V_{CC}; Chip Deselected
I_{CCS}	Standby Supply Current from V_{CC}			20	µA	$V_{CC} = 5.25$ V; $V_{in} = V_{CC}$; Output not Loaded
V_{OH}	High-Level Output Voltage	2.4 V		V_{CC}	V	$V_{CC} = 4.75$ V; $I_{OH} = -800$ µA
V_{OL}	Low-Level Output Voltage			0.4	V	$V_{CC} = 4.75$ V; $I_{OL} = 2.1$ mA
C_I	Input Capacitance			10	pF	$V_O = 0.0$ V; $T_A = 25°C$; $f = 1$ MHz†
C_O	Output Capacitance			10	pF	$V_O = 0.0$ V; $T_A = 25°C$; $f = 1$ MHz†
V_{IH}	High-Level Input Voltage	2.0		V_{CC}	V	
V_{IL}	Low-Level Input Voltage	V_{SS}		0.65	V	
I_{CC}	Operating Current from V_{CC}			20	mA	$V_{CC} = 5.25$ V; $t_{CY} = 450$ ns

Switching Characteristics ($T_A = 0$ to $+70°C$; $V_{CC} = +5.0$ V ± 5 percent; 1 Series 74 TTL LOAD; $C_L = 100$ pF)

Symbol	Parameter	Min	Typ	Max	Unit	Condition
t_{ACC}	Address Access Time			450	ns	
t_{CS}	CHIP SELECT Time			200	ns	
T_{CD}	CHIP DESELECT Time			200	ns	
t_{OH}	Previous Output Valid after Address Change	50			ns	
t_{CY}	Cycle Time	450			ns	

* Voltage values are with respect to V_{SS}.
† This parameter is periodically sampled and is not 100 percent tested.

READ CYCLE TIMING

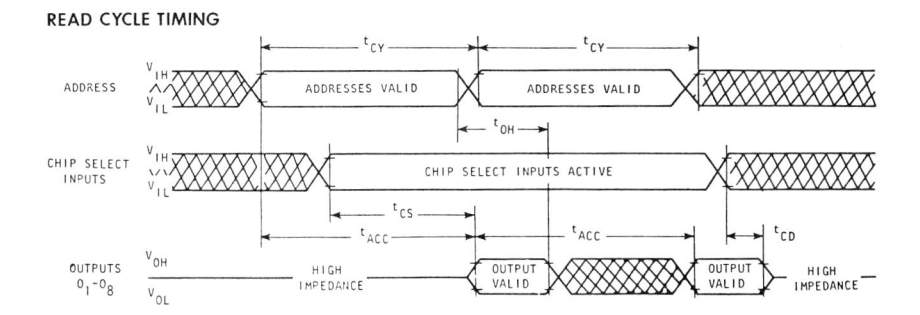

[Timing diagram showing ADDRESS, CHIP SELECT INPUTS, and OUTPUTS O_1-O_8 waveforms with timing parameters t_{CY}, t_{OH}, t_{CS}, t_{ACC}, t_{CD}, voltage levels V_{IH}, V_{IL}, V_{OH}, V_{OL}]

All of the eight outputs must be enabled by both CHIP SELECT controls before the output word can be read. Data will remain valid until the address is changed or the outputs are disabled.

The primary electrical characteristics of the CM3200 are listed in Table 4-14.

CMOS PROM/ROM Electrical Characteristics Summary

As with CMOS RAMs, there is a great deal of developmental activity taking place with CMOS ROMs/PROMs. Research and development is being conducted with ultraviolet, fusible-link, electrically alterable, and electrically programmable types.

Table 4-15 summarizes the commercially available CMOS ROMs and PROMs by memory size. And Table 4-16 presents the salient electrical characteristics of a sampling of the CMOS EPROMs and PROMs listed in Table 4-15. Many more products will be added to this list in the future.

SELECTED CMOS MEMORY APPLICATIONS

Memory System Implementation

Figure 4-50 shows a 256-word $\times$ n-b static RAM memory system. The outputs of four MCM14505 devices are tied together to form 256 words $\times$ 1 b. Additional bits are attained by paralleling the inputs in groups of four. Memories of larger words can be attained by decoding the MSBs of the address and ANDing them with the strobe input.

Fan-in and fan-out of the memory is limited only by speed requirements. The extremely low input and output leakage current (100 nA

TABLE 4-15 CMOS ROM Summary

Memory Size (b)	RCA	Motorola	Harris	Intersil	Hughes	SSS	Supertex
256 × 4		MCM14524	HM6611(P)* HM6661(P)				
256 × 8	CDP1842 (P) CDP18U42(EP)†						
512 × 8	CD40032 CDP1831 CDP1832		HM6641(P)	IM6654 (EP)	HCMP1831 HCMP1832 HNVM3004‡	SCP1831 SCP1832	
1024 × 4				IM6653 (EP)			
1024 × 8	CDP1843 CDP1833/34 CDP57U58(EP)		HM6708 (EP)		HCMP1833/ 34 HNVM 3008‡	SCP1833 SCP1834	
1024 × 12			HM6312/A	IM6312/A			
2048 × 8	CDP1835§ CDP5316§	MCM65516	HM6716 (EP)	IM6316 (EP)	HCMP1835/ 1836	SCM5316	CM1600
8192 × 8			HM6388 HM6389	IM6364			CM6400
4096 × 8 16,384 × 8 32,768 × 8							CM3200 CM1310 CM2560

NOTE: 1800 series ROMs, PROMs, and EPROMs have input latches for interface with 1802 μP, 10-V power supply range, and wider temperature range.
* P = PROM (programmable read-only memory). ‡ EEPROM (electronically erasable PROM)
† EP = EPROM (erasable programmable read-only memory). § CMOS/SOS

258

TABLE 4-16 Selected CMOS EPROM (ROM) Electrical Characteristics Summary

Part Number	Supplier	Organi- zation (b)	EPROM/ PROM	ROM	t_{AA} (ns max)	I_{CC} (μA max)	Power Supply (V max)	Package DIP (Pin)	Address Latch	TTL- Compatible I/O	Three-State Output
HM6611	Harris	256 × 4	×		250	15 μA/200 μA	12.0	16	×	×	×
HM6661	Harris	256 × 4	×		250	15 μA/200 μA	12.0	18	×	×	×
IM6654	Intersil	512 × 8	×		300, 450, 600	100 μA	5.0	24	×	×	×
HNVM3004	Hughes	512 × 8	×[a]		500 Typ	1 mA/10 μA	5	24		×	×
HM6641	Harris	512 × 8	×		200	100 mA/100 μA	5	24		×	×
HNVM3008[c]	Hughes	1024 × 8	×[a]		500 Typ	1 mA/10 μA	5	24		×	×
IM6653	Intersil	1024 × 4	×		300, 450, 600	100 μA	5.0	24	×	×	×
HM6312/A	Harris	1024 × 12		×	220 (10 V) 350 (5 V)	10 mA/800 μA	12.0	18	×	×	×
IM6312	Intersil	1024 × 12			400	100 μA	5.0	18	×	×	×
IM6312A	Intersil	1024 × 12			200	500 μA	12.0	18		×	×
IM6316	Intersil	2048 × 8	×		350 Typ	100 μA Typ	5.0	24	×	×	×
SCM5316	SSS	2048 × 8		×	450	7 mA/100 μA/10 μA	5.0	24		×	×
CM1600	Supertex	2048 × 8		×	450, 550, 800	20 mA/100 μA	5	24		×	×
MCM65516[b]	Motorola	2048 × 8		×	280 Typ	200 μA/5 μA(3V)	5	18		×	×
CM3200	Supertex	4096 × 8		×	450, 800	20 mA/20 μA	5.0	24		×	×
HM6388	Harris	8192 × 8		×	550	100 μA	5.0	24	×	×	×
HM6389	Harris	8192 × 8		×	550	100 μA	5.0	28	×	×	×
IM6364	Intersil	8192 × 8		×	350 Typ	100 μA Typ	5.0	24	×	×	×
CM6400	Supertex	8192 × 8		×	1500, 2000	15 mA/200 μA	5	24		×	×
CM1310	Supertex	16,384 × 8		×	2000	15 mA/200 μA	5	24		×	×
CM2560	Supertex	32,768 × 8		×	1500	20 mA/200 μA	5	24		×	×

NOTE: Listed limits are guaranteed at 25°C.
[a] EEPROM
[b] A discussion of the MCM65516 can be found by referring to B. Donoghue, "C-MOS read-only memory mates with a host of processors," *Electronics*, September 25, 1980, pp 127–129.
[c] Discussions of the HNVM3008 can be found by referring to G. DesRochers, "EEPROM Eclipses Other Reprogrammable Memories," *Electronics Design*, November 22, 1980, and E. K. Shelton, "Low-Power EE-PROM can be Reprogrammed Fast," *Electronics*, July 31, 1980.

259

260 CMOS Memories

maximum) keeps the output voltage levels from changing leakage significantly as more outputs are tied together. With the output levels independent of fan-out, most of the power supply range is available as logic swing, regardless of the number of units wired together. As a result, high noise immunity is maintained under all conditions.

Power dissipation is 0.1 μW/b at a 1.0-kHz rate for a 5.0-V power supply, while the static power dissipation is 2.0 nW/b. This low power allows nonvolatile information storage when the memory is powered by a small standby battery.

Figure 4-51 shows n MCM45537 256 × 1-b RAMs connected in a 256

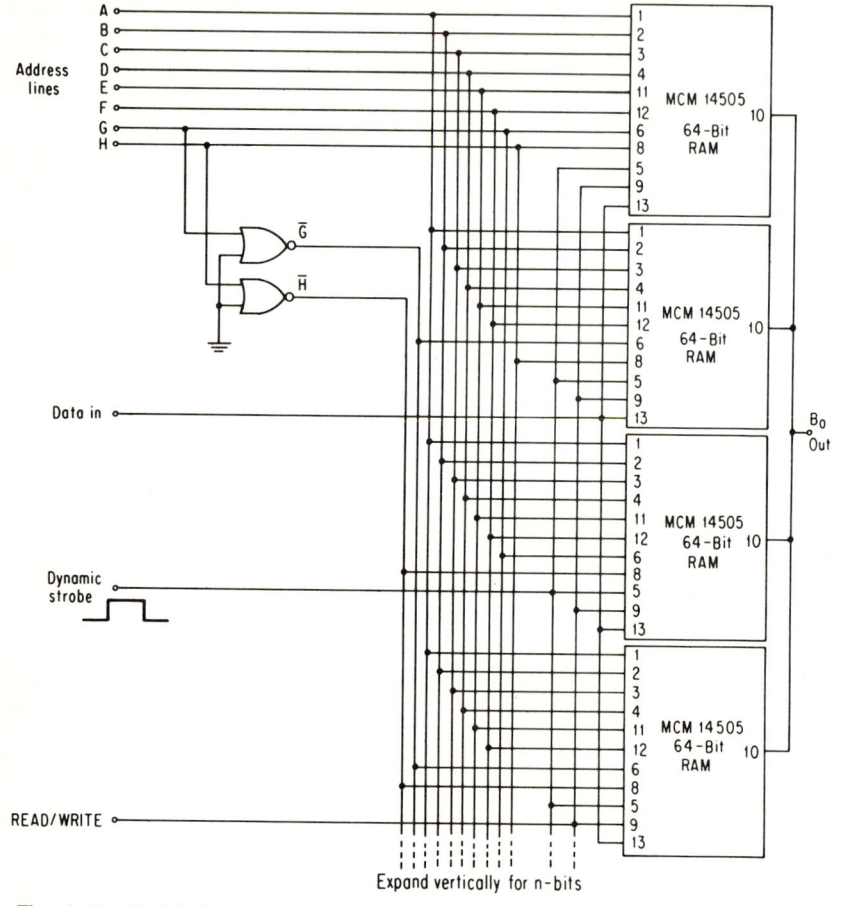

Fig. 4-50 CMOS 256-word × n-bit Static Read/Write Memory. *(Courtesy Motorola Semiconductor Products Division.)*

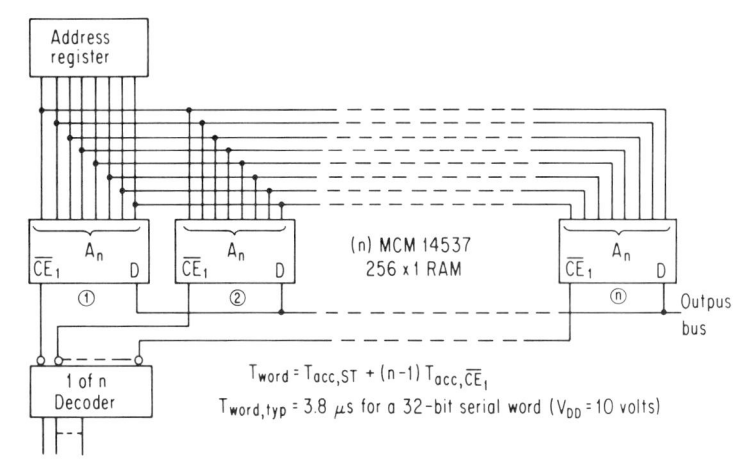

[Figure showing address register connected to multiple MCM 14537 256×1 RAM chips with CE₁ inputs, 1-of-n decoder, and output bus]

$T_{word} = T_{acc,ST} + (n-1) T_{acc,\overline{CE}_1}$

$T_{word,typ} = 3.8 \mu s$ for a 32-bit serial word ($V_{DD} = 10$ volts)

Fig. 4-51 Typical Application for Serial Words Utilizing Bus Techniques. *(Courtesy Motorola Semiconductor Products Division.)*

× n-b array. By controlling the $\overline{CE}_1$ ($\overline{CHIP\ ENABLE}$ 1) control lines with an active low 1-of-n decoder, the accessed data word present in the output memory latches is serially placed on the three-state output data bus.

CMOS Unit Controls Washing Machines with Custom Circuit

Washing machines and other similar appliances such as dishwashers and tumbler dryers are now being manufactured with CMOS ROM programmable controllers that use line frequency as the clock. If the power supply fails, all signals on the chip disappear and data is retained by holding the voltage on the chip with a capacitor.

Main functions of the controller are capacitive touch input circuit logic, program and timing ROM, timer, and control logic for a stepping motor that drives the power switches (Fig. 4-52). Power switching is performed by a 64-position motor-operated rotary disk. The disk pattern determines the sequence and timing of switch openings and closings.

A capacitive touch panel has 10 program inputs, one ECONOMY MODIFIER input, one ERASE input, and a 7-segment LED display that identifies program number. Because a touch must be maintained for 80 ms before the system accepts a VALID signal, there is immunity from electrical interference. In addition, the system will not accept a signal if more than one input is touched.

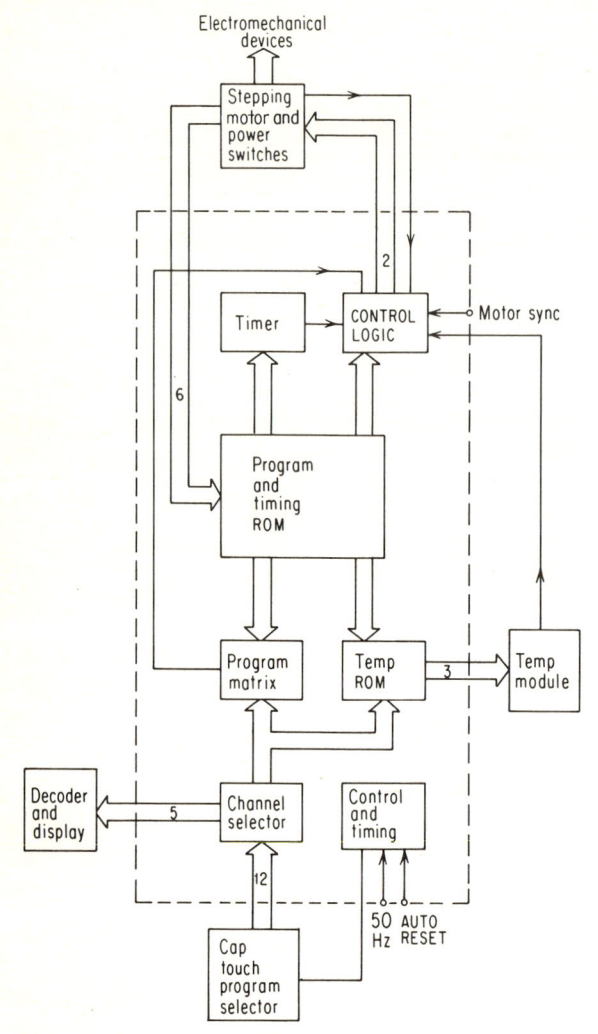

Fig. 4-52 System Architecture for Washing Machine CMOS Controller. Logic Is Simplified by Having up to 64 Program Steps Arranged Sequentially, with a Specific Time Value Assigned to Each.

Interfacing the 6800 Microprocessor*

The use of a 6800 microprocessor in an Aircraft Area Navigation System, which includes a 2-kb × 8-b and a 64-k × 4 b RAM, is shown in Fig. 4-53.

*B. Kopek, "Interfacing the 6800 Microprocessor to National CMOS MM74C910 Memory," Application Note in *Memory Applications Handbook,* National Semiconductor Corp. Santa Clara, CA, 1979. Used with permission.

Selected CMOS Memory Applications 263

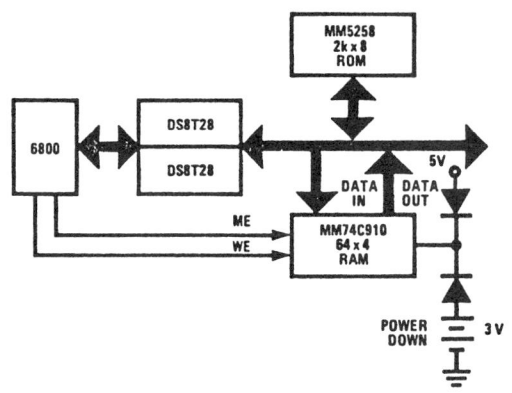

Fig. 4-53 Block Diagram of 6800 Aircraft Navigation System. *(Courtesy National Semiconductor Corp.)*

Nonvolatile memory for waypoint storage is obtained by usage of the MM74C910, 64-word × 4-b RAM.

Interfacing of the 6800 and MM74C910 is accomplished via the DS8T28 bus transceiver. However, additional hardware and software are required to extend the data time on the bus to read and write 4-b data (two MM74C910 CMOS RAMs organized by 8-b words would simplify software).

Figure 4-54 shows the clock circuit hardware required to extend the data time during WRITE. Clock phase 2 is extended from 500 to 550 ns by increasing the time constant of the monostable generator. The output

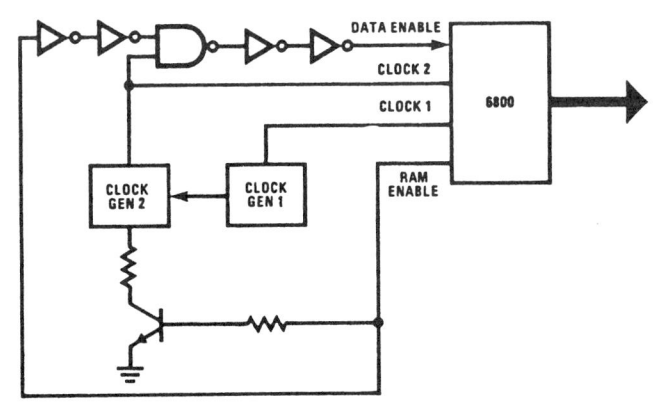

Fig. 4-54 Clock Circuit Block Diagram. *(Courtesy National Semiconductor Corp.)*

264 CMOS Memories

is passed through one NAND and two INVERTER circuits to the DATA BUS ENABLE input of the microprocessor. A data extension of approximately three gate delays is thereby accomplished, allowing a data input hold time t_{HD} (min) of 30 ns to be met before $\overline{WE}$ ($\overline{WRITE\ ENABLE}$) from the 6800 goes high. When the MM74C910 is not being accessed, the DATA BUS ENABLE input of the 6800 is high and the clock 1 and clock 2 monostable generators are a symmetrical 500 ns.

CHAPTER FIVE
Microprocessors

The true growth and application of CMOS technology lies in LSI circuits. As was mentioned in the first chapter, CMOS technology allows the fabrication and integration of functions previously not possible or practically feasible. This is especially true since the advent of producible CMOS/SOS circuits. The combination of CMOS and SOS technologies is a natural one since SOS solves the primary speed-limitation problem of CMOS.

The main practical limitation on the speed of CMOS circuits is the parasitic capacitance of the aluminum conductors that interconnect the transistors. These conductors, separated by a layer of SiO_2 (glass) from the bulk silicon semiconductor substrate, restrict switching speeds by presenting a significant capacitive load for the very low current levels being switched.

SOS technology was evolved primarily to reduce parasitic capacitance—thereby significantly increasing the speed of CMOS (by a factor of 3:1) while maintaining its low-power benefits. The CMOS circuits are fabricated in a thin silicon layer grown on a sapphire substrate. Being a nonconductive material, the sapphire virtually eliminates the problem of capacitance between the aluminized conductors and the substrate. The choice of sapphire as the nonconductor stems from two properties: First, large sapphire crystals can be grown; second, and more important, its thermal coefficient of expansion is virtually identical to that of silicon. If the thermal coefficients for the two materials were not matched, silicon circuits grown on sapphire would buckle and dislodge as the circuits heated.

Circuit capacitance is further reduced through elimination of the large horizontal junction structures used in other fabrication technologies. In SOS, these are replaced by vertical junctions, and junction capacitance is sharply diminished.

A side benefit is the elimination of a processing problem that in other technologies can cause chip failures. In SOS, as in the other technologies, the aluminum interconnects run on top of an insulating layer of SiO_2. With SOS, a pinhole in the SiO_2 layer is usually not catastrophic, because below the SiO_2 is the insulating sapphire; therefore, penetration does not result in a short circuit. In contrast, a pinhole in a non-SOS device could cause a short circuit between the conductor and the underlying silicon layer, leading to device failure.

Packing density of CMOS/SOS devices is improved by a factor of 4:1 over bulk CMOS since with SOS devices the guard bands (used in CMOS technology) are eliminated. Processing steps of the simpler SOS structure are reduced by one-third by virtue of the same reasoning—cutting costs and increasing yield.

CMOS/SOS has been faced with some serious obstacles that are only now being resolved. Despite apparent advantages of SOS technology, many years of research and the developmental efforts by many companies failed to produce significant results until 1977 when Hewlett-Packard announced the development of a 16-b single-chip microprocessor built with CMOS on sapphire. The primary obstacles have been (and are) materials and processing problems. First, significant problems are encountered in the effort to grow an epitaxial layer of silicon on a sapphire wafer; second, impurities on the surface of the sapphire present problems.

Attempts to apply conventional silicon fabrication methods to SOS met still further difficulties. For example, silicon diffusion processes were found to crack the SOS wafers. This made it necessary to develop new processing methods. Because of these and other problems, one SOS effort after another resulted in failure.

In announcing its accomplishment with this technology, Hewlett-Packard attributed its success to step-by-step application of materials and process engineering. Emphasis was placed on techniques for growing sapphire crystals and concentrated efforts were made to eliminate impurities at the silicon-sapphire boundary, which solved a series of problems, including that of transistor leakage. (This problem had caused difficulties for various manufacturers working in SOS technology; for example, RCA reportedly undertook a total encapsulation of SOS transistors to prevent such leakage.) Standard silicon processes were reengineered to adapt them to SOS; these included diffusion and ion-implant processes.

CMOS/SOS is presently being used to produce Hewlett-Packard's 16-b single-chip microprocessor and related circuits for use in the HP300 small-business system and the HP3000 series 33 computer. In addition, Hewlett-Packard is committing all small- and medium-sized computers to this technology for four reasons: high speed, low power, static logic (easier to design with), and reliability. This strong commitment to CMOS/SOS by Hewlett-Packard should be followed by a commitment on the part of other companies. Sperry Univac (Blue Bell, PA) believes that the low-capacitance and low-power characteristics of CMOS/SOS make it well suited for future VLSI design and so is beginning pilot production of such circuitry. Two large systems houses see the handwriting on the wall and are diverting their efforts to this technology.

In competition with CMOS/SOS will be the bipolar I^{2}L technology on one side and NMOS or scaled NMOS on the other. Hewlett-Packard has projected that the speed of CMOS/SOS and scaled NMOS will be approximately equal in the future. However, in the long run, it is projected that SOS will yield superior speed, especially in pure random logic circuit configurations such as microprocessor CPU designs. Perhaps this is the reason for the transfer of this technology between RCA (the largest commercial producer of CMOS/SOS) and Intel (the leader in the development of microprocessor circuitry).

Perhaps the only major disadvantage of SOS is its relatively complex processing and higher wafer cost resulting in higher finished product cost. CMOS/SOS wafers are very expensive to produce using the basic silicon ingot growing techniques. A more promising technique presently undergoing examination is sapphire ribbon technology, which produces square-type wafers cut from the long-drawn ribbon. RCA is attempting to significantly reduce sapphire ribbon costs by improving fabrication techniques.

Since one of the major applications for CMOS/SOS will be for microprocessors, it might be well to assess the increased device cost compared to that of NMOS microprocessors. The approximate cost of components in a computer is approximately 10 percent of the selling price. As such, the unit cost of a CMOS/SOS microprocessor is insignificant when compared to the total system cost, and as well it has the benefits of very high speed and throughput, high density, and very low power consumption.

Within this context it is now appropriate to present some of the LSI implementations of CMOS and CMOS/SOS technology in this final chapter on microprocessors.

MICROPROCESSOR ARCHITECTURE*

The advent of the "computer on a chip" (monolithic) microprocessor (in 1971) has brought about a dramatic change in the design of digital circuits. Instead of dealing with small individual components, designing with microprocessors involves the use of combinations of functional blocks to arrive at the desired circuit, thus immensely simplifying the design function.

Part of the microprocessor's flexibility and usefulness is the ability to implement designs by a combination of hardware and software.

It is best to begin by defining a microprocessor and proceed from there. A microprocessor may be defined as an IC component that is capable of performing arithmetical and logical operations under program control in a bit-parallel fashion. It has the same basic architecture as a larger-scale computer.

Figure 5-1 shows a generalized block diagram of a microprocessor

---

*Portions of this section excerpted from H. Tweddle, "An Introduction to Microprocessors and the RCA COSMAC COS/MOS Microprocessor," RCA Solid-State Products Division Application Note ICAN6416, RCA, Somerville, NJ, 1975. Used with permission.

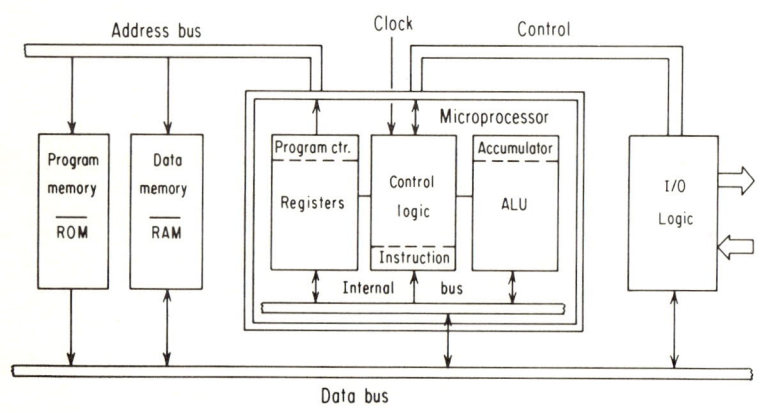

Fig. 5-1 An Elementary Microprocessor System.

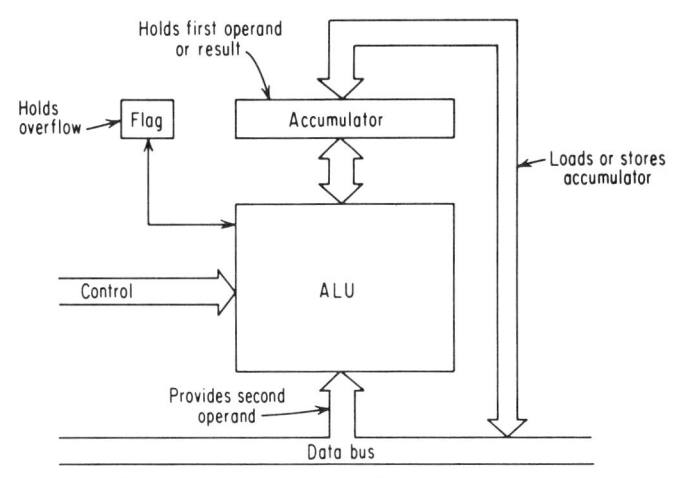

Fig. 5-2 The Arithmetic and Logic Unit.

together with the three additional functions usually required in a system: program memory, data memory, and I/O electronics. As can be seen, the microprocessor is composed of three primary blocks: the ALU (arithmetic and logic unit), the registers, and the control logic.

The Arithmetic and Logic Unit

The ALU shown in Fig. 5-2 performs arithmetic and logical operations on binary data presented to it via the internal data bus. This bus comprises a number of lines (usually 4, 8, 12, or 16) on which data words may be placed. The word length upon which the ALU operates is an important characteristic of a microprocessor: a longer word, by allowing more data to be processed at one time, provides potentially higher speed, but results in a more complex, and thus more expensive, machine. The number of different ALU functions available to the user is again a characteristic of a particular microprocessor.

Associated with the ALU is the accumulator, which is a register for temporary storage of operands and results of ALU operations. Some microprocessor architectures employ several accumulators; in the simplest case, however, one operand is loaded into the accumulator and the other is presented on the data bus, while the result appears in the accumulator, overwriting the first operand. An overflow of the accumulator sets a flip-flop called a *flag*. There may be a number of other flags, both to indicate overflow or underflow of registers and to show certain internal- and external-state information.

Registers

A microprocessor stores memory addresses and data in a number of registers, as shown in Table 5-1. One register holds the current program counter, which addresses a location in program memory. A group of registers is usually provided for temporary data storage; these registers are referred to as *scratchpad memory*. In most applications, some additional read and write data memory is required; this memory is addressed by a register containing the data pointer.

A register block may also be set aside as a "stack," which provides LIFO storage of return addresses and data. A stack is needed when the program includes subroutines. A limited stack size, however, limits the subroutine nesting capability of a microprocessor. A more versatile approach is to form a stack in external memory, in locations addressed by the stack pointer.

The word length of the address registers is important as it defines the number of memory locations that may be directly addressed. The word length of the data registers is usually defined by the word length of the machine.

The Control Logic

The control logic is responsible for defining the operation of the ALU, data movements within the microprocessor, and data transfer to and from external devices. It also provides control signals to help external logic to interface with the microprocessor and can be instructed to change the program counter, and thus the microprocessor operation sequence, in response to the state of the control inputs, flags an internal register. The control logic derives its timing from one or more clock

TABLE 5-1 Typical Microprocessor Register Usage

Program Counter	Addresses Current Instruction in Program Memory
Data Pointer	Addresses Current Location of Interest in Data Memory
Stack Pointer	Addresses Next Vacant Location in External Stack
DMA Pointer	Addresses Location in Data Memory for DMA Transfer
(Internal) Stack	LIFO Data/Address Storage
Scratchpad	Random-Access Data/Address Storage

inputs; its function is defined by the instruction, a word presented to it from program memory and stored in the instruction register.

Contained in the control logic portion of the microprocessor (Fig. 5-1) is an instruction decoder, which is typically an internal ROM that translates the machine instruction code into microinstructions which are executed by the processor. Some microprocessors permit the user to define these microinstructions on an external-control ROM chip.

Program Execution

The sequence of operations that the microprocessor is required to perform is stored as a string of instructions called a *program* in the program memory. The first of these instructions is addressed by the program counter and fetched into the instruction register. The control logic then acts upon this instruction to execute the operation it specifies. The program counter is then incremented, the next instruction is fetched, and another FETCH-AND-EXECUTE sequence is carried out. In this way, the microprocessor steps through the instructions to perform the task defined in the program.

Certain instructions, usually called *branch instructions,* change the program counter to a location other than the immediately subsequent one, allowing program jumps and the use of subroutines. Subroutines are frequently used groups of instructions which may be stored away from the main program stream and called into use whenever needed by the main program. They help conserve memory space, since a subroutine which may be used several times in the course of a program need be stored only once.

Conditional branch instructions test the state of a particular register or flag and implement the branch if the required condition is met. If the condition is not met, the microprocessor continues with the next instruction in the main program. In this way the microprocessor may be made to respond differently to different external or internal conditions.

Buses

Microprocessors generally use buses as a means for moving and transferring data, addresses, and control signals between system components. The reason is that microprocessors are packaged as conveniently as practical, which means IC packages having a minimum number of pins.

Resource Sharing

As in any computer system, a number of scarce resources in a microprocessor must be shared by different jobs or users at different times.

These resources include buses, registers, I/O pins, and memory or control programs. These resources are shared by means of time-division multiplexing. The control and synchronization section of the microprocessor defines specific time periods, or subcycles, during which certain operations are allowed to take place. These operations may be internal to the processor, or may refer to external operations such as DATA or INSTRUCTION FETCH from memory. It is the task of the system design to provide means for decoding the control and synchronization signals provided by the microprocessor to coordinate external operations.

It must be realized that as a consequence of resource sharing the microprocessor supplies information only during brief periods of time, which may not coincide with those periods when other system elements are prepared to accept and utilize that information. Auxiliary hardware, in the form of decoders and latches, must be used to capture this information and apply it to other system elements.

Input/Output

The I/O portion of the microprocessor provides information as words on the data bus, with additional controls available separately. The I/O electronics is responsible for interfacing these signals with whatever I/O devices the system uses. Broadly, the functions of the I/O electronics include synchronization of data transfer, selection and activation of one of a number of I/O devices, and the formating of data so that it is compatible with the device selected. The functions available on the control lines have a marked effect on the complexity of the I/O electronics. The I/O electronics may also be required to do logic-level conversion as part of the interfacing function if this ability is not provided within the microprocessor itself.

An I/O data transfer may be initiated either by the microprocessor in the course of its program execution (programmed mode I/O), or by one of the I/O devices.

Interrupts

Certain applications of microcomputer systems require immediate response of the processor to an external condition. In such cases the processor must interrupt the program presently being executed and begin a new program to handle the external condition or interrupt. There are a number of different types of interrupts depending on the number and priority of external devices to be serviced:

- Simple interrupt
- Vectored interrupt
- Priority interrupt

The simple interrupt merely specifies that a single external device requires servicing by the processor. The vectored interrupt provides the ability to recognize an interrupt from any one of several external devices. More particularly, the vectored interrupt specifies which device requires servicing. This specification is done by a data field, or "vector," which specifies the identity of the external device. Finally, a priority interrupt also recognizes an interrupt from any of several external devices, but specifies which device has priority over other devices.

Generally an I/O device requests an interrupt by raising the INTERRUPT input of the microprocessor. The current program execution is then halted, data and addresses required for eventual resumption of the program are stored in the stack as for a normal subroutine, and the microprocessor starts performing the interrupt service routine defined by the user. The interrupt service routine first establishes which I/O device has requested the interrupt and then performs the appropriate instructions to deal with the request.

Memory Access and Transfer

Memory access and transfer architecture are other important features that are critical for certain system design applications. High-speed operations, particularly requiring considerable data transfer between memory and peripheral devices, utilize the memory access arrangement known as *DMA (direct memory access)*. DMA is a direct transfer of blocks of data (words) from or to predetermined memory locations to or from a peripheral device without direct processor control by means of an I/O device that supplies the memory address and data for each word to be transferred. It also contains the logic to increment addresses to succeeding words, count the number of words transferred, and determine when the transfer is complete.

DMA may be externally or internally controlled. In the first case, raising the DMA REQUEST pin causes the microprocessor, after completing its current instruction, to detach itself from the data and address buses. The I/O device is then permitted to communicate with data memory at a speed limited only by the access time of the memory, rather than by the cycle time of the microprocessor. This form of DMA can be made very fast by using fast memory, but in that case the I/O electronics is required to provide ADDRESS and CONTROL as well as DATA inputs to the memory, thereby increasing interface complexity.

In the second form of DMA, the microprocessor itself provides ADDRESS and CONTROL signals in response to a DMA request; the I/O electronics need only present the data to the I/O bus. This is the more convenient form of DMA and is considerably faster than a standard programmed or interrupt mode I/O. However, the speed of data transfer is a function of the cycle time of the microprocessor, and therefore no advantage is gained from using a very fast memory.

A microprocessor, like a larger-scale digital computer, operates on the basis of synchronous sequential circuits. The processor performs certain operations during predetermined clock cycles, for example, reading and writing operations that utilize the external bus. During other clock cycles, the processor performs internal operations, thereby freeing the bus for use by the peripherals and memory. The specific times during which the bus is free is indicated by the synchronization and state information that is obtainable from the microprocessor. By decoding the synchronization and state information, an indication that the bus is free for use may be transferred to the peripherals and memory. The peripheral device then "cycle-steals" from the processor and transfers the data along the bus to the memory during the unused processor cycle time. This concept of cycle stealing is a very important one for efficient microprocessor system design using high-speed peripherals.

External Memory Requirements

Several different types of semiconductor memories are commonly used with microprocessors, as shown in Table 5-2. Data memory is written into in the course of a program; therefore it is implemented in a RAM, a read and write memory that is volatile (i.e., it loses its data when the power supply is removed). Program memory, on the other hand, must be non-

TABLE 5-2 Types of Memory Commonly Used with Microprocessors

RAM	Random-Access Memory	Volatile	Data Storage, and Program Storage in Some Applications
ROM	Read-Only Memory	Nonvolatile	Program Storage in Production Systems
PROM	Programmable ROM	Nonvolatile	Program Storage in Small-Quantity Production and Prototypes
EPROM	Erasable PROM	Nonvolatile	Program Storage in Small-Quantity Production and Prototypes

volatile and is not altered during normal microprocessor operation so that program storage, in a production system, is normally a ROM; however, if the application allows or requires reloading of the program, the program may be stored in a RAM. This memory may also be used during initial development of a system where frequent program changes are required. Another possible form of program storage is the PROM, which may be used in prototype systems or in small-quantity production systems where a mask-programmed ROM is not economical. It may be found convenient in a prototype system to use an EPROM for program storage since, although nonvolatile, an EPROM may be reprogrammed as the system is modified. Chapter 4 discussed the topic of CMOS memories in detail.

Mircoprocessor Software

A microprocessor operates by means of a sequence of instructions which constitute the "software," as opposed to the physical hardware on which the instructions are executed. Software, or machine instructions, are essentially replacements of hardware components: Instead of using two particular hardware components, for example, an instruction will specify that the operation be performed twice on a single component. Once the basic logical functions are supplied in hardware, any arithmetic or logical calculation may be done by a program of instructions.

The use of microprocessor software as a replacement for hardware components is one of the key advantages of microprocessor systems. Any hardware arrangement can be modeled or simulated by software, and implemented in a microprocessor system. Cost considerations that make a hardware design impractical for a given application are not applicable to software design, and many current microprocessor applications consist of the replacement of high-cost, random-logic hardware systems by a lower-cost programmed microprocessor system.

Three kinds of software are offered by IC manufacturers: operating, diagnostic, and program development.

Operating software is a group of programs that run on the microcomputer under normal use, i.e., they directly control the hardware operation. In a finished system the programs reside in ROMs or PROMs. The user must write her or his own operating software because it represents the logic design of the system or product being built. A manufacturer may supply some prepackaged items, such as mathematical subroutines, but the rest must be created to suit the application.

Diagnostic software, on the other hand, is a fixed package of programs supplied by the microprocessor manufacturer. These test the microcomputer hardware and verify that the system is operating properly. There are also software diagnostic programs, such as simulators and debuggers, that test for proper program sequencing and functioning.

Program development software represents the largest investment on the part of the microprocessor supplier. It is this type of software that is usually referred to when one speaks of a vendor's "software support."

For a designer, much of the start-up (developmental) effort is linked to the coding phase. Coding converts system programs, or algorithms, into instructions that can be loaded directly into memory. The basic tools, themselves programs, typically require the use of time-sharing services or other computer facilities.

Of all the available software tools, few are more important to designers than the assembler, a program that converts symbolic mnemonic commands into the binary form needed by a microcomputer. The mnemonic commands themselves form an assembly language that offers a shorthand way of writing the binary instructions.

Generally, a single assembly statement generates a single storable command. The shorthand statements are grouped into fields designated by the following four names: label, operator, operand, and comments.

The four elements, when combined on a single line, are separated from one another by some form of delimiter, such as one or more blank spaces, a slash, or a comma. The comments field is used only to help others understand what the programmer intends; it will not generate any instructions for the microcomputer.

For example a sample assembler statement might appear as follows:

Label	Mnemonic	Operand	Comment
UPDAT	LDAA	NB	Begin the Loop

Labels help the programmer use branch commands; one can direct the program to go backward or forward to a specific statement in an assembly listing just by giving the statement's label. The mnemonic command LDA A instructs the microprocessor to load the accumulator known as A with the data that will come from the location described by operand NB. The operand tells the microprocessor to fetch data from the location called NB.

How data is addressed affects computing efficiency. Too long an address can slow the microprocessor down. Too short an address can limit the number of words that can be accessed readily.

COMMERCIALLY AVAILABLE CMOS MICROPROCESSORS

The microprocessors to be discussed make use of CMOS technology with its well-known advantages of low power consumption, high noise immunity, wide power supply tolerances, and reliable operation over a wide temperature range. As such, CMOS microprocessors find usage in battery-powered applications such as electronic toys and games (the CDP1802 and TMS1000), automotive applications such as RCA's 1803 for Chrysler's spark control computing system, and hobby and home computers.

The CMOS microprocessors available today fall into two major categories: those developed specifically as CMOS microprocessors and those that are CMOS versions of machines originally designed with other LSI technologies.

At present two processor families dominate the CMOS microprocessor area: RCA's 1800 series and Intersil's 6100 series, while a host of others are rapidly going into production. Both manufacturers offer compatible memories and expanding support circuits for these microprocessors (Table 5-3). This situation will change, however, with the proliferation of the CMOS versions of the 8-b NMOS microprocessors, as well as many new 4-b microprocessors.

As will be seen, an impressive armament of CMOS microprocessors is available, or will be so shortly, to handle a host of tasks in conjunction with many of the support circuits discussed already in the text (A/D and D/A converters, memories, etc.). However, these support circuits, which in a typical system outnumber the microprocessors used, are not limited to use only with CMOS microprocessors—they work with all types of microprocessors.

The Motorola MC14500 1-b Microprocessor

Designed as a programmable logic controller, the single-chip MC14500 1-b CMOS industrial control unit (Fig. 5-3) replaces multibit processors or hand-wired logic in decision-oriented tasks. Housed in a 16-pin package, the device accepts sixteen 4-b instructions. Each instruction performs logical operations on data appearing on a 1-b bidirectional data line. The main attribute of the device is its simplicity in repetitive control system applications.

A minimum system consists of the processor, an external memory, a program counter, an 8-channel data selector, and an 8-b addressable latch. Instruction lines are TTL-compatible and the 1-b bidirectional data line has three-state capability. On-chip there are three 1-b registers that are directly addressable, and the clock oscillator.

Sixteen 4-b instructions comprise the entire instruction set. There are seven logic, five program control, two output, and two no operation instructions. All operations are performed at the bit level. Looping control structure is used where the program counter feeding the external memory that inputs instructions to the processor wraps around after reaching its highest value and repeats the program.

Time-invariant software is the important feature of this processor. This means that the processor can effect a conditional jump without parallel loading of the program counter with the jump address. Whole

TABLE 5-3 Two Families Currently Dominate the CMOS microprocessor field. But forthcoming Introductions of CMOS Equivalents for NMOS Microprocessors Will Undoubtedly Change this Market Picture

Device Family	1802	6100

Suppliers	RCA
	Solid State
	  Scientific
	Hughes
	  Semiconductor
		Intersil
		Harris
		  Semiconductor

Word Size	8 b	12 b
Supply Range	4–12 V	4–11 V
Clock Frequencies	DC–6.4 MHz	DC–8 MHz
Current at 5 V	100 μA Max (Quiescent)	2.5 mA Max
Comments	16 Scratchpad Registers; Can Address up to 64 kb	DEC PDP-8/E In-Chip Form; Can Address up to 32 kb

Support Chips Available	UART
	Video-Display
	  Controller
	TV Interfaces
	Multiplier/Divider
	I/O Port
	Buffers, Bus Drivers
	Color Generator
	Latches, Decoders
		Peripheral Interface
		  Element
		UART
		DMA Memory, Timer,
		  Controller
		Bit-Rate Generator
		Manchester
		  Encoder/Decoder
		Parallel I/O Port
		Buffers, Bus Drivers
		Latches, Decoders

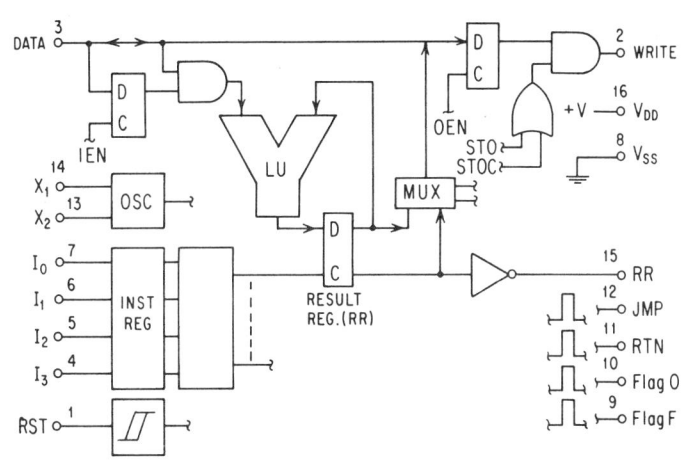

Fig. 5-3 MC14500 Block Diagram.

blocks of instructions can be either turned on or off. This leads to the looping control structure, in which the same sequence of commands is encountered with certain blocks of code being selectively enabled or disabled. In conventional systems the execution time of the program varies with the state of the input signals.

No prototyping hardware is available for this circuit. The design cycle for implementing a working system is intended to be so short as not to require any additional hardware or software. Particular emphasis is made of the fact that this processor is much easier to use than 4-, 8-, or 16-b models.

Available support hardware consists of the MC14099 8-b addressable latch, the MC14512 8-channel DATA SELECT circuit, and the MC14599 8-b READ/WRITE addressable latch with master reset.

The TI TMS1000C 4-b Microcomputer

The TSM1000C series consists of CMOS versions of the TMS1000/1200 4-b microcomputers. Fully software-compatible with the TMS1000 series, these new versions are intended for applications requiring low power consumption, higher speeds, and low systems cost. The TMS1000C series offers a flexible supply voltage (3 to 6 V); the power drain is typically 5 mV at 5 V. The clock rate of the device is 1 MHz. It also features three subroutine levels and eight data inputs in the TMS1200C 40-pin package, offering double the number of data inputs in the original P-channel TMS1000 family.

280 Microprocessors

The CMOS versions also provide a power-down mode, controlled by the HALT pin, that drops typical power consumption to 5 μW, while preserving memory data and the current machine state.

Each TMS1000C microcomputer has ALU, control, clock, I/O circuits and memory on a single silicon chip (Fig. 5-4). The RAM is organized as 64 × 4-b words, with the ROM organized as 1024 × 8-b words. The ROM is mask-programmed for the user's requirements.

The TMS1000C instruction set contains a total of 54 commands that are divided into five basic groups: 12 register reference instructions, 27 arithmetic and logic operations, three bit-manipulation commands, five I/O instructions, and seven memory addressing instructions.

Software support for the TMS1000 family consists of an assembler, a simulator, a high-level language compiler (TIML) and a variety of utility programs. There is no program library available.

Special features of the software include the wide variety of register and accumulator operations possible as well as the individual bit set, reset, and test operations on the contents of a memory location.

Hardware support for the TMS1000 family starts with the 64-pin system evaluator chips that permit external RAM and ROM interfaces and prototype boards for the evaluation chips. Also available is the AMPL

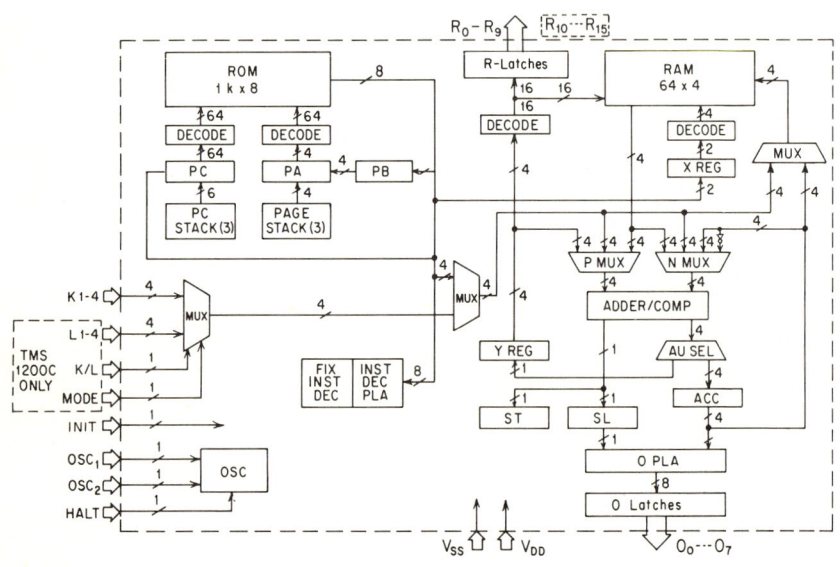

Fig. 5-4 TMS1000C Block Diagram.

TABLE 5-4 Key TMS1000C Specifications

	TMS 1000C	TMS 1200C
Package pin count	28 pins	40 pins
Instruction ROM	1024 × 8-b	
Data RAM	64 × 4-b	
"R" Individually Addressed Output Latches	10	16
Maximum Rated Voltage (O, R, and K)	6 V	
Data Inputs	4	8
Instruction Set	43	
Power Supply	5 V	
Typical Dissipation	5 mW	
Clock Frequency	1 MHz Max	
Subroutine Levels	3	
HALT (Power-Down)	Yes (5 μW)	

(amplifier) development system which permits assembly language programming simulation and in-circuit emulation.

The salient features of the 1000C family are summarized in Table 5-4.

The RCA CDP1802 8-b Microprocessor*

The widely used CDP1802 is an 8-b register-oriented CPU designed as a general-purpose computing or control element in a wide range of stored program systems. It includes a dedicated DMA pointer on-chip, which is used to generate timing functions with accuracies limited only by the crystal-clock oscillator tolerance.

Packaged in a 40-lead DIP (Fig. 5-5), the CDP1802's operating condi-

* Portions of this section excerpted from RCA Solid-State Products Division Application Note ICAN6416 (see reference for the section titled "Microprocessor Architecture"). Portions also excerpted from RCA file No. 1023, Microprocessor Products CDP1802D/CDP1802CD, RCA Solid-State Products Division. Used with permission.

282 Microprocessors

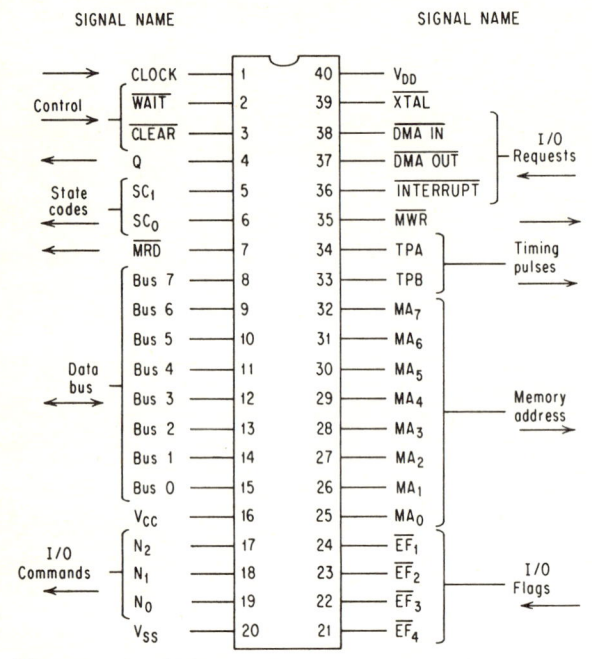

Fig. 5-5 The CPD1802 Microprocessor Pin Connection Diagram. *(Courtesy RCA Solid-State Division.)*

Pin connections:

Signal Group	Signal	Pin		Pin	Signal	Signal Group
Control	CLOCK	1		40	V_{DD}	
Control	$\overline{WAIT}$	2		39	XTAL	
Control	$\overline{CLEAR}$	3		38	$\overline{DMA\ IN}$	I/O Requests
	Q	4		37	$\overline{DMA\ OUT}$	I/O Requests
State codes	SC_1	5		36	$\overline{INTERRUPT}$	
State codes	SC_0	6		35	$\overline{MWR}$	
	$\overline{MRD}$	7		34	TPA	Timing pulses
Data bus	Bus 7	8		33	TPB	Timing pulses
Data bus	Bus 6	9		32	MA_7	
Data bus	Bus 5	10		31	MA_6	
Data bus	Bus 4	11		30	MA_5	
Data bus	Bus 3	12		29	MA_4	Memory address
Data bus	Bus 2	13		28	MA_3	Memory address
Data bus	Bus 1	14		27	MA_2	
Data bus	Bus 0	15		26	MA_1	
	V_{CC}	16		25	MA_0	
I/O Commands	N_2	17		24	$\overline{EF_1}$	I/O Flags
I/O Commands	N_1	18		23	$\overline{EF_2}$	I/O Flags
I/O Commands	N_0	19		22	$\overline{EF_3}$	I/O Flags
	V_{SS}	20		21	$\overline{EF_4}$	

tions at 25°C include 5 μs clock input rise or fall time, 400- to 800-kbyte DMA transfer rate, 80- to 160-ns clock pulse width, and 150- to 300-ns clear pulse width. Absolute maximum ratings include temperature ranges of -55 to 125°C operating, -65 to 150°C storage; 500-mW power dissipation per package (for -55 to 100°C operating temperature range); and 100-mW device dissipation per output transistor (for -55 to 125°C operating temperature range).

The CDP1802 can operate with any combination of standard RAM and ROM from a single voltage supply. It requires no minimum clock frequency, is TTL-compatible, has an 8-b parallel organization with bidirectional data bus, and has a flexible programmed I/O mode, program interrupt mode, and programmable output port. In addition, it has a single-phase clock and simple control of RESET, START, and PAUSE operations. Four I/O flag inputs are directly tested by branch instructions.

The architecture of the 1802 revolves around the on-chip 16 × 16-b register file, scratch-pad RAM that is used to simplify addressing and memory reference commands (Fig. 5-6). Because of the on-chip register

Commercially Available CMOS Microprocessors **283**

file and the built-in clock, a minimal system can be built around the processor and a single ROM.

The instruction set of the CDP1802 consists of 91 single-byte commands grouped into five basic types: register, memory, and logic; arithmetic; branch, skip, and control; and I/O byte transfer instructions.

Most instructions require 2 machine cycles (one instruction period). The only exceptions are the LONG BRANCH and LONG SKIP instructions, which require 3 cycles. Each machine cycle is internally divided into eight equal time intervals T, so the instruction time is $16T$ for 2 machine cycles and $24T$ for 3 cycles.

Each CPU instruction is fetched on the first machine cycle and executed during the second cycle, except for LONG BRANCH and LONG SKIP instructions that require the first machine cycle to fetch the instruction and the second and third cycle to fetch the address (execute).

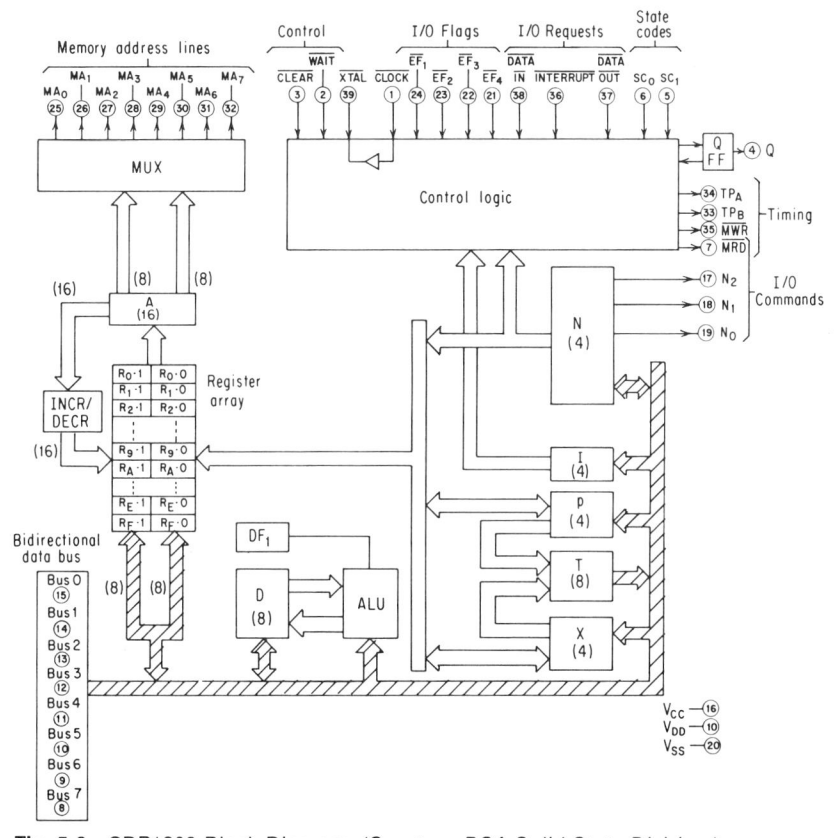

Fig. 5-6 CDP1802 Block Diagram. *(Courtesy RCA Solid-State Division.)*

Each instruction is broken into two 4-b hexadecimal digits, designated as I (the higher-order digit) and N (the lower-order digit). The I word specifies the instruction type, and the N word either designates the scratchpad register to be used or acts as a special code.

Register operations include instructions that count or move data between internal 1802 registers. Memory reference commands provide directions to load or store a memory byte. Branching operations provide conditional and unconditional branch instructions that can either work in the current memory page or go to any location.

Arithmetic and logic instructions provide many of the common operations: ADD, SUBTRACT, AND, OR, EX-OR, and SHIFT, while control and I/O commands take care of all the timing and data transfer operations. The control functions facilitate program interrupt, operand selection, and branch and link operations, and control the Q flip-flop. The I/O functions handle memory loading and all data transfer operations into and out of the 1802.

Software support includes arithmetic, resident editor and assembler,

TABLE 5-5 1802 Primary Specifications

Data Word Size	8 b
Address Bus Size	16 b
Direct Addressing Range	65,536 Words
Instruction Word Size	1–3 Bytes
Number of Basic Instructions	91
Shortest Instruction/Time (MOST)	2.5 μs
Longest Instruction/Time (LONG BRANCH)	3.75 μs
Clock Frequency (Min/Max)	DC/6.4 MHz
Clock Phases/Voltage Swing	1/supply voltage
Dedicated I/O Control Lines	9
Package	40-Pin DIP
Power Requirements	4–12 V/1.6 mA (5 V)

TABLE 5-6 1802 Support Circuits

Model	Description
CDP1802	CPU
CDP1851	Programmable I/O
CDP1852	8-b I/O Port
CDP1853	N-bit-of-8 Decoder
CDP1854/A	Universal Asynchronous receiver-transmitter
CDP1855	Multiply/Divide Unit
CDP1856	4-b Bus/Buffer Separator
CDP1857	4-b Bus/Buffer Separator
CDP1858	4-b Latch Decoder
CDP1859	4-b Latch Decoder
CDP1861	Video Display Controller
CDP1862	Color Video Circuit
CDP1863	Video Sound Circuit
CDP1864	P.A.L. System Video Interface
CDP1866	Latch Decoder for 1-kb × 4-b RAMs
CDP1867	Latch Decoder for 1-kb × 4-b RAMs
CDP1868	Latch Decoder (for Video Games)
CDP1869	Address and Sound Generator
CDP1870	Video Generator
CDP1871	Keyboard Encoder
Memories	Listed in Tables 4-2 and 4-15

cross assembler/simulator, and firmware debug packages as well as a full floppy-disk-based program development system. Also available is a high-level interpretive language.

Features of the software include simple 1-, 2-, or 3-byte instructions and simple timing loops for debugging. There is also a wide variety of branch and skip instructions from which to choose to permit rapid selection of a subroutine or program jump. A programmable serial port is also included on the chip to permit simple serial I/O without any specialized communications circuits.

A summary of the primary specifications for the 1802 microprocessor is presented in Table 5-5. Table 5-6 lists some of the available support circuits developed for the 1802. The CDP1802A operates at 4 MHz, up from 3.2 MHz for the basic CDP1802.

Designing a simple system based on the 1802 is straightforward. A parallel I/O system can be built from only four ICs: the microprocessor, two I/O ports, and a RAM or ROM (Fig. 5-7). A family of ICs, designed for interface and memory use, supports the 1802 (Table 5-6).

286 Microprocessors

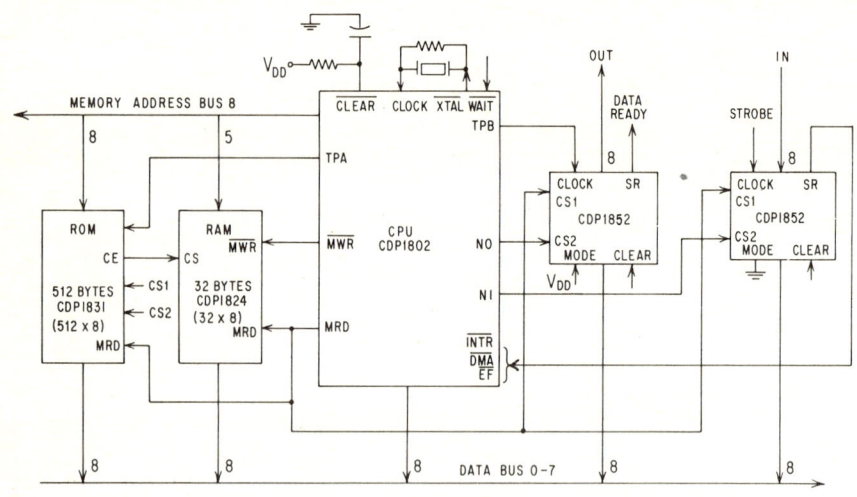

Fig. 5-7 A Minimal Operating System for the 1802 Includes an I/O Port and a Small Amount of RAM or ROM. If a Large Amount of Memory Is Needed, an Extra Set of Data Latches Must Be Used to Deliver a 16-b Address. *(Courtesy RCA Solid-State Division.)*

Almost any RAM or ROM circuit can be added to an 1802-based system. A small system with 32 bytes of RAM (CDP1824) and 512 bytes of ROM (CDP1831) requires no additional devices. One determines the address space for the memory when the program to be stored in ROM is written.

The 7-b user-specified sector address (the 7 MSBs of the 16-b multiplexed address bus) locates the ROM in one of the 128 memory sectors, each of which contains 512 bytes. When the proper sector address is input to the ROM, the CHIP ENABLE output goes high and disables the RAM. The RAM is enabled when the ROM is not selected.

Expanding the memory is simple. The circuit in Fig. 5-8 shows how to increase the ROM space from 512 bytes to 65 kbytes. The system needs no address decoding and is directly compatible with the 1802. A system with 4 kbytes of RAM uses the 1858 address latch-decoders and the 1822 RAMs (Fig. 5-8*b*). The 1824, 1822, and 1831 memories are directly compatible with the 1802, and the 1831 has the address latches built in. If larger RAM or ROM circuits, made by other suppliers, are used, the latches must be added externally.

The simplest way to bring data into the microprocessor is through the four external flag inputs, EF_1 to EF_4. Each input can perform as an independent serial data link, under program control. Branch instructions can test the flag inputs and divert the program on the appropriate

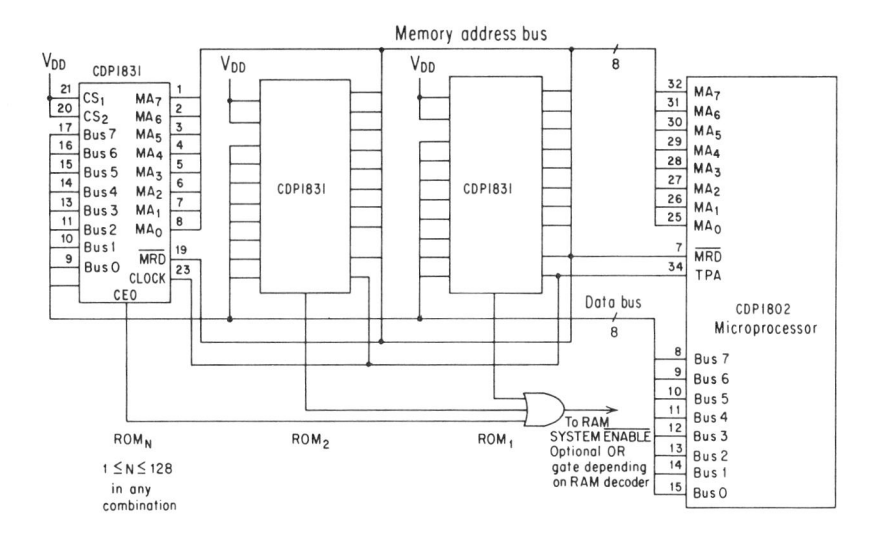

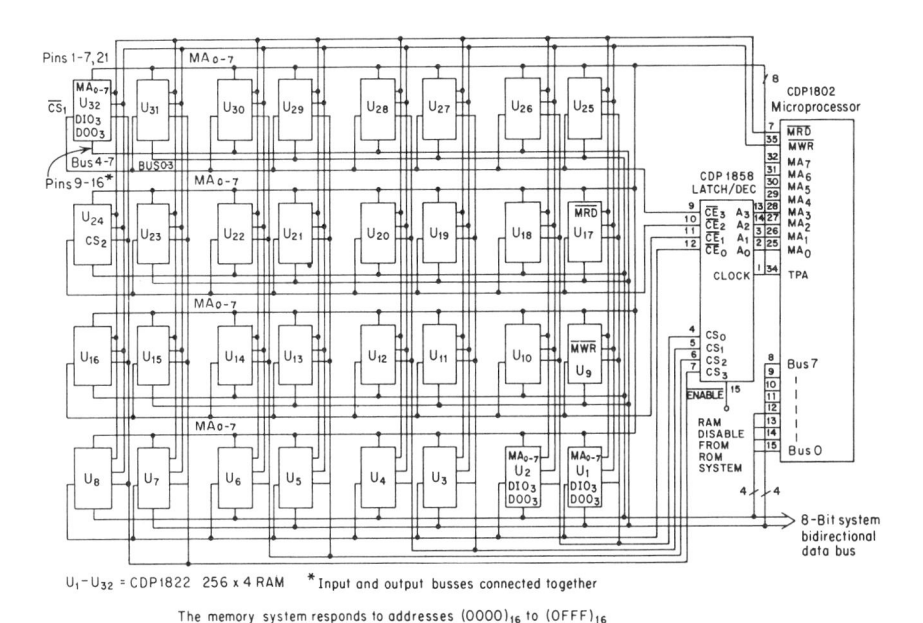

Fig. 5-8 An All-ROM-Controlled Microprocessor System Is Possible without Any Additional Circuits (a). Adding ROMs Increases Memory from 512 bytes to 65 kbytes. Each of the 1831 ROMs Contains the Necessary Data Latches to Hold the First Byte of the Address. For a 4-kb RAM System, however, Control Circuitry Must Be Added (b). *(Courtesy RCA Solid-State Division.)*

condition. Loops can be constructed to continually test the flag lines and manipulate the data as required.

The Q output of the 1802 serves as a control port or serial output. The Q bit can be set, reset, and tested by branch instructions or it can be used either as a pulse-width-modulated output or a variable-frequency output.

Programmed I/O data transfers can be used to directly select any of three possible peripherals. The instructions use a hexadecimal format of 6X, where the 3 LSBs of the code for the X digit are routed to the N_0, N_1, and N_2 pins of the microprocessor to activate a device.

Even more peripherals (up to 14) can be selected if an 1853 N-bit decoder is used. If a two-level decoding scheme is used, the number of external devices is unlimited.

CDP1802 APPLICATIONS

The CDP1802 microprocessor has been extensively used in automotive and consumer applications (video games* and television[†],[‡] circuits). Some of these applications will now be discussed.

Process Controller

Figure 5-9 shows the use of the CDP1802 microprocessor and peripheral circuits as a process controller. The processor consists of 4 kbytes of EPROM for storing instructions, 256 × 4-b of working RAM, one input port, two output ports, a 3.14573-MHz crystal clock, an ac power loss RESET circuit, and a counter that sets the basic timing of the instrument.

Automobile Engine Controller

A major automotive application for microprocessors and digital converters is as an on-board car and truck controller to monitor such things as spark advance, carburetor gas flow, alternator failure, and the like. The automobile engine controller shown in Fig. 5-10 is designed to extract maximum energy from the fuel while minimizing pollutants. All pertinent parameters (such as engine pressure, temperature, speed, etc.) are monitored in real time in order to arrive at the optimum gasoline-air mixture and the precise instant of ignition. This process must be repeated at least 100 times a second.

* P.K. Balter and J.A. Weisbeker, "Fun and Games with Cosmac," Session 41, *Electro 77 Proceedings,* San Francisco, CA, August 1977.
† K. Karstad, "Microprocessor Adds Flexibility to Television Control System," *Electronics,* Nov. 22, 1979, pp. 132–138.
‡ *IEEE Spectrum,* January 1980, pp. 66–67.

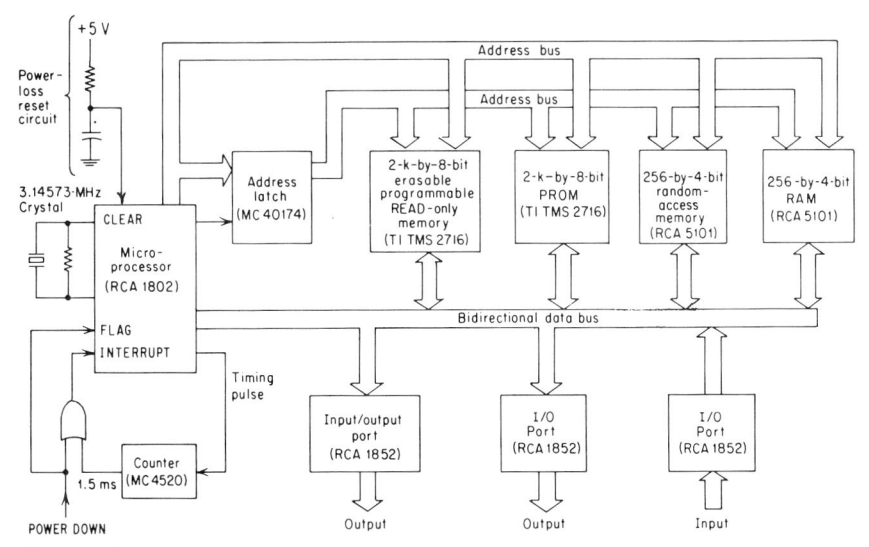

Commercially Available CMOS Microprocessors   289

Fig. 5-9 Process Controller.

The controller can be implemented using a CDP1802 microprocessor. To achieve minimum cost, the processor chips must assume A/D conversion, all computation, timing tasks, and D/A conversion. The microprocessor can perform the computation effortlessly, and major savings in A/D and D/A circuits can be realized if pulse width is used as the intermediate analog signal at the inputs and outputs. The most difficult task is keeping the timing under control.

The controller first multiplexes the analog inputs to permit the sharing of an ADC. The digitized inputs go to the algorithmic portions for real-time solutions. One of these sections is for fuel injection, another for ignition timing, and the third for optimum spark energy (dwell time). All three blocks share the use of a single arithmetic unit.

These activities are coordinated by the logic control section and sent to the elapsed-time modifier, which provides the correct real-time distribution of the control functions. Then the signals are converted to forms suitable for the driven devices: A pulse of precise timing and width is required for engine ignition; an analog signal of proper amplitude must be generated for fuel-injection control, and several digital lines for communications and display must be available.

Many functions must be timed simultaneously. The microprocessor must maintain elapsed time, perform a complete I/O cycle in less than 10

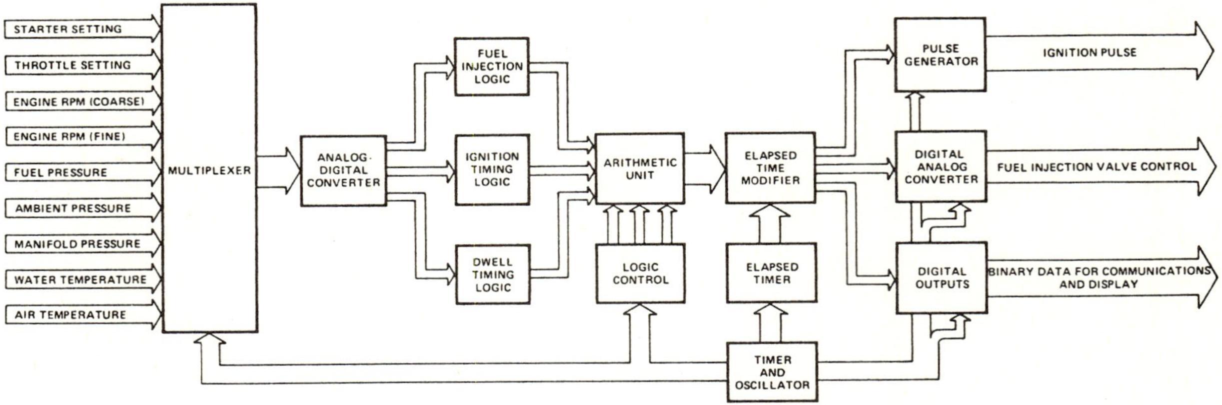

Fig. 5-10 Automotive Controller.

ms, monitor the engine rotation, sample the analog input pulse widths, and generate the correct ignition time and spark pulse width. Perhaps the most stringent timing requirement is that of measuring the pulse widths of the analog inputs—in particular, the 200-μs engine pressure pulse—to an accuracy of ± 1 μs, a formidable task.

The temperature extremes encountered in automotive applications are well suited to the use of stable CMOS circuits.

BMW (Munich, Germany), manufacturer of automobiles, uses the RCA 1802 microprocessor as the heart of a digital electronics system that integrates the circuitry for high accuracy and stable control of both the fuel injection and ignition. Various engine sensors pick up information that is compared with data on engine revolution and gas pedal position stored in the microcomputer memory. Using this data, the computer calculates the injection and ignition timing and the amount of fuel to be injected for a certain engine revolution value and gas pedal position. It then feeds its output in the form of command signals to the transistorized ignition system and to the fuel-injection nozzles. This system has been shown to save substantial fuel, and it facilitates cold starting.

The RCA CDP1804/1804C CMOS/SOS 8-b Microcomputer

The CDP1804 is a high-speed 8-b CMOS all-in-one microcomputer. Not only does the 1804 contain a full software-compatible 1802 CPU, but 64 bytes of RAM, 2048 bytes of ROM, and a timer-counter. Built from a combination of CMOS and SOS technologies, the processor provides operating speeds that are significantly higher than equivalent devices (such as the CDP1802) built from bulk CMOS processing. And there is no power penalty since the SOS technology reduces power requirements. The register-oriented 1804 provides easy I/O access and is geared for control applications.

The salient specifications for the CDP1804 are summarized in Table 5-7.

The architecture of the 1804 is based on that of its predecessor, the 1802. Along with the CPU is a 16×16-b register file, a 64×8-b RAM, a 2048×8-b ROM, and an 8-b counter-timer (Fig. 5-11). Except for the V_{CC} pin of the 1802, both the 1804 and 1802 are pin-compatible. On the former V_{CC} pin, the 1804 has an external memory select function. The counter-timer can operate in several modes—event counter, timer-counter, pulse duration.

However, the CDP1804 is more than just faster than the CDP1802.

TABLE 5-7 CDP1804 Specifications

Data Word Size	8 b
Address Bus Size	8 b
Direct Addressing Range	256 bytes
Instruction Word Size	1, 2, or 3 bytes
Number of Basic Instructions	102
Shortest Instruction/Time (Most)	2 μs (10 V)
Longest Instruction/Time (Long Branch)	3 μs
Clock Frequency (Min/Max)	DC/8 MHz (10 V)
Clock Phases/Voltage Swing	$1/V_{cc}$
Dedicated I/O Control Lines	13
Package	40-pin DIP
Power Requirements	4.75–10.5 V/ 0.5 mA (5 V)

The CDP1802's load mode is replaced in the CDP1804 by a test mode. And with an $\overline{EMS}$ signal replacing V_{cc}, there does not exist split-voltage operation.

In addition, Schmitt trigger inputs are provided on CLOCK, RESET, and WAIT lines and the CDP1804 also contains a PLA which provides the designer access to the processor's internal architecture for customized instructions.

The instruction set of the CDP1804 processor is upwards compatible with the instruction set of the CDP1802 microprocessor. An additional 15 instructions have been added to the processor to permit control of the on-chip counter-timer. The basic instruction set common to both processors includes 10 for control, seven for memory reference, seven for register operation, 12 for logic operations, 12 for arithmetic operations, 20 short branch commands, eight long branch operations, nine skip instructions, and 14 I/O commands.

Software support includes arithmetic packages, resident editors and

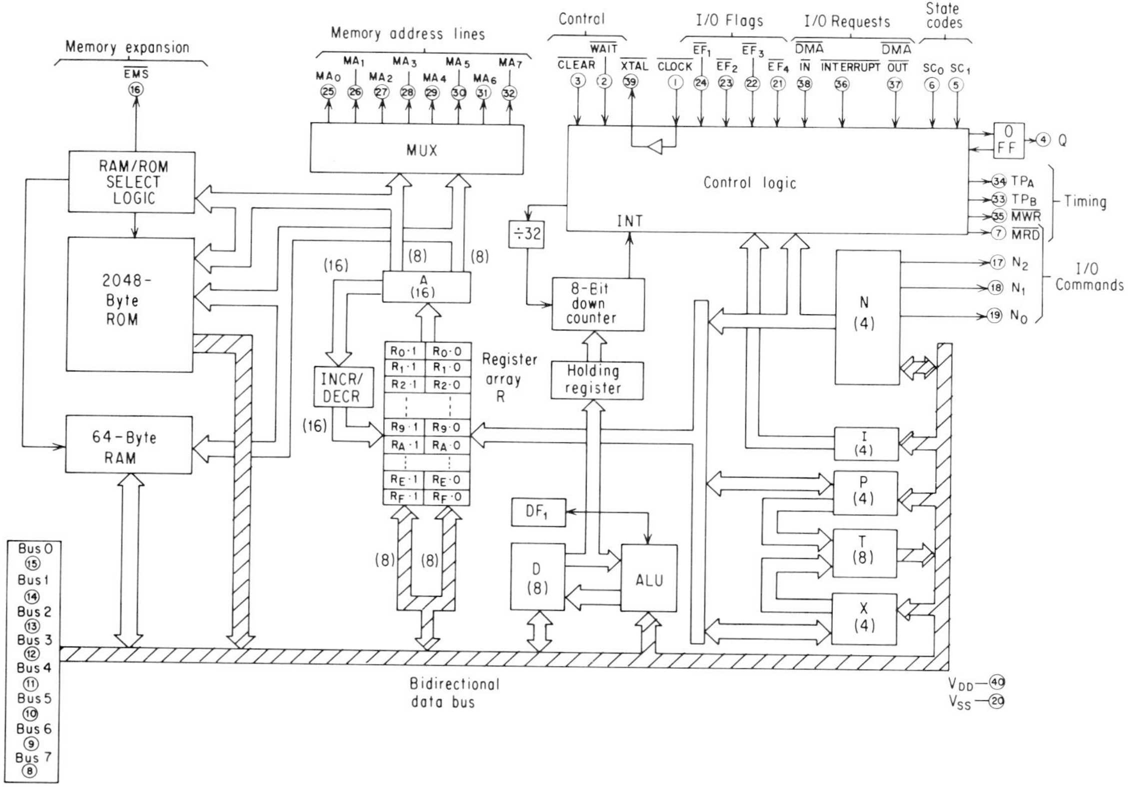

Fig. 5-11 CDP1804 Block Diagram.

assembler, cross assembler-simulator, and firmware debug packages as well as a full floppy-disk-based program developmental system. Also available is a high-level interpretive language.

Software features include the special instructions used to control the counter-timer. Most instructions are single-byte commands and there are simple timing loops available for debugging. There is also a wide variety of branch and skip instructions. In addition, software can be used to manipulate the flag lines to act as a simple serial I/O port.

The hardware support circuits for use with the CDP1804 are the same as those listed for the CDP1802 microprocessor.

RCA will also supply the CMOS/SOS CDP1805, a ROM-less version of the 1804.

CMOS and CMOS/SOS Versions of 8-b Single-Chip NMOS Microprocessors

The development of CMOS versions of various microprocessor and microcomputer designs is proceeding at a rapid pace in hopes of capturing high-power-consumption applications as well as extremely energy conscious applications: garage openers, stereo receivers, TV, home security systems, telephones, toys, games, dishwashers, and selected automotive applications.

The widespread popularity of Intel's second generation 8-b NMOS microprocessors and the adoption of the same as the industry standard has led to the announced fabrication of these devices using CMOS and CMOS/SOS technologies. In addition, other NMOS microprocessors are being fabricated using CMOS technology as well.

In order to provide both 8-b and 12-b microprocessors, Intersil is fabricating a large number of CMOS microprocessors: the Industry Standard 8049, 8741, 8748, and 8035 using CMOS technology. Intersil's IM87C48 is functionally identical to Intel's 8748 except that it consumes only 10 mA at 5 V while running at 6 MHz compared to 135 mA at 5 V for the 8748. The 87C48 consists of 1 kbyte of EPROM, 64 bytes of RAM, 27 I/O lines, and an on-chip counter-timer for real-time application. The CMOS IM87C41—a slave processor that can be tied to any bus and is best suited for use in intelligent peripherals—is also available. The IM80C48 and IM80C41 are masked ROM versions of the 8748 and 8741 parts, respectively. Forthcoming will be a proprietary 2-kbyte version of the 80C41. The Intersil IM80C49 is equivalent to Intel's 8049 one-chip microcomputer, with 2 kbytes of ROM and 128 bytes of RAM. The

IM80C35, equivalent to Intel's 8035, has 64 bytes of RAM but uses outboard ROM for program storage. Some of the available peripheral circuits to complete these microprocessors are the IM82C43 I/O expander, a CMOS version of the NMOS 8243, and an IM87C41 PPIM (programmable peripheral interface microcomputer) with 1-kb × 8-b EPROM/64-byte architecture. Also planned is the IM80C42 2-kb × 8-b/128-byte RAM PPIM microcomputer, which has no current NMOS counterpart.

National Semiconductor will also offer the 80C48, which will feature 8748 performance with ultra-low power for 6-MHz operation. This will be the initial entry for an entire family of CMOS microcomputers.

Harris Semiconductor also plans to introduce a CMOS version of the 8748 8-b microprocessor with a 1.35-μS instruction cycle time. In addition Harris, using an oxide-isolated double polysilicon CMOS process, will become an alternate supplier of Intel's 16-b 8086 microprocessor, the 80C86. Using N-channel dynamic logic circuits that are run from clocks independently of the system clock in conjunction with scaling, Harris hopes to obtain speed and die size comparable to that of N-channel devices: 500 nS cycle time for the 80C86.

Nippon Electric's version of the 8048, the μPD80C48, draws a maximum of 10 mA at a full 6 MHz. In the HALT mode, the on-chip oscillator continues running and the chip draws only 1 mA.

Table 5-8 summarizes the characteristics of the 8048 and 8085 microprocessors that use NMOS, CMOS, and CMOS/SOS technologies.

It has been predicted that by the mid-1980s from 35 to 40 percent of the total 8048 market will be for CMOS technology.

National Semiconductor's NSC800 8-b microprocessor combines the low-power advantages of CMOS and the speed of existing NMOS microprocessors. Architecturally, the NSC800 also combines the best features of Intel's 8085—its multiplexed data and address bus—with those

TABLE 5-8 CMOS/NMOS Microprocessor Comparison

Device	Process	Clock Rate (MHz)	Power (mW)	Die Size (mil × mil)
8048	NMOS	4	675	200 × 200
	CMOS	3–4	65	240 × 240
	CMOS/SOS	3–5	50	240 × 240
8085	NMOS	5	850	164 × 222
	CMOS/SOS	4–6	50	225 × 225

of the Z80—its register structure and instruction set—and is pin-compatible with the Z80. National uses a unique oxide-isolated two-level polysilicon process (dubbed P²CMOS) to attain the 1-μs instruction time of the 4-MHz Z80. A system using the NSC800 dissipates less than 125 mW.

The NSC800 executes the Z80 instruction set even though its internal architecture differs from the Z80's. As shown in Fig. 5-12, the NSC800 incorporates an 8-b data and 16-b address bus that can directly address 65,536 bytes of memory plus 256 locations in a separate memory space for I/O.

Within the CPU, an 8-b bus provides communication among the register array, the ALU, and the status registers, as well as between the data and address buffers and the instruction register. The processor's 22 registers, all of which can be used by the systems designer, are divided into three main groups: the first and second groups each comprise eight 8-b

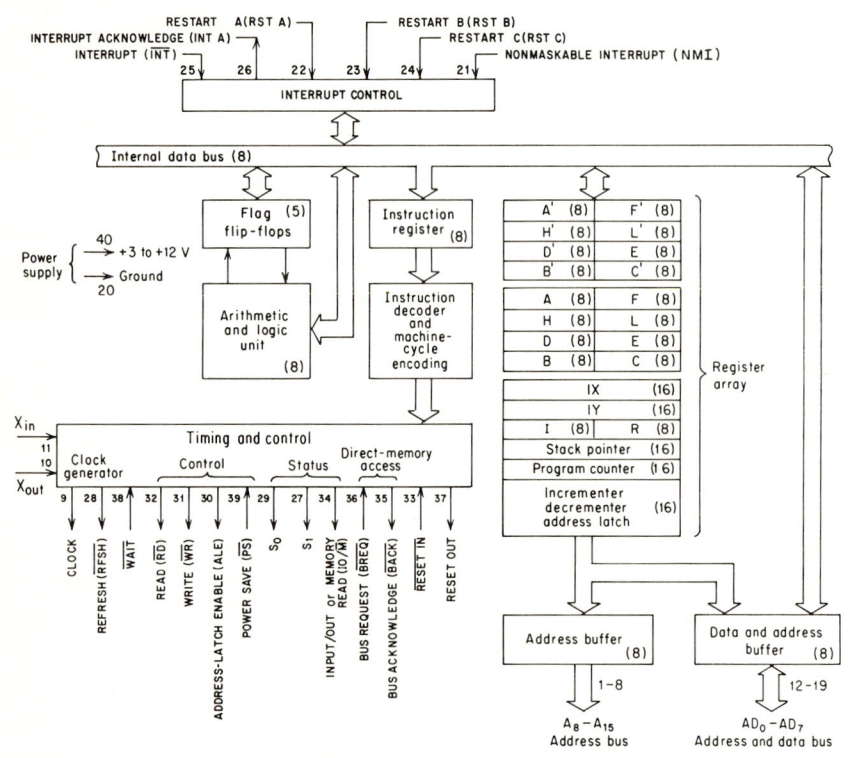

Fig. 5-12 The NSC800 Central Processing Unit Combines the Best Features of the Intel 8085 and the Zilog Z80 CPUs. It Borrows the Multiplexed Address-Data Bus Scheme of the 8085, Yet Has the Register Structure and Instruction Set of the Z80.

registers (A, B, C, D, E, H, L, and F, and their primes); the third group has two 16-b index registers, one 16-b stack-pointer register, and one 16-b program-counter register. In addition, two 8-b registers store an interrupt vector and the next refresh address for dynamic RAMs.

The processor's internal clock rate—2.5 MHz for the 800 and 4 MHz for the 800A—is half the basic oscillator frequency. This buffered signal is available on a pin (CLK) for use as a clock for peripheral circuits; as such, it cannot be interrupted by the CPU.

The NSC800 instruction set is completely compatible with that of the Z80 down to such details as the block-I/O and memory transfers; bit sets, resets, and tests; and indexed addressing. The 800, however, breaks down instruction cycles into several basic subcycles.

The NSC800 provides a unique low-power mode through its $\overline{PS}$ (POWER-SAVE) pin. The $\overline{PS}$ input is sampled at the last T state of the last M_1 cycle of an instruction. If a low level is detected on $\overline{PS}$, the CPU stops its internal clocks, thereby reducing power dissipation, yet maintaining all register data and the internal control status. The only power consumed in the POWER-SAVE mode is that required by the oscillator and the system clock.*

Table 5-9 compares the salient features of the NSC800 with those of the 8085 and Z80.

Motorola's 146805 is a CMOS version of the single-chip 6805 NMOS microcomputer. It offers an instruction set that is slightly trimmed down and modified. Both processors will eventually have 1 kbyte or 2 kbytes of ROM; 16, 20, or 32 I/O lines; 64 or 112 bytes of RAM; a flexible 8-b timer; and a prescaler. The 6805 comes in a 28-pin package and has 16 I/O lines and 1 kbyte of ROM. The first 146805 will come in a 40-pin, ROM-less version. The architecture of the 146805 is similar to that of the 6800 microprocessor, with the addition of some specialized memory-mapped registers.†

The units differ only slightly. One programming difference is that the NMOS versions use mask options to program the timer functions; the CMOS units will use software. Because the processors will probably be used in many low-cost control and game applications, their instructions are designed for bit operation and for test and branch on a bit.

---

* WESCON 1980 Session 24 Paper 24/5 discusses the NSC800 and its peripheral support circuits in detail.

† Support circuits for the 146805 are already in development. A detailed discussion of the 146805 can be found by referring to P. Smith et al., "CMOS Microprocessor Wakes Itself Up," *Electronics,* Sept. 25, 1980. See also Wescon 1980 Session 24, Paper 24/1.

TABLE 5-9 Comparing the NSC800, 8085, and Z80 Processors

	NSC800	8085	Z80
Power Supply Requirements			
Voltage Range (V)	3–12	5	5
Power Consumption at 5 V (mW)	50	850	750
Bus Drive Capacity (100-pF TTL Load)	1	1	1
Dynamic RAM Refresh Counter	8-b	No	7-b
Automatic WAIT State on I/O	Yes	No	Yes
Number of Instruction Types	158	80	158
Number of Registers Accessible			
to Programmer	22	10	22
Block I/O and Search	Yes	No	Yes
On-Chip Clock Generator	Yes	Yes	No
Minimum Instruction			
Execution Time (μs)	1	0.8	1
Number of On-Chip Vectored			
Interrupts	5	5	2
Early READ/WRITE Status	Yes	Yes	No
Bus Structure (Lines)			
Address	8	8	16
Data	8	8	8
Memory Refreshing	Yes	No	Yes
Minimum System Comparison*			
Chip Count	4	3	6
P_{DISS} (W)	0.1 @ 5 V	2.8	2.5
Memory Address (kbytes)	64	64	64
I/O Address	256	256	256

* Minimum system contains 256 bytes of RAM, 2 kbytes of ROM, and 32 I/O lines.

TABLE 5-10 MC6805 and MC146805 Comparison

	MC6805 (NMOS)	MC146805 (CMOS)
Power Dissipation	300-500 mW	0.5-50 mW
Standby Power	No Standby Mode	0.1-5 mW
Voltage Range	4.75–5.75 V dc	3–8 V dc
Cycle Range	100 kHz to 1 MHz	0 Hz to 1 MHz
Cycles/Instruction (Average)	5.36	4.02
Expansion Bus	Not Planned	Yes
Noise Immunity	Good	Better

Table 5-10 compares the salient features of the 6805 with its CMOS counterpart, the 146805.

In addition, General Instruments will offer CMOS versions of the 1655 and 1670 8-b NMOS microcomputers; Mostek will offer a CMOS 3870 (the 38C70); AMI will offer CMOS counterparts for the S2200 and 6809 NMOS microprocessors, including CMOS support circuits; and Hitachi will offer CMOS versions of the 6801, 6805, and 6805R2.

All of these developments will present the design engineer with a wide variety of parts with which to effect a design solution.

The Intersil IM6100 12-b Microprocessor*

The IM6100 is a single-address, fixed-word length, parallel transfer microprocessor using 12-b 2's complement arithmetic. Able to emulate the Digital Equipment Corp. PDP-8 minicomputer's instruction set, the IM6100 processor provides the user with a wealth of readily available software. The CMOS processor can also operate from a single supply and, since it draws only about 2 mA (at 5 V and 4 MHz), it is ideal for portable equipment design. The processor's architecture is also very similar to that of the PDP-8, the only major difference being in the bus—the IM6100 does not use the Digital Equipment Corporation's patented Unibus structure and timing.

The internal circuitry is completely static and designed to operate at any speed between dc and the maximum operating frequency. Two pins are available to allow for an external crystal, thereby eliminating the need for clock generators and a level translator. The crystal can be removed and the processor clocked by an external clock generator. A 12-b memory accumulator ADD instruction, using a +5-V supply, is performed in 5 μs by the IM6100, in 6 μs by the IM6100C, and in 2.5 μs by the IM6100A using a +10-V supply. The device design is optimized to minimize the number of external components required for interfacing with standard memory and peripheral devices. The IM6100 is packaged in a 40-pin DIP.

The IM6100 has six 12-b registers, a programmable logic array, an ALU, and associated gating and timing circuitry. A block diagram of the processor is shown in Fig. 5-13.

*Portions of this section excerpted with permission from Intersil IM6100 *CMOS 12-Bit Microprocessor Handbook*.

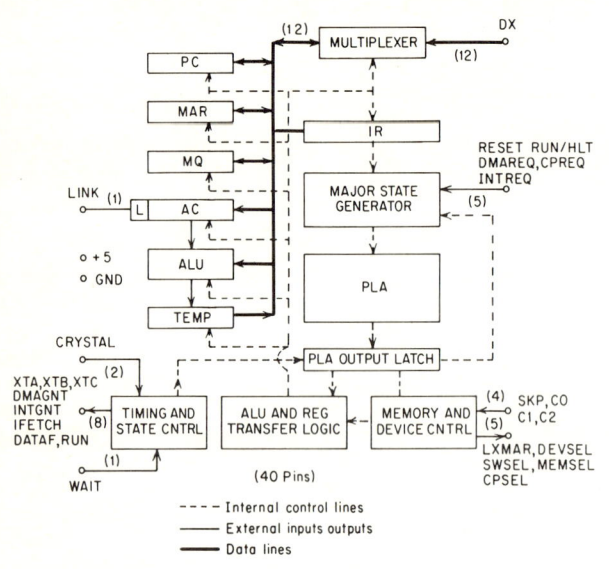

Fig. 5-13 IM6100 Block Diagram. *(Courtesy Intersil.)*

The IM6100 instructions are 12-b words stored in memory. The IM6100 makes no distinction between instructions and data; it can manipulate instructions as stored variables or execute data as instructions when it is programmed to do so. There are three general classes of IM6100 instructions. They are referred to as MRI (memory reference instruction), OPR (operate instruction), and IOT (input/output transfer instruction). There are six MRIs with three addressing modes each. Another 62 commands are OPRs, and the remaining 12 instructions are IOTs.

Software support offered includes an extended package containing loaders, editors, assemblers, debuggers, and a floating-point arithmetic program. Also available is FOPAL, a Fortran cross assembler, FOCAL (FOrmula CALculator), a high-level interpreter, and the DECUS (Digital Equipment Corporation User's Society) program library with over 1000 programs.

The most important software features include code compatibility with the PDP-8 and the flexible addressing modes. And, some of the OPRs can be combined with other OPERATE commands to make multifunction instructions.

PDP-8/E COMPATIBILITY

Since the IM6100 and the PDP-8/E (a trademark of Digital Equipment Corporation) are software-compatible, the basic PDP-8/E paper-tape software system supplied by the Digital Equipment Corporation will operate properly with the IM6100. This basic software package includes binary loaders, a PAL III assembler, a symbolic editor, a DDT (dynamic debugging technique), an ODT (octal debugging technique), a 23-bit floating-point package, and FOCAL. The IM6100 will execute the complete set of CPU diagnostics for PDP-8/E.

Since the bus structure of the IM6100 can be adapted to provide a subset of the PDP-8/E OMNIBUS signals, all programmed I/O interfaces for the PDP-8E—for example, teletype, paper-tape reader-punch, etc.—will operate with the IM6100 without any hardware or software modification.

The DMA structure of the IM6100 and PDP-8/E are different; the IM6100 DMA structure is similar to the PDP8- 1-CYCLE BREAK, but not compatible.

The IM6100 handles 4-kb words of memory directly. Like the PDP-8/E, an external extended memory control element can be used to extend the addressing space up to 32 kb. All necessary control and timing signals to implement the memory extension controller are generated by the IM6100.

The EAE (extended-arithmetic element) and the UF (user flag) processor options of the PDP-8/E cannot be used with the IM6100. The EAE is used for hard-wired MULTIPLY and DIVIDE and the UF for time sharing.

Tables 5-11 and 5-12 summarize the key specifications and available support circuits for the IM6100 respectively.

IM6100 APPLICATIONS
ALL-CMOS System

The IM6100 microprocessor family provides for the capability of building an all-CMOS system with no additional support components. The CMOS RAM devices are organized 1024 × 1-b (IM6508/18) or 256 × 4-b (IM6551/6561). They have internal address latches and operate synchronously with an address strobe. A 1024 × 12-b mask-programmable CMOS ROM (IM6312) is also provided. The IM6402/6403 is an industry standard UART with the option of operating directly from a high-frequency crystal. The IM6101 parallel interface element provides all signals necessary to communicate with an external device, including a vectored priority interrupt chain. For example, a parallel teletype interface can be designed with only two logic elements—the IM6101 for control and the IM6403 for data handling. The dynamic power dissipation of the CMOS system will be less than 60 mW at +5 V (Fig. 5-14).

Microprocessors

TABLE 5-11 IM6100 Key Specifications

Data Word Size	12 b
Address Bus Size	12 b
Direct Addressing Range	4096 Words
Instruction Word Size	12 b
Number of Basic Instructions	80
Shortest Instruction/Time (AND, OR, JUMP, etc.)	2.5 μs
Longest Instruction/Time (Autoindexed Increment and Skip if Zero)	5.5 μs
Clock Frequency (Min/Max)	DC/8MHz (10 V)
Clock Phases/Voltage Swing	1/CMOS or TTL
Dedicated I/O Control Lines	24
Package	40-pin DIP
Power Requirements:	5 V/2.5 mA or 10 V/10 mA

TABLE 5-12 IM6100 Support Circuits

Model	Description
IM6100	12-b CPU Commercial
IM6101	Programmable Parallel Interface Element
IM6102	Memory Extender/DMA Controller/Timer
IM6103	Multimode Latched Port
IM6402	CMOS UART (16× Clock)
IM6403	CMOS UART (xtal Clock)

Memories listed in Tables 4-2 and 4-15

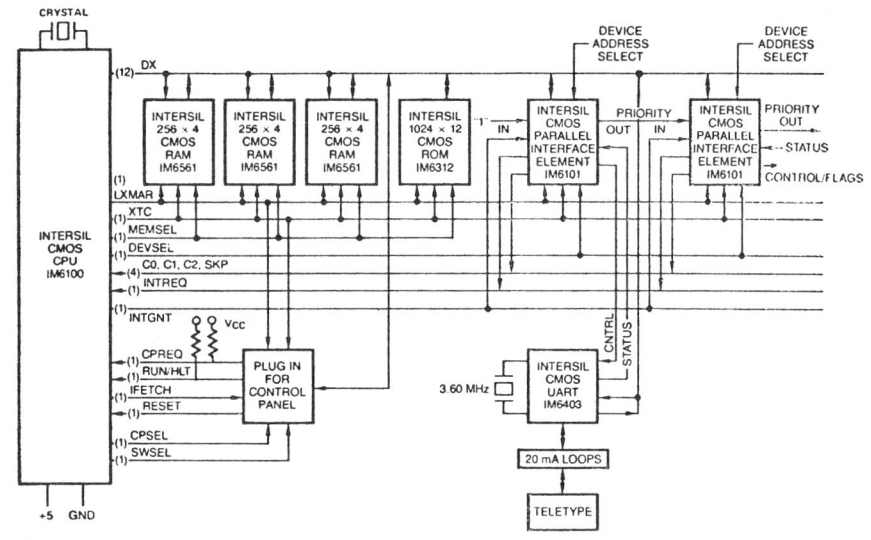

Fig. 5-14 All CMOS Microprocessor System. *(Courtesy Intersil.)*

256 × 12-b RAM, 2-kb × 12-b PROM Memory System

A low-power nonvolatile memory system with extremely low standby power requirements can be constructed as shown in Fig. 5-15. A 256 × 12-b RAM, 2-kb × 12-b PROM organization seems to be sufficient for typical microprocessor applications. Provisions are made, however, to expand the RAM/PROM capacity up to 4-kb words. The 2-kb 6306 bipolar PROMs are power-strobed with HD6605s. The CMOS RAMs have extremely low quiescent power requirements, less than 300 μW for a 256 × 12-b array, and they can be made nonvolatile with an inexpensive battery back-up. The system designer can reduce memory power dissipation considerably with CMOS RAMs and power-strobed PROMs since the memory utilization of microprocessors is typically less than 30 percent. The power dissipation of the system shown in Fig. 5-15 is less than 0.5 W at 5 V.

Transparent Control Panel

A unique feature of the IM6100 is the provision for a dedicated, completely independent control panel with its own memory separate from the main memory, as shown in Fig. 5-16. The concept of a "transparent" control panel is an important one for microprocessors since microprocessor-based production systems normally do not have a full-fledged panel and the system designer would like to use the entire capac-

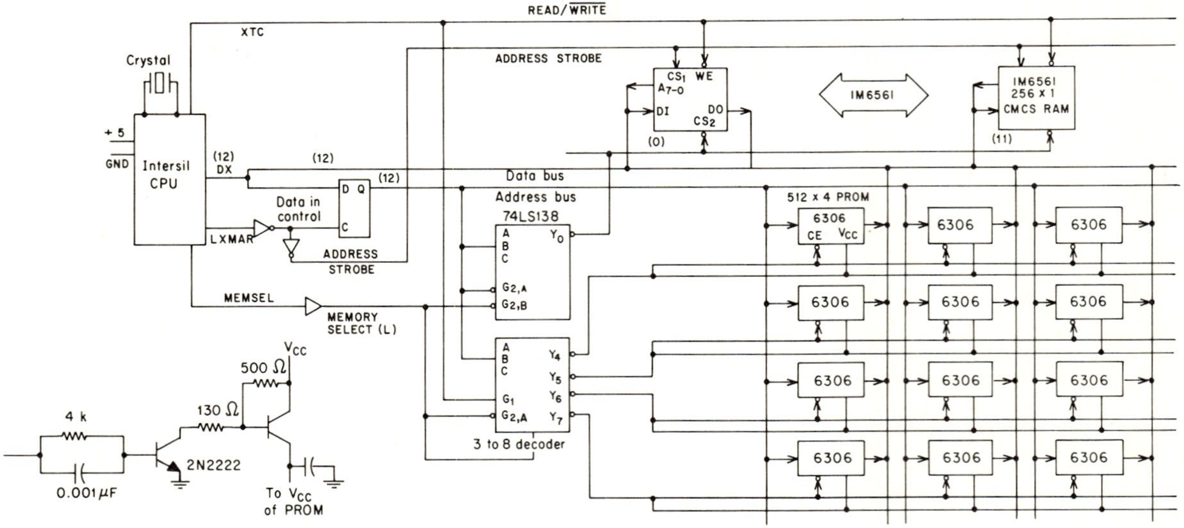

Fig. 5-15 Low-Power Nonvolatile Memory System. *(Courtesy Intersil.)*

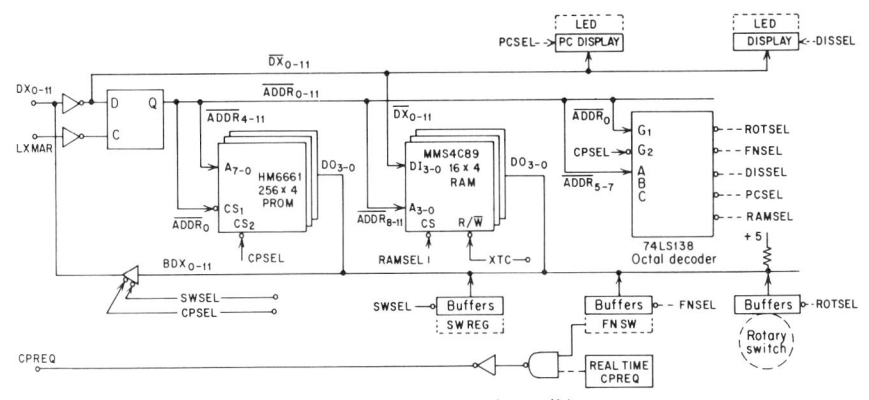

Fig. 5-16 Transparent Control Panel. *(Courtesy Intersil.)*

ity of the main memory for the specific system applications. A number of panel options which can greatly increase the usefulness, flexibility, and reliability of the system, such as testing, maintenance, and diagnostic routines, bootstrap loaders, etc., can be incorporated just by increasing the size of the panel memory to handle more software. The panel can be considered as a portable device which can be plugged into a socket on the CPU board whenever the panel functions are needed, and disconnected when not needed, without disturbing any part of the user program.

IM6100-to-CMOS-RAM Interface

The IM6100 provides all the control signals to interface directly with standard CMOS RAMs. Since the CMOS RAMs have internal address latches, the address information on the D_X lines is latched internally with the address strobe. ADDRESS, DATA IN and DATA OUT can be multiplexed on the D_X lines without any degradation in performance (Fig. 5-17).

Microprocessor-Counter Interface

Figure 5-18 shows the circuitry necessary to interface the ICM7227 4-digit UP/DOWN counter-decoder-driver to an Intersil IM6100 microprocessor. The ICM7227 in the processor environment can perform many accessory functions that are inefficient or impossible for the processor to perform. For simple systems, the ICM7227 can provide a cost-effective display-latch-decoder-driver. By adding a time base such as an ICM7213, and using an ICM7227, an inexpensive real-time clock-display directly accessible by the processor can be implemented.

306 Microprocessors

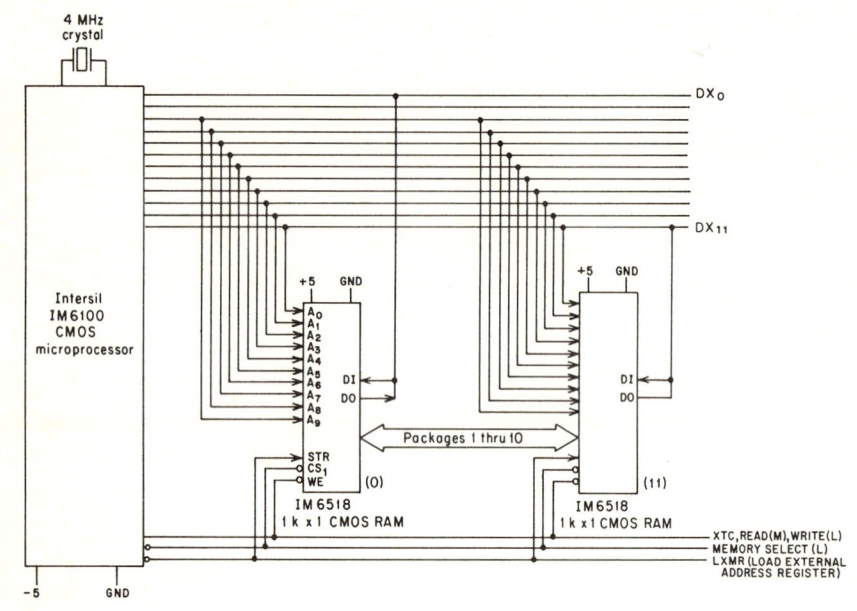

Fig. 5-17 CMOS Microprocessor. *(Courtesy Intersil.)*

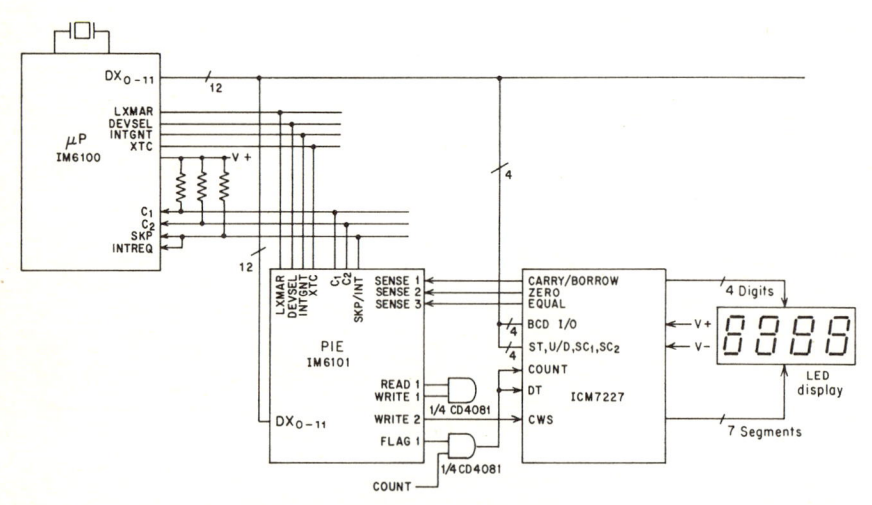

Fig. 5-18 UP/DOWN Counter as Peripheral to IM6100 Microprocessor. IM6101 Parallel Interface Element Allows Addition of One or More Counters as Generalized Peripherals to Any IM6100 System, with a Minimum of External Components. (*Courtesy Intersil*.)

Hewlett-Packard CMOS/SOS Microcomputer Chip Set*

A microcomputer chip set (Fig. 5-19) is one of the first applications for Hewlett-Packard's SOS process. Components in the set include a 16-b CPU chip [MCC (microcomputer chip)], a 8192-b ROM, a 2048-b RAM, and an interface chip (PHI) to couple the processor to the IEEE-488 standard interface bus. The chips communicate with one another over a parallel asynchronous bus that can run at any speed up to 4 MHz. The MCC can use this bus to directly address up to 65,536 16-b memory locations and 2048 external I/O registers used for peripheral control. Memories tied to the bus can run at any mixture of access and cycle times within the same product.

The MCC is the heart of the microcomputer chip set and is tailored for control applications, with an instruction repertoire covering 34 groups of instructions. These instructions take from 750 ns to 1.4 μs to execute (a full 16-b register-to-register addition requires only 875 ns). While operat-

* A. Capell, et al., "Process Refinements Bring CMOS on Sapphire into Commercial Use," *Electronics*, May 26, 1977, pp. 99–110. Copyright © 1977 by McGraw-Hill, Inc. Reprinted with permission.

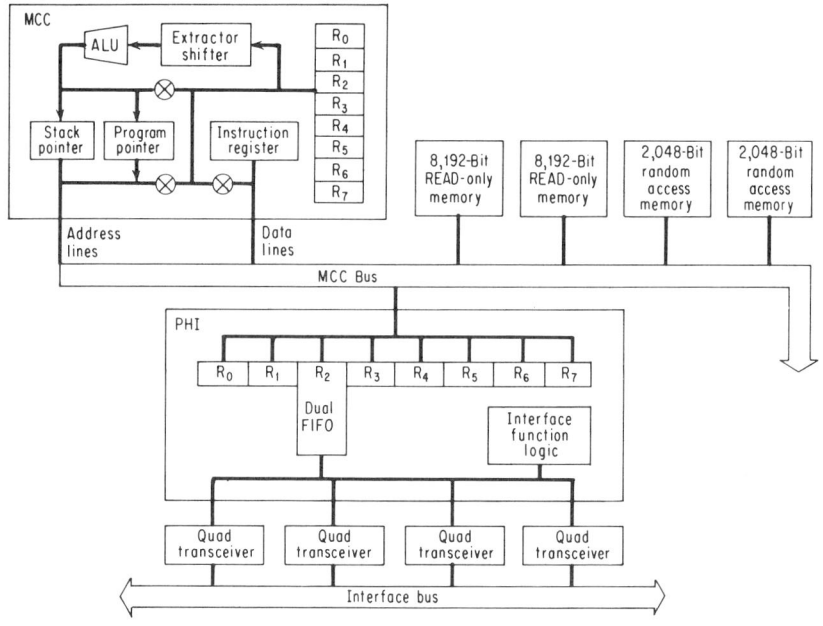

Fig. 5-19 Hewlett-Packard's 16-b SOS Microcomputer System. *(Reprinted from Electronics, May 26, 1977. Copyright © 1977 by McGraw-Hill, Inc. Used with permission.)*

ing at its maximum clock rate of 8 MHz, the CPU chip typically dissipates only 350 mW drawn from a single 12-V power supply.

The bulk of its instructions operate directly on 16-b registers, which can be selected either from the eight general-purpose registers within the MCC or from a "bank" of eight I/O registers external to the MCC. Because external-register interrogation is so important in control-oriented environments, the associated access time is optimized and approaches that of the internal registers.

MEMORY REFERENCE instructions can access directly or with indexing any of the 65,536 available memory locations. A system stack residing in external RAM is used by subroutine calls and interrupts for return address storage. Stack-oriented PUSH and POP instructions are included to allow the programmer to use this stack.

Since the MCC was developed for control rather than processing applications, emphasis was placed on manipulation and interrogation of small clusters of bits as well as of full 16-b words. Word-oriented instructions such as ADD or AND can operate on extracted fields of 4, 8, or 16 b from within a specified register, and bit-oriented instructions can operate on any selected bit of the register. Since all of these instructions can operate directly on an external I/O register, they can be very sophisticated in their manipulation of a peripheral device that the bits in that register control.

Built-in priority encoding is provided in the MCC, allowing it to make quick decisions based on the result of a previously executed instruction. After reading from the fault-indication register of a peripheral device, for example, the MCC could use this feature to branch immediately to a code sequence corresponding to the highest-priority fault detected.

Simplified debugging is stressed in the architecture of the MCC. In addition to allowing its clock input to stop so that the system state can be examined, the MCC also provides an IDLE mode during which all its internal registers (including the program pointer, stack pointer, and general-purpose registers) can be interrogated by external circuitry as if they were in two external register banks. IDLE mode is entered either by external request or when an instruction from a special group is executed. In the latter case, external logic can decode and execute the instruction while the MCC is idle and then return control back to the MCC when execution is complete.

Hewlett-Packard, when participating in the development of the IEEE-488 instrument interface standard, began a strong commitment to make its new products compatible with this standard. However, to interface to the standard bus requires a significant amount of logic—up to 200 standard TTL ICs if all features of the IEEE-488 are used. The PHI is a single-chip alternative which provides a complete high-performance bus interface for both instruments and bus controllers and performs all the interfacing functions defined in IEEE-488. It ties directly into a MCC system, appearing to the MCC as a bank of eight external I/O registers.

All data bytes that pass through the PHI chip are buffered in one of two FIFO buffers—one for inbound data and one for outbound data. These buffers appear to the MCC as one of the eight registers in the PHI bank and improve performance by compensating for data transfer irregularities such as MCC interrupt response delays or temporary slowdown of the device at the other end of the bus. If the PHI is being used within a bus controller, the outbound FIFO also plays an important part in sequencing and synchronizing bus activities.

The PHI interfaces to the 16 IEEE-488 bus lines through four bipolar quad transceivers with which it communicates at standard low-power, Schottky-TTL, logic levels. Its interface to the MCC bus, however, can be configured to operate either at these logic levels or at the 12-V levels compatible with the MCC. To achieve 1-MHz data transfer rates over the bus, the PHI responds to standard handshakes in typically 50 ns. This performance, in an 8000-transistor circuit requires the ability to perform large-scale integration of high-speed asynchronous circuits to a degree obtainable only with CMOS-on-sapphire.

Program and variable-storage capabilities are provided by an 8192-b ROM chip and a 2048-b RAM chip, which were each designed in CMOS-on-sapphire for use in MCC systems. Both provide 8 b of parallel output and are used in pairs to achieve the 16-b word length of the MCC. Typical access times are 50 ns for the ROM and 80 ns for the RAM. Since the RAM is static, it requires no refreshing and consumes less than 50 μW during battery standby operation. Both the RAM and the ROM are contained in 24-pin DIPs and operate from a single 12-V power supply.

Another development with a CMOS/SOS microprocessor demonstrates the viability of this approach. Mikros Systems Corp. also offers a SiGate CMOS/SOS 16-b microcomputer—the MK-16. The 16-b CMOS/SOS microcomputer results in low power consumption (5 to 7 W for the CPU board); comparable PUSH and POP operations to the PDP-11/70 (1 μs); indirect indexing twice as fast as a Z8000 NMOS microprocessor; and it can operate in a byte mode.

A 16-b Parallel CMOS Microprocessor*

A 16-b parallel microprocessor using C²MOS (clocked CMOS) circuitry in LSI form has been developed using dynamic logic circuits. In particular, by using a dynamic ROM for the instruction decoder and control circuit, a reduction of the number of basic instructions can be avoided.

A circuit diagram of the dynamic ROM used in the instruction decoder and control circuit is shown in Fig. 5-20 where P-channel MOSFETs are used in-parallel to reduce delay time in the circuit. The timing of the input data of such a dynamic ROM must be synchronous with the precharge pulses of the ROM. The output of an instruction register, whose timing is ½-b time ahead of the basic clock and synchronous with the WRITE pulse of the register, serves as the input to the ROM. By providing a ½-b delay circuit whose timing is synchronous with a READ pulse of the register at the output side of the ROM, unnecessary pulses caused by precharging of the output nodes of the ROM can be eliminated and the delay circuit can operate as a driver circuit.

* K. Manabe et al., "A C²MOS 16-bit Parallel Microprocessor," *1976 IEEE ISSCC Digest of Technical Papers*, Philadelphia, PA, February 1976. Copyright © 1976 by the Institute of Electrical and Electronics Engineers, Inc. Reprinted with permission.

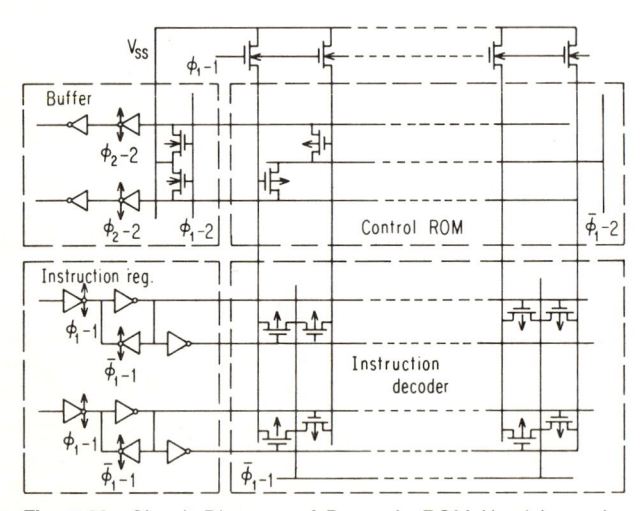

Fig. 5-20 Circuit Diagram of Dynamic ROM Used in an Instruction Decoder and Control Circuit. *(Reprinted from 1976 IEEE ISSCC Digest of Technical Papers, Philadelphia, PA, February 1976. Copyright © 1976 by the Institute of Electrical and Electronics Engineers, Inc. Used with permission.)*

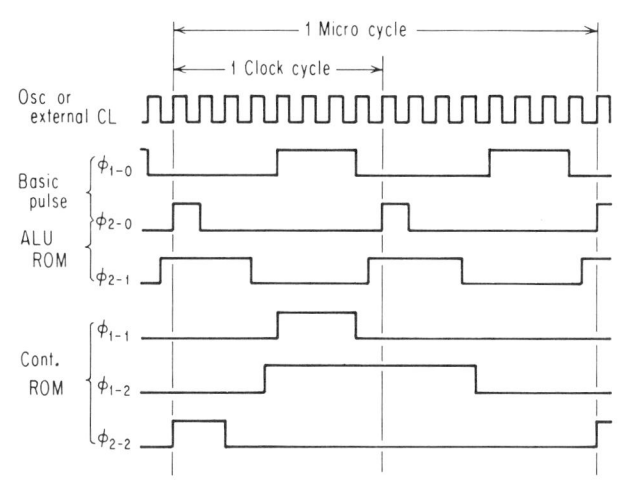

Fig. 5-21 Basic Output Waveforms of Timing Generator Included in the LSI. *(Reprinted from 1976 IEEE ISSCC Digest of Technical Papers, Philadelphia, PA, February 1976. Copyright © 1976 by the Institute of Electrical and Electronics Engineers, Inc. Used with permission.)*

Since many dynamic logic circuits are used in the microprocessor, some clock pulses are required to operate those dynamic circuits efficiently. Therefore, the microprocessor has an internal timing generator which generates several clock pulses by dividing a basic input pulse by eight and using some logic gates. Moreover, as the LSI circuit has a built-in inverter for use with a quartz oscillator, it is not necessary to use any external pulse generators. Figure 5-21 shows some basic output waveforms of the timing generator and especially those used in the two ROMs, i.e., one is for arithmetic logic circuits and the other is for an instruction decoder and control circuit.

A block diagram of the microprocessor is shown in Fig. 5-22. About 6000 devices are integrated on the chip. This LSI circuit can operate with a single power supply: from 3 to 12 V. It can be seen that the ALU consists of an 8-b parallel logic circuit in which there are eight internal bus lines and eight I/O ports. By considering 2 clock cycles to be an instruction cycle, it can be seen that the microprocessor operates as a 16-b parallel system.

The merits of the method are the following:

- Chip area for the ALU, internal bus lines, and I/O ports can be reduced to one-half.

- Transient characteristics of the dynamic ROM are generally not as good as those of other random-gate circuits. By using this method, the instruction cycle time can be twice as long as the other circuit cycle time and the operation of the ROM can be considered to provide as much margin as that of the other circuits.
- Since 16-b parallel I/O and address data can be divided into upper and lower parts each consisting of 8-b parallel data, the number of pins for the IC can be reduced. Consequently, this LSI circuit requires only 28 pins.

Metal lines are preferentially used for clock pulse lines to decrease additional capacitance and series resistance, which would reduce the circuits' transient performance. MOSFETS placed in the ROM are automatically generated at required positions on the basis of the data from a logic simulation of the LSI system.

Characteristics of the power dissipation of the microprocessor illustrated in Fig. 5-23 show that it can operate with very little power dissipation.

The main features of the processor are

- The number of basic instructions is 72, and instruction execution time is 10 μs to 26 μs at 8-MHz input pulse rate.

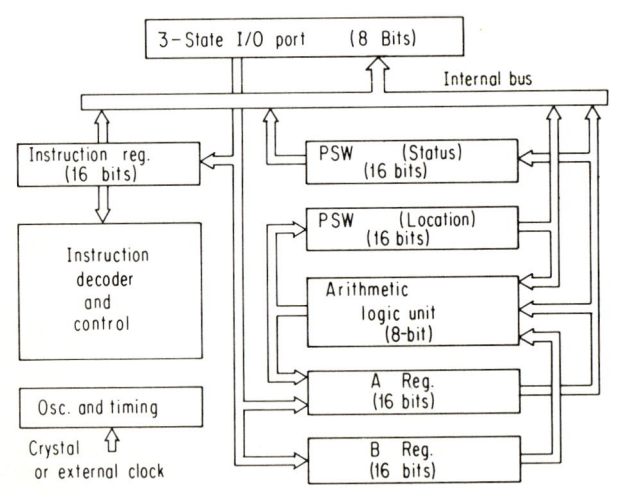

Fig. 5-22 Block Diagram of the Microprocessor. *(Reprinted from 1976 IEEE ISSCC Digest of Technical Papers, Philadelphia, PA, February 1976. Copyright © 1976 by the Institute of Electrical and Electronics Engineers, Inc. Used with permission.)*

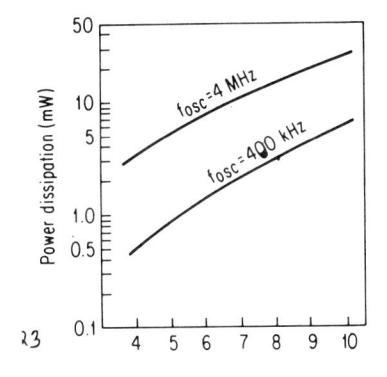

Fig. 5-23 Characteristics of the Power Dissipation of the Microprocessor LSI. *(Reprinted from 1976 IEEE ISSCC Digest of Technical Papers, Philadelphia, PA, February 1976. Copyright © 1976 by the Institute of Electrical and Electronics Engineers, Inc. Used with permission.)*

- The processor has three levels of INTERRUPT REQUEST inputs, and multiple interrupt can be handled very easily.
- Addressing modes are (1) direct addressing, (2) indirect addressing, (3) indexed addressing, (4) immediate addressing, and (5) pointer addressing.
- Program status words (PSW) include condition code, stack pointer, interrupt mask, location, and other flags.

Since many CMOS memories are commercially available (Tables 4-2 and 4-15) in addition to this microprocessor, it is possible to realize a microcomputer system with CMOS devices only, which affords an extremely low-power system and permits even battery operation.

New 4-Bit µP Developments

With the emphasis on high-performance 8-b μPs and μCs and 16-b μPs, one might assume that the application and interest as well as design and development effort in 4-b μPs and μCs has ceased. This, however, is inaccurate. 4-b μC development is very active with National Semiconductor, OKI Semiconductor, and Applied Microcircuits, Inc., all introducing devices and support circuits for use in telecommunications, automotive, consumer, and security applications.

National Semiconductor's COP (control-oriented processor) family of single-chip microcontrollers are the first members to be fabricated using complementary-MOS technology. One result is what is believed to be the lowest power dissipation available for 4-b microcomputers.

Designated the COP420C, COP421C, COP320C, and COP321C, the additions are complete microcomputers containing all system timing, internal logic, read-only memory (8 kb organized as 1024×8 b), random-access memory (256 b organized as 4 × 64 b), and the input/output facility necessary for dedicated control functions. (See Table 5-13 for salient device characteristics.)

In addition to the power-saving idle state, the devices also include software control of POWER UP and POWER DOWN. In a power-loss situation, the devices can turn off, then turn back on, taking up where they left off, using a backup battery.

The vectored INTERRUPT, plus RESTART, and three-level subroutine stack featured in the new devices allow the user to achieve program efficiency unattainable in other microcontrollers. The CMOS units employ the same instruction set, pinout, and architecture as other members in the COP family, enabling users to incorporate them into systems without the need for extensive retraining. Also, the programmability of the I/O ports, featured in the entire COP family, reduces the amount of external hardware required and provides drive characteristics compatible with the user's external circuitry.

As with other COP members, the new devices are Microbus-compatible, allowing them to be used as peripheral microprocessor devices, receiving and transmitting data and commands from and to a host microprocessor in less than 1 μS. They are also compatible with Microwire, National's standard three-wire bus interface, a feature that enables the end user to add several peripheral devices.

The four microcontrollers operate from a single power supply (2.4 to 6.0 V), and contain an internal time-base counter for real-time processing.

TABLE 5-13 NSC COP Microcomputer Family

P/N	COP 420C	COP 320C	COP 421C	COP 321C
Size	4-b	4-b	4-b	4-b
ROM	1 kb × 8	1 kb × 8	1 kb × 8	1 kb × 8
RAM	4 × 64	4 × 64	4 × 64	4 × 64
I/O lines	23	23	19	19
Temp.	0 to 70°C	−40 to +85°C	0 to 70°C	−40 to +85°C
P_{DISS} idle (typ.) and 32-kHz clock and 2.4-V V_{cc}	50 μW	50 μW	50 μW	50 μW
Operating (typ.) 500-kHz clock and 5-V supply	4 mW	4 mW	4 mW	4 mW
Package	28-pin DIP	28-pin DIP	24-pin DIP	24-pin DIP

To support these control-oriented processors, National has a family of support circuits in development. The first circuit is the COP498 random-access memory and timer peripheral circuit that turns off power to the host microcontroller during idle periods. Volatile data is held in the peripheral's 64- × 4-b memory until a preset time interval or external interrupt signals the chip to wake up the microcontroller and transfer back the stored data.

OKI Semiconductor's OLMS-40 family provides five different variations in ROM, RAM and I/O to allow one to address different design priorities (for example, a fast cycle time or compact package). These devices can call upon nearly a dozen support circuits as well as several powerful development tools.

Table 5-14 summarizes the salient features of the OLMS-40 family and Table 5-15 lists the CMOS peripheral, circuits, development, and software systems.

CMOS Microprocessor Summary

Tables 5-16 and 5-17 summarize the key features of the commercially available general-purpose and complete-chip CMOS microprocessors, respectively.

Microprocessor Support Circuits

UART

UARTs are single-chip subsystems for interfacing microprocessors to an asynchronous serial data channel. The receiver converts serial start, data, parity, and stop bits to parallel data, verifying proper code transmission, parity, and stop bits. The transmitter converts parallel data into serial form and automatically adds start, parity, and stop bits. The data word length can be 5, 6, 7, or 8 b. Parity may be odd or even. Parity checking and generation can be inhibited. The stop bits may be 1 or 2 or $1\frac{1}{2}$ when transmitting 5-b code.

The Intersil IM6402/6403 are CMOS/LSI UARTs that can be used in a wide range of applications including modems, printers, peripherals, and remote data acquisition systems. CMOS/LSI technology permits operating clock frequencies up to 4 MHz (250 kbaud) an improvement of 10 to 1 over previous PMOS UART designs. Power requirements, by comparison, are reduced from 300 to 10 mW. Status logic increases flexibility and simplifies the user interface.

TABLE 5-14 OLMS-40 μC Family

Series 40 Single-Chip CMOS Microcomputers	OLMS 40 MSM5840/5840E	OLMS 40 MSM5840-XX	OLMS 45 MSM5845-XX	OLMS 42 MSM5842-XX	OLMS 421 MSM58421-XX	OLMS 423 MSM58423-XX
Package	42-pin DIP (plastic)	42-pin DIP (plastic)	42-pin DIP (plastic)	28-pin DIP (plastic)	60-pin FLAT (plastic)	40-pin DIP (plastic)
Supply Voltage	3-6 V	3-6 V	3-6 V	3-6 V	3-6 V	3-6 V
Typical/Max Supply Current @ 5 V	0.8/3 mA	0.8/3 mA	2/3 mA	1.5/4 mA	2/5 mA	1 mA (typ)
Clock Rate	2 MHz Max.	2 MHz Max.	4 MHz Max.	4 MHz Max.	4 MHz Max.	2 MHz Max.
Basic Instruction Execution Time	16 μs at 2 MHz	16 μs at 2 MHz	8 μs at 4 MHz	8 μs at 4 MHz	8 μs at 4 MHz	8 μs at 2 MHz
Instruction Set	98 Instructions (Single Byte 93)	98 Instructions (Single Byte 93)	49 Instructions (Single Byte 46)	52 Instructions (Single Byte 48)	52 Instructions (Single Byte 48)	52 Instructions (Single Byte 48)
ROM Size	4096 × 8 (External)	2048 × 8 (Internal) 2048 × 8 (External)	1280 × 8	768 × 8	1536 × 8	1280 × 8
RAM Size	128 × 4	128 × 4	64 × 4	32 × 4	40 × 4	32 × 4

Working Register	4	4	1	1	1
Input Port	1 × 4 1 × 2 (Latched)	1 × 4 1 × 2 (Latched)	1 × 4 1 × 2 (Latched)	1 × 4 1 (Latched)	1 × 4 1 (Latched)
Output Port	4 × 4	4 × 4	4 × 4	5 × 7 (Segment) 1 × 5 (Discrete)	12 × 12 Multi-plexed VF High Voltage Outputs
Input/Output Port	2 × 4	2 × 4	2 × 4	2 × 4	1 × 4
Stack	4	4	2	1	2
Timer/Counter	1 (8 Bit) R/W	1 (8 Bit) R/W	1 (8 Bit)	1 (12 Bit)	1 (8 Bit)
Interrupt	2 Levels	2 Levels	1 Level	—	1 Level
Evaluation Chip	MSM5840	MSM5840	MSM5840E	MSM5840	MSM5840
Evaluation Board	MPB 201/202/203	MPB 201/202/203	MPB 201/202/203	MPB 201 with LCD Attachment	MPB 201 with External Adapter
Remarks	Evaluation Chips for Series 40. External 4-kbyte ROM.	Additional 2-kbyte External ROM		With 5-Digit LCD Driver	12 × 12 Matrix VF Outputs

317

TABLE 5-15 OLMS-40 Support Circuits and Development Tools

CMOS I/O Products	
MSM5837RS	Serial to 12-bit parallel out converter and LED driver.
MSM5870RS	4 line to 16 line output expander.

CMOS Peripheral Circuits	
MSM5201RS	Analog display driver for radio recvr freq.
MSM58281RS	Display driver, 4-digit VF.
MSM58282RS	Display driver, 4-digit VF/LED.
MSM5829GS	Display driver, 5-digit LCD.
MSM58291GS	Display driver, 5-digit VF.
MSM5832RS	Microprocessor real time clock/calendar.
MSM5838GS	Row scanning controller for LCD DOT matrix driver.
MSM5839GS	Column data register for LCD DOT matrix driver.

Development and Evaluation Systems	
MPB201	PC board emulator Series 40 MCU operation. Stand-alone or download. ROM based self-assembler, debug monitor.
MPB202	PC board emulator MSM5840 operation. Stand-alone final program verification.
MPB203	PC board provides MPB201 with EIA interface, 20-mA loop interface or TTL compatible interface.

System Software	
SDP40-I	ISIS® based software development package.
SDP40-C	CP/M® based software development package.
SDP40-S	OKI Series 40 program development package, stand-alone.

Development System Packages	
MPSP-I	Evaluation boards, software, complete User's Manuals—ISIS®.
MPSP-C	Evaluation boards, software, complete User's Manuals—CP/M®.
MPSP-S	Eval. boards, software, complete User's Manuals—stand-alone.

The IM6402 differs from the IM6403 on pins 2, 17, 19, 22, and 40, as shown in Fig. 5-24. The IM6403 utilizes pin 2 as a control and pins 17 and 40 for a inexpensive crystal oscillator. TBR_{empty} and D_{ready} are always active.

Figure 5-25 depicts a block diagram of the IM6402/6403 UART.

As shown in Fig. 5-26, the transmitter section accepts parallel data, formats it, and transmits it in serial form on the TR_{output} terminal. At Ⓐ, data is loaded into the transmitter buffer register from the inputs TR_1

through TR_8 by a logic low on the TBR_{load} input. Valid data must be present at least t_{set} prior to and t_{hold} following the rising edge of TBR_L. If words less than 8 b are used, only the least-significant bits are used. The character is right-justified into the least-significant bit TR_1. At Ⓑ, the rising edge of TBR_L clears TBR_{empty}. One-half to one and one-half clock cycles later data is transferred to the transmitter register and TR_{empty} is cleared. One-half cycle later transmission starts. Output data is clocked by TR_{clock}. The clock rate is 16 times the data rate. One-half clock cycle later, TBR_{empty} is reset to a logic high. At Ⓒ, a second pulse on TBR_{load} loads data into the transmitter buffer register. Data transfer to the transmitter register is delayed until transmission of the current character is complete. At Ⓓ, data is automatically transferred to the transmitter register, and transmission of that character begins 1 clock cycle later.

Referring to Fig. 5-27, data is received in serial form at the R_{input}. When no data is being received, R_{input} must remain high. The data is clocked through the RR_{clock}. The clock rate is 16 times the data rate. At Ⓐ, a low level on DR_{reset} clears the D_{ready} line. At Ⓑ, during the first stop bit data is transferred from the receiver register to the $RB_{register}$. If the word is less than 8 b, the unused most-significant bits will be a logic low. The output character is right-justified to the least-significant bit RBR_1. A logic high on O_{error} indicates overruns. An overrun occurs when D_{ready} has not been cleared before the present character is transferred to the $RB_{register}$. At Ⓒ, $\frac{1}{2}$ clock cycle later, D_{ready} is reset to a logic high, P_{error} and F_{error} are evaluated. A logic high on F_{error} indicates an invalid stop bit was received, a framing error. A logic high on P_{error} indicates a parity error.

Table 5-18 summarizes the key characteristics of the IM6402/6403 as compared with PMOS UARTs, and Fig. 5-28 shows how the IM6402 UART is interfaced with the IM6100 microprocessor.

Harris Semiconductor's HD6402/A UART (second source of the IM6402/A) and HD4702/6405 bit-rate generators provide the system designer with architectural flexibility and allow one to convert parallel data to serial and back again asynchronously, substantially reducing the amount of interconnect in a data acquisition system (Fig. 5-29), printer, modem, or peripheral.

THE MOTOROLA MC14469 PERIPHERAL CIRCUIT

The MC14469 CMOS peripheral circuit adds as many as 256 I/O ports to many microprocessors—increasing their interface capability. In addition to tying 128 remote inputs or outputs to a central processor, it also puts remote address-recognition logic, a slave UART, and two 8-b parallel I/O

TABLE 5-16 CMOS Microprocessor Summary (General Purpose)

Supplier	Processor P/N	Word Size (Data/ Instruction) (b)	Direct Addressing Range (Words)	Number of Basic Instructions	Maximum Clock Frequency (MHz)/ Phases	Instruction Time, Shortest/ Longest (μs)	TTL-Compatible	BCD Arithmetic	On-Chip Interrupts/ Levels
Motorola	MC14500	1/4	0	16	1/1	1/1	Yes	No	Yes/1
National Semicond.	NSC800	8/8	64K	150+	8/1	0.5/2.88	Yes	Yes	Yes/5
RCA	CDP 1802	8/8	64K	91	6.4/1	2.5/3.75	Yes	Yes	Yes/1
	8085AC	8/8	64K	80	5.5/1	0.8/5.2	Yes	Yes	Yes/4
Intersil	IM 6100	12/12	4K	81	4/1	2.5/5.5	Yes	No	Yes/1
Intersil	80C35/80C39	8/8	64K	96	6/1	2.5/5	Yes	Yes	Yes/1
Toshiba	T88000†	16/16	16M	151	10/1	0.4/1.6	Yes	N/A‡	N/A

* Standard TTL or MOS circuits will suffice.
† CMOS/SOS
‡ N/A = not available.

ports on one chip. Up to now, each remote point would have needed a UART of its own and perhaps 6 MSI chips—and several more wires would have been needed for the "party line."

Moreover, this chip's twisted line can carry power for this new receiver-transmitter chip, as well as data to and from a 4800-baud UART at the central location.

In operation, the chip's transmitter-receiver serially receives two 11-b words, one for address, one for command. It transmits two words, each with 8 data bits, a start bit, a stop bit, and 1 even-parity bit.

Seven address pins on the MC14469 can be programmed by tying them to ground or leaving them open. Once an address word that matches the address programmed on its pins arrives, the chip starts to decode and execute the command, which can either select the data to be transmitted or provide data for a remote peripheral device.

Up to 128 MC14469s tied to one pair of wires——which can be several thousand feet long—can operate in either simplex or full duplex fashion. Any microprocessor can serve as the central processor, and any UART can form the interface to the line, tying the MC14469s together.

Commercially Available CMOS Microprocessors

Number of Internal General-Purpose Registers	Number of Stack Registers	On-Chip Clock	DMA Capability	Specialized Memory and I/O Circuits Available	Prototyping System Available	Package Size (Pins)	Voltages Required (V)	Comments
1	0	Yes	No	No*	No	16	3–18	Needs External Program Counter
4	RAM	Yes	Yes	Yes	Yes	40	3–12	Executes Z80 Instructions and Has 8085 Bus Structure
6	RAM	Yes	Yes	Yes	Yes	40	3–12	Superseded Two-Chip Version
3	RAM	Yes	Yes	Yes	Yes	40	5	CMOS Equivalent and Pin-Compatible with 8085A
0	RAM	Yes	Yes	Yes	Yes	40	4–11	Emulates PDP-8 Instruction Set. Second sourced by Harris
4	RAM	Yes	Yes	Yes	Yes	40	5	CMOS Equivalent of Intel 8035/8039, P_{DISS} = 50 mW; "ROM-less" Version of 8048/8049
A	RAM	N/A	N/A	N/A	N/A	N/A	5	On-chip Floating point and decimal-processing hardware

A diode and a capacitor at each location assure that each chip has a fairly steady drain voltage supply. One external transistor used as a line driver can amplify the CMOS output for each remote MC14469 that must reply to the central processor.

The inherent noise immunity of CMOS, and the ability of CMOS to operate over wide ranges of temperature and voltage, make the MC14469 suitable for factory environments and other nonideal locations. In addition, the new power requirements of CMOS mean that voltage drop along the line is not a problem.

The MC14469 operates over a supply range of 4.5 to 18 V and is encased in a 40-pin DIP for ease of use.

Current SOS/CMOS Status

Becoming less bullish about the prospects of silicon-on-sapphire technology, Hewlett-Packard Company has backed off from further developing SOS 2, its polysilicon-gate 2.5–3-μm third-generation process. But Hewlett Packard is pursuing an earlier process, SOS 1.5, a second-generation aluminum-gate process with 3–4-μm features, with which it expects to build most if not all SOS circuits for at least another decade.

TABLE 5-17 Commercially Available CMOS Microprocessor Summary (Complete Chip)

Supplier	Device P/N	Word Size (Data/Instruction) (b)	On-Chip RAM Size	On-Chip ROM/PROM Size (Words)	Off-Chip Memory Expansion	Number of Basic Instructions	Maximum Clock Frequency (kHz)	On-Chip Clock	Instruction Time, Shortest/Longest (μs)
Hitachi	HMCS43/43C	4/10	80 × 4	1024 × 10	No	74	780/500	Yes	10
National Semicond.	COP420C	4/8	64 × 4	1024 × 8	Yes	49	250	Yes	16/32
	COP421C	4/8	64 × 4	1024 × 8	Yes	49	250	Yes	16/32
NEC	UPD650	4/8	96 × 4	2000 × 8	No	80	440	Yes	4.5/9
	UPD651	4/8	64 × 4	1000 × 8	Yes	58	440	Yes	4.5/9
	UPD652	4/8	32 × 4	1000 × 8	No	58	440	Yes	4.5/9
Panasonic	MN1450	4/8	64 × 4	1024 × 8	No	75	500	Yes	6/12
	MN1453	4/8	16 × 4	512 × 8	No	50	500	Yes	6/12
	MN1454	4/8	16 × 4	512 × 8	No	48	500	Yes	6/12
	MN1435	4/8	128 × 4	2048 × 8	No	75	500	Yes	6/12
Texas Inst.	TMS1000C	4/8	64 × 4	1024 × 8	No	43	1000	Yes	6/6
Intersil	87C41	8/8	64 × 8	1024 × 8	Yes	90	6000 (5 V)	Yes	2.5/5
	80C48/ 87C48	8/8	64 × 8	1024 × 8	Yes	96	6000 (5 V)	Yes	2.5/5
	80C49	8/8	128 × 8	2048 × 8	Yes	96	11,000	Yes	1.4/2.8
Motorola	146805	8/8	64 × 8	1100 × 8	Yes	61	3580	Yes	2/4
RCA	CDP1804	8/8	64 × 8	2048 × 8	Yes	113	8000	Yes	2/3
ITT Semi	SAA6001	4/8	← Not Available →						
OKI	See Table 5-14								

Hewlett-Packard concluded recently that SOS's earlier advantages over other MOS technologies will not be as pronounced for fabricating devices with smaller geometries as they are for those with larger ones.

Though some semiconductor manufacturers are dropping efforts to develop standard SOS circuits or are emphasizing other processes, HP is increasing its SOS circuit development activity. According to HP, SOS will continue to make a contribution for many years wherever very high performance with low power dissipation is needed.

Commercially Available CMOS Microprocessors

TTL-Compatible	BCD Arithmetic	On-Chip Interrupts/Levels	Subroutine Nesting Levels	General-Purpose Internal Registers	Number of I/O Lines	Additional Special Support Circuits	Package Size (DIP Pins)	Voltages Required (V)	Comments
No/Yes	Yes	Yes/2	RAM	RAM	32	No	42	−10/+5	
Yes	Yes	Yes/1	3	RAM	20	Yes	28	2.4–6.3	
Yes	Yes	No	3	RAM	16	Yes	24	2.4–6.3	
Yes	Yes	Yes/1	3	RAM	35	No	42	+5	
Yes	Yes	Yes/1	1	RAM	35	No	42	+5	
Yes	Yes	Yes/1	1	RAM	21	No	42	+5	
Yes	No	Yes/1	2	RAM	30	Yes	40	4.25–6	
Yes	No	Yes/1	2	RAM	13	Yes	18	4.25–6	
Yes	No	Yes/1	2	RAM	10	Yes	16	4.25–6	
Yes	No	Yes/2	2	2+ RAM	34	Yes	40	4.25–6	
Yes	Yes	No	3	2+ RAM	22/32	Yes	28/40	3–6	
Yes	Yes	Yes/1	8	16+ RAM	18	Yes	40	5–10	CMOS Equivalent to Intel 8041; 50-mW P_{DISS}
Yes	Yes	Yes/1	8	RAM	27	Yes	40	5–10	CMOS Equivalent to Intel 8048/8748; 50-mW P_{DISS}
es	Yes	Yes/1	8	RAM	27	Yes	40	5–10	CMOS Equivalent to Intel 8049; 50-mW P_{DISS}
es	Yes	Yes/1	0	RAM	20	Yes	28	5	CMOS Version of 6805
es	Yes	Yes/1	RAM	RAM	13	Yes	40	5–10	Compatible with 1802 Software
←———— Not Available ————→							60	3	135-μW Power Consumption; 45-μW standby power; can directly drive 8-digit LCD

At present, HP is using SOS to build many 16-b microprocessors, as well as bus-controlled chips for the HP-IB (IEEE-488) interface bus and other communications chips.

Even as HP continues with the pursuit of SOS for commercial applications, the enthusiasm of some devotees is waning.

EFCIS (Société pour l'Etude et la Fabrication de Circuits Intégrés Speciaux), the French MOS manufacturer jointly owned by Thomson-CSF and the French Atomic Energy Agency, has all but dropped its efforts

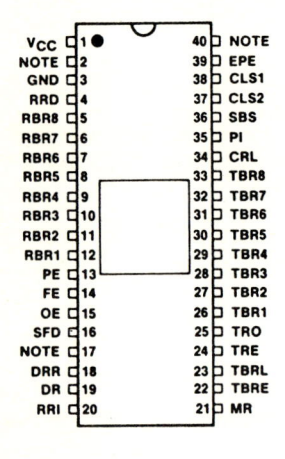

Fig. 5-24 IM6402/6403 Pin Connection Diagram.

to develop standard SOS parts. The firm's market strategists estimate that substrate costs will keep them from being competitive.

Interestingly, RCA, the biggest proponent of CMOS/SOS, has decided to deemphasize SOS and shift its focus to a new oxide-isolated bulk CMOS process now in development for commodity parts. The high cost of sapphire and concurrent complex processing required plus the development of very high-speed bulk CMOS processes have resulted in this decision and a change of direction.

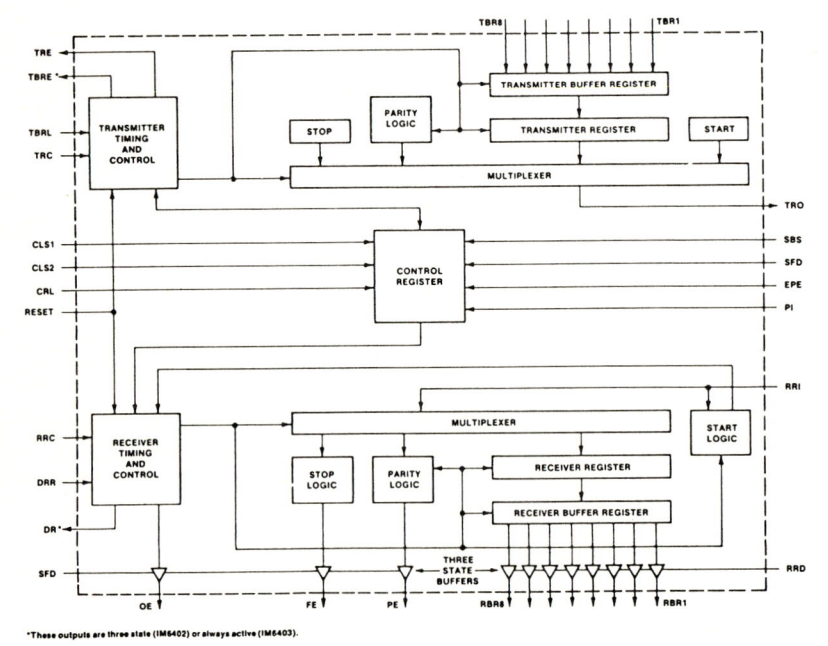

Fig. 5-25 IM6402/6403 Block Diagram.

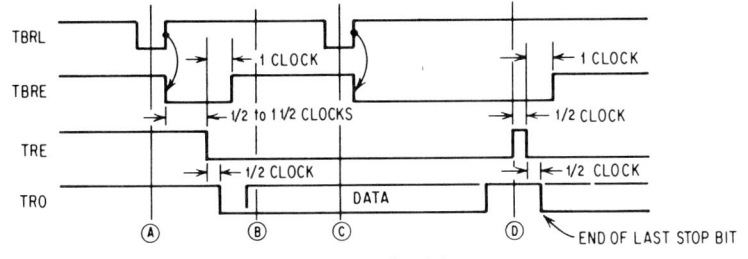

Fig. 5-26 Transmitter Timing (Not to Scale).

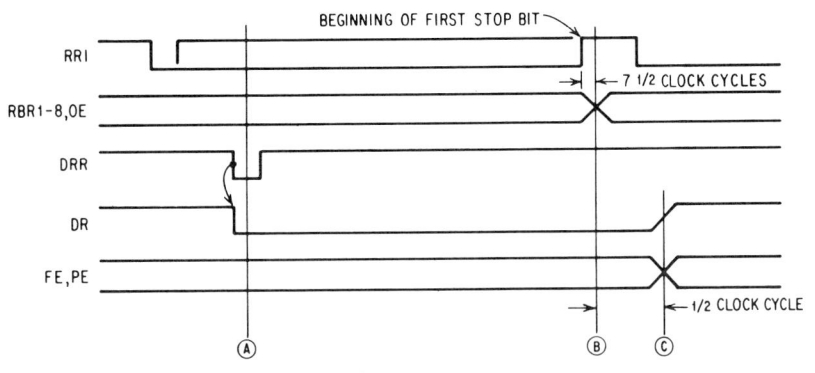

Fig. 5-27 Receiver Timing (Not to Scale).

TABLE 5-18 CMOS/PMOS UART Comparison

	IM6402	PMOS UART
Speed	Up to 4 MHz	200 kHz
Power Consumption		
Active	36 mW	675 mW
Standby	2 mW	675 mW
Power Supplies	Single	Two or Three
Noise Immunity	50%	30% or Less
Operating Temp.	−40 to +85°C*	0–70°C

* Military temperature range device also available.

INTERFACING WITH THE IM6100 MICROPROCESSOR

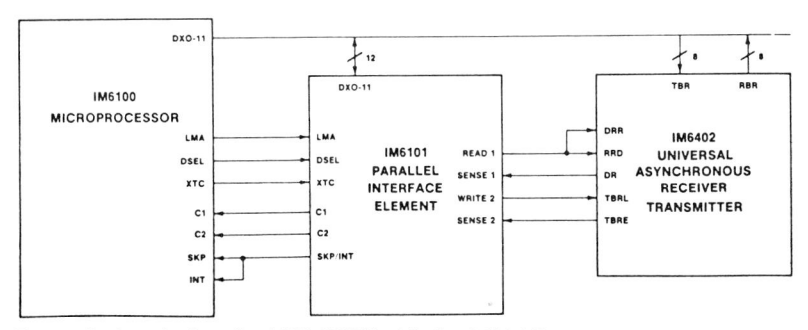

Fig. 5-28 Interfacing the 6402 UART with the 6101 Microprocessor.

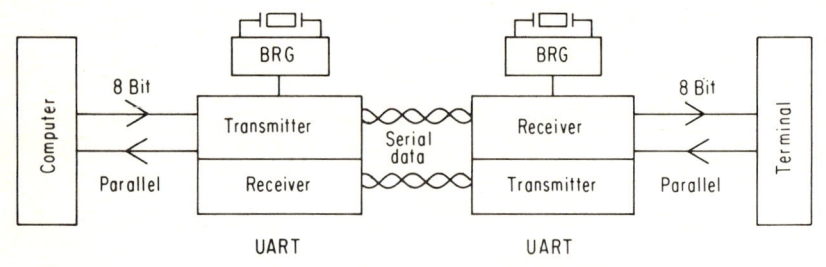

Fig. 5-29 Use of Monolithic CMOS UART and Bit-Rate Generator in Data Acquisition System.

Rockwell International Corporation's Microelectronics division in Anaheim, California, perhaps most active in building SOS circuits for military applications, has yet to set a timetable to move the technology into the commercial market. The projects Rockwell is working on are as follows.

- A VLSI chip that is an 8200-transistor Viterbi error-correction decoder implemented in CMOS/SOS technology to offer a small, lightweight package that consumes little power and can be used to assure data integrity in communications satellites. This device, with power measured in milliwatts, replaces a former relay rack of PC boards and ICs, where power is measured in tens of watts.
- A VLSI functional module, which demonstrates analog capabilities, that is a 12-b A/D converter now being implemented in 4-μ-m CMOS/SOS. This module offers a conversion speed of 2.5 μs for 12-b, with an average power dissipation of only 15 mW and operates from ±10 V, accepting input signals ranging from 0 to 10 V.
- A CMOS/SOS frequency synthesizer that offers up to 120-MHz performance in a 3-μm version, yet consumes only 25 mW of power. Individual experimental prescaler devices have 2-μm channel lengths and provide a divide-by-$^6/_7$ function of 300 MHz, with 15 mW of dissipation at 5-V bias.

Toshiba, Ltd., is ready to begin production on a high-performance CMOS/SOS microprocessor and static RAM.* The prototype 4-kb × 1-b RAM has a typical access time of 20 nS with a power dissipation of only 250 mW when operating from a single 5-V supply, whereas a 4-kb RAM

---

*See 1981 IEEE ISSCC Digest of Papers, New York, N.Y. Paper 16.1 and 1.1 respectively for details on these devises.

fabricated with a scaled NMOS process typically dissipates twice the power (500 mW) to achieve the same access time (22 nS).

The new design features a negligible standby power drain without incurring the additional chip-select time typical of N-channel MOS devices operated in a power-down mode. Moreover, tests indicate that the lower resistance of the silicide word lines reduces the propagation delay and thus access time by about 6 nS compared with similar RAMs using polysilicon word lines.

Photolithography and dry etching are used to define the silicide gates and word lines. The effective channel length in these devices is 1.5 μm, which is large enough for devices operated from a 5-V power supply despite short-channel effects.

The cell size for this RAM is 36 by 36 μm (14 by 14 mils) and chip size is 3.23 by 4.19 mm (127 by 165 mils).

The basic SOS process will also be used by Toshiba to fabricate microprocessors. But polysilicon gates will then be used, because long connections in these chips are made with aluminum rather than with the gate material, as in memories.

Fujitsu reasearchers say they do not think SOS will be used for memories in the future. However, Fujitsu has developed a 10,000-gate CMOS processor with a 400-ns micro instruction cycle and 130-mW power dissipation.

While the future of SOS for commercial applications remains cloudy, primarily because of the high processing costs involved, the strong performance advantages of this technology are certain to make their weight felt in military electronics by 1985. As the speed and density of computer systems continue to increase and the size of this equipment continues to decrease, a clear-cut advantage will emerge for CMOS/SOS in military systems where size, weight, and power dissipation are extremely critical.

The Department of Defense's very high-speed integrated circuit (VHSIC) program is currently sparking CMOS/SOS development work at several IC companies as well as military equipment suppliers that have captive semiconductor manufacturing capabilities. Because of this, CMOS/SOS could become the predominent VLSI technology of the decade (edging out NMOS, PMOS and I^{2}L, for example), particularly if the Department of Defense gives SOS its blessing.

The availability of the aforementioned products plus the research efforts of companies like General Electric, McDonnell-Douglas, Sanders Associates, Hewlett-Packard, Hitachi, OKI, Toshiba, Fujitsu, NEC, Rockwell, Raytheon, Lockheed Missiles and Space Company, and

Hughes Aircraft Corporation will set the stage for the adoption of CMOS and CMOS/SOS as VLSI technology. However, many of these developments will be for in-house use and not outside sales. It has been predicted by Rockwell International that CMOS will eventually become the main microcomputer technology. Whether SOS has any commercial future will depend on its becoming a cost-effective producible process and upon the action of Intel.

Intel holds the key to the commercial success or failure of SOS. If Intel chooses not to exercise its cross-licensing rights from RCA (signed in April 1978), SOS will remain in its present role as a specialty technology for military and aerospace applications, notwithstanding other uses at Rockwell, HP, and elsewhere.

As of this writing (1981), Intel had decided not to pursue the technology.

There is no doubt that CMOS/SOS could emerge as the leading VLSI technology for military devices by the mid-1980s; its commercial success will depend on the number of other IC manufacturers besides Intel that enter the market, as well as on the cost of sapphire. Increased volume will naturally bring such costs down, but the demand must be there first.*

Only by measuring the progress of the aforementioned product announcements and the practical applications of the same will one be able to determine the trend for the next generation of integrated circuits— the realm of VLSI.

LSI Trends

The trend of future LSI and VLSI circuits is very clear. These circuits will be complete subsystems or systems that contain both analog and digital functions on the same chip, using a wide variety of technologies in order to obtain desired performance objectives in various sections of the chip.

Several particular application areas currently forwarding this trend are those of data acquisition systems, signal-processing systems, and speech-synthesis/voice-recognition systems. The increasing sophistication of memories, microprocessors, and digital signal-conditioning devices obviates the need for complex data conversion and processing of analog signals, which still do not readily lend themselves to digital-signal

---

* Further discussion of CMOS/SOS and bulk CMOS for VLSI use is contained in IEDM 1980, Dec. 8–10, Washington, D.C., Paper 10.1.

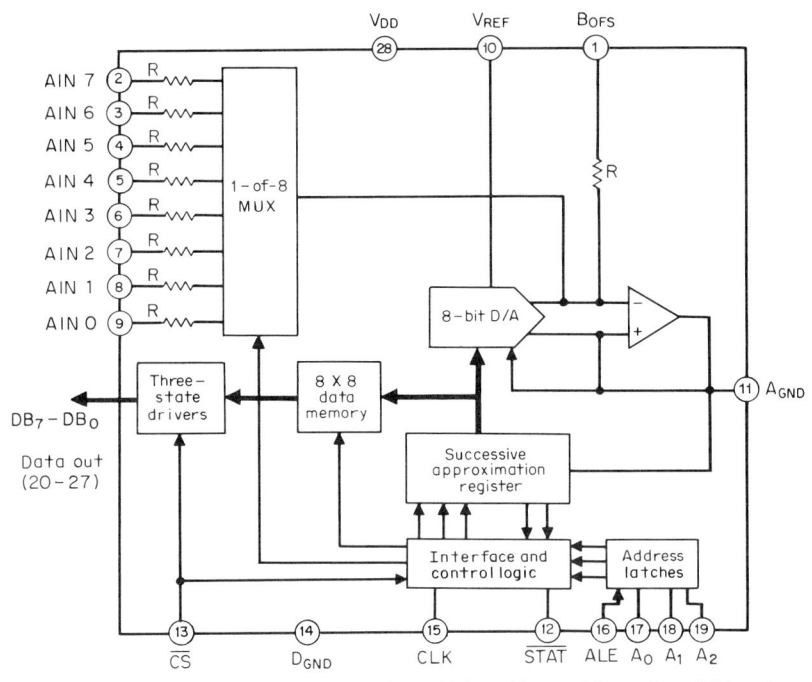

Fig. 5-30 Block Diagram of AD7581 Monolithic 8-Channel Data Acquisition System.

processing because of the speed and resolution required or because of environmental conditions. Several examples serve to illustrate the point.

The first Monolithic commercial implementation of this philosophy was Intel's 2920 analog microprocessor. The 2920 is a special-purpose microprocessor which is intended to operate as a real-time single-chip system for analog signal processing applications. It contains analog-to-digital and digital-to-analog conversion capability as well as program storage (EPROM), scratch-pad RAM, a binary scaler, and a 25-b microprocessor.

Analog Devices' AD7581 monolithic CMOS microprocessor-compatible data acquisition chip is another telling example of this trend. The AD7581 (Fig. 5-30) contains a dual-port 8-byte RAM, an 8-channel multiplexer, an 8-b successive approximation A/D converter, address latches, interface logic, and three state buffers for direct connection to a microprocessor bus. With this on-board memory, the system microprocessor views the AD7581 as a memory not as a converter; thus, there is no need to interrupt software, since data are taken from the A/D converter in a normal microprocessor READ cycle. The AD7581 is treated as an interleaved DMA device.

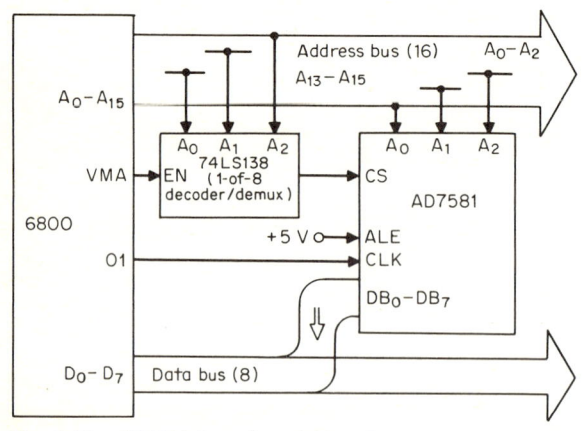

Fig. 5-31 AD7581 Data Acquisition System Interface with 6800 Microprocessor.

Using precision-multiplying D/As, AD7581 microprocessor-compatible DAS ICs are in two temperature ranges: 0 to 70°C and −25 to 85°C. Versions can be specified with either ±1⅛, ±¾, or ±½ LSB relative accuracy over either available temperature range. Differential nonlinearity is ±1⅛, ±⅞, or ±¾ LSB. The chip can be set up for a choice of unipolar (0 to 10 V) or bipolar (−5 to +5 V) analog input operation. The AD7581 is fast, accomplishing A/D conversion in 15 μs. Operation is from common +5-V logic supplies. A −10-V reference is also required. Fig. 5-31 depicts the ease of interfacing the AD7581 with a typical microprocessor.*

American Microsystems Incorporated's AMI 2210 is another complete single-chip system containing a microprocessor and A/D and D/A converters.

These few examples are only the tip of the iceberg and portend the large monolithic system chips (data acquisition, controller, telecommunications, speech processing, etc.) that will be forthcoming in the future and will be fabricated using the new generations of CMOS processes, which provide both high-speed operation and energy-conscious low power consumption.

---

*Further discussion of the AD7581 may be found by referring to H. Tucholski, "DAS Packs Memory for μP Interface," *Electronics Design*, Sept. 13, 1980, pp. 79–83.

Index

A Law transfer characteristic, 155, 158–160, 162, 163, 166, 167
Access time, 195, 199, 308, 326–327
  PROMs, 246, 248–250, 259
  RAMs, 209, 217, 223, 224, 226, 231, 234–236, 238–239
  ROMs, 252, 253, 256, 259
Accumulator, 269
Active filter, 163
A/D (analog-to-digital) coverter display interface, 119–123, 129–132
A/D (analog-to-digital) converters (ADCs), 1, 12, 85–150
  dual-slope integration (see Dual-slope integration A/D converter)
  flash, 87
  microprocessors, 289–290, 326, 329
  pulse duration modulation, 124, 125, 127
  simultaneous-integration, 106–108
  successive-approximation, 87, 90, 91–95, 127, 329
  telecommunications circuits, 155, 156, 167
  3½-digit, 97, 112–132
  video, 87
  (See also Commercially available A/D converters)
Address decoder, 206, 207, 211
Address latch, 207, 248, 249
Aircraft Area Navigation System, 262–264
ALU (arithmetic and logic unit), 269
American Microsystems, Inc. (AMI), 171, 173, 299, 330
  codecs: S3501 coder, 12, 162–165
  S3502 decoder, 12, 162–165
  S3505, 166–167
  RAMs, 214, 238
  repertory dialer, 189–190
Analog Devices, Inc.:
  A/D converters (see Commercially available A/D converters, Analog Devices)
  D/A converters, 88, 89
  AD7520, 68–77, 88
  AD7522, 76–84, 88

Analog Devices, Inc. (Cont.):
  data acquisition system, AD7581, 329–330
Analog microprocessor, 329
Analog-to-digital converters (see A/D converters)
Arithmetic and logic unit (ALU), 269
Asynchronous memory, 199–201, 254
Asynchronous RAM, 199–201
Automobile engine controller, 288–291
Autozeroing converter, 114–124

Battery-supported backup, 2, 3, 14, 15, 121, 223, 231, 260, 313
Beckman Corp., 89
Bipolar offset binary conversion, 73, 74, 78, 79, 81
BMW (Munich, Germany), 291
Branch instructions, 271, 284, 288
Buffered logic, 7–10
Buried contact memory cell, 226–229
Bus driver, 61–62

C^2MOS (see Clocked CMOS)
CCITT (International Telephone and Telegraph Consultative Committee), 160, 163, 166
Channel banks, 155
Channel cross talk, 161, 165, 166
Channel noise, 161, 166
Clocked CMOS (C^2MOS), 98, 103–106, 229–231, 310–315
CMOS (complementary metal-oxide semiconductor) construction, 3, 4, 7–11
  logic functions, 19–20
  PROM, 246
  RAMs, 226, 228–232, 234, 235
  telecommunications circuits, 174, 185–187
CMOS (complementary metal-oxide semiconductor) technology, 1–11

331

CMOS (complementary metal-oxide semiconductor) technology *(Cont.):*
  A/D converters, 95, 101–106, 114–118, 124–129, 133–136
  D/A converters, 68–73, 78, 79, 84, 85
  logic circuits: 14400/14500 series, 6, 12, 13, 55–62
    4000 series, 6, 17, 31, 33
    54C/74C series, 6
  logic performance characteristics, 18–45
  LSI circuit implementation using, 63–65
  memories, 191, 194, 199–202
    PROMs, 242, 246–250
    RAMs, 223, 224, 226, 228–232, 235
  microprocessors, 265–268, 312, 321–324, 326–330
  telecommunications circuits, 163–177, 181, 182, 185–187
Codec filters, 155, 156, 160, 163, 164, 166–168
Codecs, 1, 2, 12, 155–173
  commercially available, 12, 161–173
Coherent detection, 154
Commercially available A/D converters, 87, 90–151
  Analog Devices, AD7550, 108–112, 138–139
    AD7570, 91–96, 138–139
    AD7574, 87, 90–91, 138–139
  Intersil, 140–141
    ICL7106/7107, 112–114, 140–141
    ICL7109, 95, 97–102, 138–139
  microprocessor-compatible, 98, 103–106
  microprocessor-interface applications, 142–151
  National Semiconductor, 138–141, 147–151
    ADC0800 ("Naked 8"), 87
    ADC0816, 132–139
    ADD3501, 124–132, 140–141
  Nippon, 138–139
    simultaneous integration, 106–108
  RCA, 87, 90, 138–139
    high-speed 6-b, 87
  Siliconix, 140–141
    LD130, 114–124, 140–141
  summary of, 138–141
  Teledyne Semiconductor, 8700 series, 138–147

Commercially available codecs, 12, 161–173
Commercially available communication circuits, 12, 13
  data acquisition, entry, and manipulation, 55–62
  data conversion circuits: A/D converters *(see* Commercially available A/D converters)
    D/A converters *(see* Commercially available D/A converters)
  microprocessors, 301, 303, 305, 318, 319, 324–326
  telecommunications circuits, 161–190
Commercially available D/A converters, 65–86, 88–89
  Analog Devices, 88, 89
    AD7520, 68–77, 88
    AD7522, 76–84, 88
  discrete, 65–68
  summary of, 88–89
  Teledyne Semiconductor, 8640/8641, 84–86, 88
Commercially available display circuits, 45–54, 305, 306
Commercially available memories, 14–15, 53–54, 191–193
  (See also Commercially available PROMs; Commercially available RAMs; Commercially available ROMs)
Commercially available microprocessors, 13–14, 222, 252, 277–306
  A/D converters, 95–102, 137, 142–151
  display circuits, 53, 54
  Intersil, 294–295, 320–323
    IM6100, 13, 98, 102, 277, 278, 299–306, 320–321
  Motorola: MC14469 peripheral circuit, 319–321
    MC14500, 279–281
    MC146805, 14, 297, 298, 322–323
    MC6800, 97, 98, 101
  National Semiconductor: COP family, 313–315, 322–323
    NSC800, 14, 295–298, 320–321
  new 4-b μP developments, 313–317
  OKI Semiconductor, 317–320
  RCA, 320–321, 324, 326, 328
    CDP1802, 277, 278, 281–291, 320–321

Commercially available microprocessors, RCA *(Cont.)*:
  CDP1804/1804C, 291–294, 322–323
  summaries of, 320–323
  support circuits, 315, 318–319, 324–325
  Texas Instruments, TMS1000C, 279–281
Commercially available PROMs, 242–250
  display circuits, 53, 54
  Harris Semiconductor, 258, 259
    HM6611, 242–244, 258, 259
  Intersil, 258, 259
    IM6603/6604, 244–250
  microprocessors, 274–275, 304–305
  summaries of, 258, 259
Commercially available RAMs, 14, 15, 199, 212–229, 231–235, 238–239, 252
  display circuits, 53, 54
  Harris Semiconductor, 199, 215, 238–239
    HM6514, 215, 217–223, 239
  Hitachi, 215, 239
    HM6147, 15, 215, 231–235, 239
    HM6148, 215, 232
    HM6116, 215, 232, 234
    HM6167, 234
  Intel, 212, 219, 224
    5101/5101L, 213, 214, 216, 217
  Intersil, 215, 238
    IM6504, 215, 222–224, 239
    IM6508/6518, 15, 199, 213, 215, 216, 238
  memory applications, 257, 260–264
  microprocessors, 274–275, 285–289, 301, 303–307, 309, 326–327
  National Semiconductor, 214
    MM54C920, 202–208, 214
    MM54C930, 202–208, 214
    MM74C921, 208–211, 214
  RCA, 199, 214, 238–239
    MWS5104, 214, 226–229
    MWS5114, 214, 224–228, 239
  summaries of, 214–215, 238–239
  types and processes, 191–193
Commercially available ROMs, 252–257
  Harris Semiconductor, 258, 259
    HM6312A, 252–254, 258, 259
  memory applications, 262
  microprocessors, 271, 274–275, 285–287, 301, 303, 307, 309
  summaries of, 258, 259

Commercially available ROMs *(Cont.)*:
  Supertex, CM3200, 252, 255–257
Commercially available telephone ICs, 161–190
Common data I/O systems, 217
Communication circuits, 2, 12–14
  data acquisition, entry, and manipulation, 55–62
  data conversion circuits: A/D converters, 85–87, 90–151
    D/A converters, 65–86, 88–89
  microprocessors, 289, 290, 301, 303, 318, 319, 324–326, 328–330
  telecommunications circuits, 151–190
  (See also Commercially available communications circuits)
Compander, 159–161
Companding D/A converter, 159–161
Complementary metal-oxide semiconductor *(see* CMOS construction; CMOS technology)
Computer-controlled telephone, 188
Control logic, 270–271
Controllers, 261–262, 277–279, 288–291
Conversion accuracy, 87, 90, 104–106, 112
Conversion cycle, 103, 104, 106, 107, 128, 129
Conversion rate and time, 87, 90, 91, 97, 103, 105, 106, 126, 132, 143
Converter parallel loading, 80–83
Converter serial loading, 83, 84
Current-fed R/2R ladder network D/A converter, 66, 67
Cycle stealing, 274

D flip-flop, 37, 38
D/A (digital-to-analog) converters (DACs), 1, 12, 65–85, 88, 89
  companding, 159–161
  current-fed R/2R ladder network, 66, 67
  microprocessors, 289–290, 330
  multiplying, 68–86, 91, 92, 330
  telecommunications circuits, 155, 156, 160, 161, 166, 168–170, 173–175, 185, 186
  (See also Commercially available D/A converters)
Data acquisition systems, 94–95, 98, 111, 132–137, 328–330
Data bussing, 55–56

Index

Data converters, 1, 12, 63–150
(*See also* A/D converters; D/A converters)
Data processing, 12–13, 55–62
Data retention, 218, 222, 224, 226, 228, 235, 250, 251
Data routing, 57–61
Data selector, 57–59
Digital communication system, 151–159
Digital dwell tachometer, 122, 124
Digital Equipment Corp., 299–301
Digital linearization, 110–112
Digital panel meter (DPM), 119–132
Digital thermometer, 122, 123
Digital-to-analog converters (*see* D/A converters)
Digital voltmeter (DVM), 119, 124–132
Direct memory access (DMA), 273, 274, 281, 282, 301, 329
Display circuits, 45–55, 119–132
  commercially available, 45–54, 305, 306
Display counter, 46–51
DMA (*see* Direct memory access)
DPM (digital panel meter), 119–132
DTMF (dual-tone multiple-frequency) receiver, 177, 180–182
Dual-polarity converter, 148
Dual-slope integration A/D converter, 95, 97–104, 106, 108, 113–114, 124, 125, 127
Dual-tone multiple-frequency (DTMF) receiver, 177, 180–182
DVM (digital voltmeter), 119, 124–132
Dynamic logic, 11, 42–44, 63–65, 229, 310–313

EAROM (electrically alterable ROM), 250–252
EEPROM (electrically erasable PROM), 258, 259
Electrically alterable ROM (EAROM), 250–252
Electrically erasable PROM (EEPROM), 258, 259
Erasable PROMs (EPROMs), 14, 244–250
  display circuits, 53, 54
  microprocessors, 274–275, 288, 289
  summary of, 258, 259

Fairchild Camera and Instrument Corp., 163, 214

FAMOS (floating-gate avalanche-injection metal-oxide semiconductor) memory, 251
Flag, 269, 270
Flash A/D converter, 87
Flip-flops, 33–34, 37–40
Floating-gate avalanche-injection metal-oxide semiconductor (FAMOS) memory, 251
Floating-gate fuse technology, 237, 246, 250
Four-phase logic, 43–44
Four-quadrant multiplication, 73, 74, 78, 79
Frequency counter, 50–52
Frequency synthesizer, 326
Fujitsu America Corp., 215, 327
Fused-link PROMs, 237, 240–244, 303, 304

General Instruments, Inc., 299

Harmonic distortion, 173, 175, 176
Harris Semiconductor Corp., 13, 295
  encoders-decoders, 167
  microprocessor support circuits, 319
  PROMs, 258, 259
    HM6611, 242–244, 258, 259
  RAMs, 199, 215, 238–239
    HM6514, 215, 217–223, 239
  ROMs, 258, 259
    HM6312/A, 252–254, 258, 259
Hewlett-Packard Company, 14, 266–267, 321–324
  CMOS/SOS microcomputer chip set, 307–309
Hitachi Semiconductor:
  HM6148, 215, 232
  HM6116, 215, 232, 234
  HM6167, 234
  microprocessors, 299, 322–323
  RAMs, 215, 236, 239
    HM6147, 15, 215, 231–235, 239
Hughes Solid State Products, 215

I^2L (integrated injection logic), 2, 3, 329
IEEE-488 bus, 308, 309, 323
Input capacitance, 24, 25
Input/output (I/O), 272–274
Integrated injection logic (I^2L), 2, 3, 329
Intel, 14, 15, 97, 98, 100, 142, 255, 257
  microprocessors, 222, 267, 294, 328
  2920 analog, 329

Index

Intel *(Cont.)*:
  RAMs, 212, 219, 224
    5101/5101L, 199, 213, 214, 216, 217, 238
Interdigit blanking, 48, 50, 121, 122
Intermetall, GMBH, 166
International Telephone and Telegraph Consultative Committee (CCITT), 160, 163, 166
Interrupts, 272, 273, 297–299, 313
Intersil, 13–15
  A/D converters, 140–141
    ICL7106/7107, 112–114, 140–141
    ICL7109, 95, 97–102, 138–139
  D/A converters, 88–89
  display circuits, 45–54
  encoder, ICM7206, 172–179
  microprocessors, 294–295, 320–323
    6100 series, 13, 98, 102, 277, 278, 299–306, 320–321
    support circuits, IM6402/6403 UART, 315, 318–319, 324–325
  PROMs, 258, 259
    IM6603/6604, 244–250
  RAMs, 215, 238
    IM6504, 215, 222–224, 239
    IM6508/6518, 15, 199, 213, 215, 216, 238
  ROMs, 258, 259
Inverter, 3, 4, 19–33, 42–44, 65, 174, 196
I/O (input/output), 272–274
ITT Semiconductor, 322–323

JK flip-flop, 37, 39–40

Large-scale integration (LSI) circuits, 1, 11, 12, 14, 18
  data conversion and telecommunications circuits, 63, 65
  data storage, manipulation, and transfer, 194
  microprocessors, 265, 268, 311
    future trends, 328–330
    UARTs, 315
LCD *(see* Liquid crystal device displays)
Leakage current, 23, 24, 196
Light-emitting diode (LED) displays, 45–54, 112, 114, 119, 121, 123, 126, 129–132
Linearity, 70, 78

Linearity errors, 70, 73, 90, 97, 132
Linearization, digital, 110–112
Liquid crystal device (LCD) displays, 45, 52, 112, 114, 120, 121
Logic circuits, 6–11, 17–62
LSI *(see* Large-scale integration circuits)

Manchester encoder, 167
Matsushita, 15
Memories, 1, 13–15, 191–264
  A/D converters, 87, 90–91, 95, 96, 112, 126
  applications, 257, 260–264
  asynchronous, 199–201, 254
  FAMOS, 251
  microprocessors, 274–275, 285–289, 301, 303–307, 309, 326, 327
  RAMs, 202–239
  ROMs/PROMs, 211–212, 237, 240–259
  scratchpad, 270, 329
  telecommunications circuits, 190
  (See also Commercially available memories)
Memory access and transfer, 273, 274
  (See also Direct memory access)
Memory cells, 40–41, 195–203, 235, 241, 248
  architecture, 196–202
  buried contact, 226–229
Memory expansion, 286–288
Memory mapping, 91, 153
Memory organization, 202–209, 211–212
Memory systems, 54, 254, 287–289, 301, 303–305
  applications, 257, 260–264
  nonvolatile, 14, 15, 303
Memory timing, 208–211, 218–221, 225–227, 233, 234, 251
Metal-oxide semiconductor (MOS) characteristic, 19–21
Metal-oxide semiconductor field-effect transistor *(see* MOSFET)
Microcomputers, 13, 14, 54, 294–295, 309
  CDP1804/1804C, 291–294
  COP family, 313–315
  OLMS-40 family, 315–317
  summary of, 322–323
  TMS1000C, 279–281
Microcontrollers:
  COP family, 313–315

Microcontrollers *(Cont.)*:
    MC14500 microprocessor, 277–279
    OLMS-40 family, 315–318
    summary of, 322–323
    TMS1000C microcomputer, 279–281
Micropower, Inc., 89
Microprocessor architecture, 268–275
    CDP1802, 282–284
    CDP1804/1804C, 291–293
    Hewlett-Packard CMOS/SOS microcomputer chip set, 307–309
    IM6100, 299, 300
    NSC800, 295–297
    16-b parallel CMOS microprocessor, 310–313
    TMS1000C, 280
Microprocessor-compatible converters, 76, 87, 90, 91, 95–112, 329–330
    ADC0816 data acquisition system, 132–137
    applications, 142–151
    single-chip C^{2}MOS 12-b A/D converter, 98–106
Microprocessor-converter interface, 95, 96, 98–102, 136, 137, 329, 330
    applications, 142–151
Microprocessor instructions, 270, 271, 275, 276
    CDP1802, 283–284
    CDP1804, 292, 294
    Hewlett-Packard CMOS/SOS microcomputer chip set, 307–308
    IM6100, 300
    MC14500, 278
    16-b parallel CMOS microprocessor, 310, 312
Microprocessor peripheral circuits, 12, 13, 62, 153, 278, 279, 295, 324–326
    A/D converters, 95, 96, 98–102, 148–150
    CDP1802, 285, 286
    COP family, 314–315
    Hewlett-Packard CMOS/SOS microcomputer chip set, 307, 309
    IM6100 family, 301, 303, 305, 306, 325
    MC14469, 319–321
    MC14500, 278, 279
    OLMS-40, 318
Microprocessor software, 275, 276, 279, 280, 284, 285, 292, 294, 300, 318
Microprocessor software support, 280, 284, 285, 318

Microprocessor systems, 54, 303–316, 319, 325, 330
    all-CMOS, 301
    based on CDP1802, 285–291
    elementary, 268–269
    microprocessor-compatible A/D converter, 98–106
    microprocessor-interface applications, 142–151
    telecommunications circuits, 152, 153
Microprocessors, 1, 2, 13, 14, 262–264, 265–330
    A/D converters, 87, 90, 91, 95–112, 132–137, 142–151
    analog, 329
    D/A converter, 76
    display circuits, 48, 50, 53, 54
    memory cell architecture, 201–202
    parallel, 310–313
    ROMs/PROMs, 247, 252, 254
    telecommunications circuits, 160
    (*See also* Commercially available microprocessors; *entries beginning with term:* Microprocessor)
Mikros Systems Corp., 309
Mitel Corp., 14, 182
MOS (metal-oxide semiconductor) characteristic, 19–21
MOS Technology, Inc., 98, 101
MOSFET (metal-oxide semiconductor field-effect transistor), 3, 4, 15, 19–21, 30, 229, 310, 316
Mostek Corp., 173, 299
    A/D converters, 136, 138–139
    codecs: MK5116, 12, 162, 163
    MK5150, 168–171
    MK5151, 12, 162, 163
    MK5156, 162, 163
    telephone dialers, 182
    MK5087 tone dialer, 185–188
    MK5098 pulse dialer, 183–185
Motorola Semiconductor Products, Inc., 6, 12–14
    codecs, 167
    MC14406, 12, 162, 163
    MC14407, 12, 162–164
    data acquisition, entry, and manipulation, 55–62
    display circuits, 45, 46
    microprocessors (*see* Commercially available microprocessors, Motorola)

Motorola Semiconductor Products, Inc. (Cont.):
  RAMs, 214, 238
    memory system implementation, 257, 260–264
Mu (μ) Law transfer characteristic, 155, 157, 159, 160, 162, 163, 165–167
Multiplexers, 60, 61, 62, 94, 95, 103, 132, 133
Multiplying D/A converters, 68–86, 91, 92, 330

N-channel metal-oxide semiconductor (see NMOS technology)
National Semiconductor Corp., 6
  A/D converters (see Commercially available A/D converters, National Semiconductor)
  codecs, 161, 167
    LF3700, 12, 162
    MM58100, 12, 162
  D/A converters, 88, 89
  microprocessors, 295
    COP family, 313–315, 322–323
    NSC800, 14, 295–298, 320–321
    6800, interfacing, 262–264
  RAMs, 199, 214, 238
Nippon Electric Microcomputers, Inc. (NEC):
  A/D converters, 106–108, 138, 139
  microprocessors, 295, 322–323
  RAMs, 215, 239
Nitron, 166
NMOS (N-channel metal-oxide semiconductor) technology, 2, 3, 14
  memory systems, 191–194, 212, 231–232, 234, 235
  microprocessors, 267, 294–295, 297–299, 327
Noise filtering, 129, 132
Noise immunity, 2, 4, 5, 9, 26, 27, 43, 68, 212, 321
Noncoherent detection, 154
Nonlinear function generator, 75
Nonlinearity, 70, 73, 90, 97, 132
Nonvolatile memory system, 14, 15, 303

OKI Semiconductor, Inc., 215, 313, 315–318

Output data latch-buffer, 208, 209, 211
Oxide-isolated CMOS, 6, 11, 65, 223, 224, 295, 296, 324

PAGE MODE memory organization, 203
Panasonic, Inc., 322, 323
Parallel microprocessor, 310–313
PCM (pulse code modulation), 155–160, 163, 165, 166
Phase-locked loop (PLL), 165, 167
PLA (programmable logic array), 189
PLL (phase-locked loop), 165, 167
Positive-polarity converter, 149
Power dissipation, 2–4, 12, 14, 15
  A/D converters, 91, 97, 106, 108, 132
  D/A converters, 68, 69
  display circuits, 52
  dynamic shift registers, 43
  inverters, 22–24, 26
  memories, 192–193, 199, 200
    RAMs, 213, 217, 223, 224, 231, 232, 235–239, 260
    ROMs/PROMs, 237, 250
  microprocessors, 295–298, 301, 303, 308, 312–315, 325–327
  reduction through CMOS circuitry, 17–18
  telecommunications circuits, 161–169
Power distribution, 129
Power series generation, 74–76
Power supply noise, 129
Process controller, 288, 289
Program execution, 271
Programmable logic array (PLA), 189
Programmable logic controller, 277–279
Programmable read-only memories (PROMs), 14, 237, 240–253, 258, 259
  display circuits, 53, 54
  fused-link, 237, 240–244, 303, 304
  microprocessors, 274–275, 288, 289, 303–305
  programming, 237, 240–243, 246, 247, 251
  (See also Commercially available PROMs)
Programing circuitry, 237, 240–242, 247, 251
Programming cycle, 241, 242, 246, 247
PROMs (see Programmable read-only memories)
Propagation delay, 24, 25, 28, 30–32, 202

Index

Pulse code modulation (PCM), 155–160, 163, 165, 166
Pulse dialer, 183–185
Pulse duration modulation A/D converter, 124, 125, 127

Quad slope conversion, 108
Quantized feedback conversion, 114, 117

Random-access memories (RAMs), 13–15, 190, 191–239, 252
   A/D converter, 90–91
   asynchronous, 199–201
   display circuits, 53, 54
   memory applications, 257, 260–264
   microprocessors, 274–275, 285–289, 301, 303–307, 309, 326–327
   organization, 202–211
   synchronous, 199–202, 218–224, 301
   (*See also* Commercially available RAMs)
Ratiometric converters, 91, 103, 136, 137
RCA Solid State Products Division, 6, 13
   A/D converters, 85–87, 90, 138–139
      high-speed 6-b, 87
   codecs, 167
   D/A converters, discrete, 65–68
   microprocessors, 267, 320–321, 324, 326, 328
      CDP1802, 277, 278, 281–291, 320–321
      CDP1804/1804C, 291–294, 322–323
   RAMs, 199, 214, 238, 239
      MWS5104, 214, 226–229
      MWS5114, 214, 224–228, 239
   ROMs/PROMs, 258
      floating-gate avalanche injection 2-kb EAROM, 250–252
READ cycle, 209–210, 218, 219, 225, 226, 233, 257
READ-MODIFY-WRITE cycle, 220, 221
Read-only memories (ROMs), 13, 195, 252–259
   A/D converters, 90–91, 95, 96, 112, 126
   memory applications, 261, 262
   microprocessors, 217, 274–275, 285–287, 301, 303, 307, 309
   organization, 211–212
   (*See also* Commercially available ROMs)

Refresh, 194, 196, 212, 298
Registers:
   microprocessor, 270, 271
   CDP1802, 282–289
   CDP1804, 291–293
   Hewlett-Packard CMOS/SOS microcomputer chip set, 307–308
   IM6100, 299
   MC14500, 278, 279
   NSC800, 295–297
   TMS1000C, 280
   shift, 42–44
   storage, 56–58
Repertory dialers, 188–190
Rockwell International, Inc., 326, 328
ROMs (*see* Read-only memories)
Row decoder, 202–204, 216, 230, 231

SAMPLE-AND-HOLD (S/H) circuit, 160–161, 170
Sapphire ribbon technology, 267
Scratchpad memory, 270, 329
SCU (subscriber channel unit), 163, 164
Sense amplifier, 207–209, 229, 230
Series 14400/14500 CMOS, 6, 12, 13, 55–62
Series 4000 CMOS, 6, 17, 31, 33
Series 54C/74C CMOS, 6
SET/RESET (S/R) flip-flop, 33–34
S/H (SAMPLE-AND-HOLD) circuit, 160–161, 170
Sharp Corp., 13
Shift registers, 42–44
Sign magnitude, 74, 75
Signal processing, 328
Silicon-on-sapphire (SOS) fabrication methods, 1, 6, 11, 14, 18, 65
   memory systems, 191–193, 224–229, 234–237, 239, 250–252
   microprocessors, 265–268, 291–295, 307–309, 321–324, 326–328
Siliconix, Inc., 12, 45, 114–124, 140, 141, 162
Simultaneous-integration A/D converter, 106–108
Single-channel A/D port, 148–150
SLIC (subscriber-loop interface circuit), 164
Solid State Scientific, Inc., 45
SOS (*see* Silicon-on-sapphire fabrication methods)
Speech synthesizer, 2, 13, 328

Index 339

Sperry Univac, 267
S/R (SET/RESET) flip-flop, 33–34
Stack, 270
Storage registers, 56–58
Subscriber channel unit (SCU), 163, 164
Subscriber-loop interface circuit (SLIC), 164
Successive-approximation A/D converter, 87, 90–95, 127, 329
Supertex, 252, 255–257
Switched capacitor filter, 166, 167, 171, 173, 177
Synchronous detection, 154
Synchoronous RAMs, 199–202, 218–224, 301

Tachometer, 50–51
  dwell, digital, 122, 124
Telecommunications circuits, 1, 2, 12, 13, 151–190
  commercially available, 161–190
  digital, 151–159
Teledyne Semiconductor, Inc.:
  A/D converters, 8700 series, 138–147
  D/A converters, 8640/8641, 84–86, 88
Telephone dialers, 182–190
Telephone exchange, 167, 168
Telephone handset, 177, 178
Telephone system, 151, 154–159, 164, 166–169, 173–190
Temperature coefficient, 69, 71, 73
Temperature variation ofsCMOS parameters, 21, 22, 24–26, 28–30, 32, 33, 70
Texas Instruments, Inc., 14, 222, 257, 279–281, 322–323
Thermometer, digital, 122, 123
3½-digit A/D converter, 97, 112–132
Three-state logic, 45
Threshold voltage, 27–29
Time delay, 75, 77
Tone dialer, 185–188
Tone encoder, 12, 173–176
Tone generator, 13, 177, 179
Tone ringer, 190

Toshiba, Ltd., 215, 236, 237, 239, 320–321, 326, 327
Transfer characteristics, 5, 9, 21–23, 25–27, 30
  (*See also* A Law transfer characteristic; Mu Law transfer characteristic)
Transistor-transistor logic (TTL), 2, 3, 5, 17, 27, 28, 192, 196, 197, 213
Transmission gate, 34–37, 40, 42–44
Transparent control panel, 303, 305
TTL (*see* Transistor-transistor logic)
TTY keyboard, 95, 96
Two-quadrant multiplication, 73, 74, 78
2's complement code, 74, 78, 109

UARTs (universal asynchronous receiver-transmitters), 12, 13, 62
  A/D converters, 95, 96
  microprocessors, 301, 303, 315, 318, 319, 324–326
Unbuffered logic, 7–10
Unipolar D/A conversion, 73, 74, 78–80, 103
Universal asynchronous receiver-transmitters (*see* UARTs)

Very high-speed integrated circuit (VHSIC), 327
Very-large-scale integration (VLSI), 17, 267, 326–328
VHSIC (very high-speed integrated circuit), 327
Video A/D converter, 87
Viterbi error-correction decoder, 326
VLSI (very-large-scale integration), 17, 267, 326–328
VMOS, 197
Voltage-fed R/2R ladder network D/A converter, 66, 67

Washing machine controller, 261–262
WRITE control logic, 207–208
WRITE cycle, 201, 210, 211, 219, 220, 225, 227, 234